Applied Mathematical Sciences
Volume 80

Applied Mathematical Sciences

(continued following index)

Derek F. Lawden

Elliptic Functions and Applications

With 23 Illustrations

Springer-Verlag
New York Berlin Heidelberg
London Paris Tokyo Hong Kong

Derek F. Lawden
University of Aston in Birmingham
Birmingham B4 7ET
United Kingdom

Editors

F. John
Courant Institute of
  Mathematical Sciences
New York University
New York, NY 10012
USA

J. E. Marsden
Department of
  Mathematics
University of
  California
Berkeley, CA 94720
USA

L. Sirovich
Division of Applied
  Mathematics
Brown University
Providence, RI 02912
USA

Mathematics Subject Classifications (1980): 33A25

Library of Congress Cataloging-in-Publication Data
Lawden, Derek F.
  Elliptic functions and applications/Derek F. Lawden.
    p. cm.—(Applied mathematical sciences; v. 80)
  Bibliography: p.
  Includes index.
  ISBN 0-387-96965-9
  1. Functions, Elliptic. I. Title. II. Series: Applied
mathematical sciences (Springer-Verlag New York Inc.); v. 80.
QA1.A647 vol. 80
[QA343]
510 s—dc20
[515′.983]   89-6171

Printed on acid-free paper.

Phototypesetting by Thomson Press (India) Limited, New Delhi.
Printed and bound by R. R. Donnelley & Sons, Harrisonburg, Virginia.
Printed in the United States of America.

9 8 7 6 5 4 3 2 1

ISBN 0-387-96965-9 Springer-Verlag New York Berlin Heidelberg
ISBN 3-540-96965-9 Springer-Verlag Berlin Heidelberg New York

Preface

The subject matter of this book formed the substance of a mathematical seam which was worked by many of the great mathematicians of the last century. The mining metaphor is here very appropriate, for the analytical tools perfected by Cauchy permitted the mathematical argument to penetrate to unprecedented depths over a restricted region of its domain and enabled mathematicians like Abel, Jacobi, and Weierstrass to uncover a treasurehouse of results whose variety, aesthetic appeal, and capacity for arousing our astonishment have not since been equaled by research in any other area. But the circumstance that this theory can be applied to solve problems arising in many departments of science and engineering graces the topic with an additional aura and provides a powerful argument for including it in university courses for students who are expected to use mathematics as a tool for technological investigations in later life. Unfortunately, since the status of university staff is almost wholly determined by their effectiveness as research workers rather than as teachers, the content of undergraduate courses tends to reflect those academic research topics which are currently popular and bears little relationship to the future needs of students who are themselves not destined to become university teachers. Thus, having been comprehensively explored in the last century and being undoubtedly difficult · to master, the theory of elliptic functions has dropped out of most university courses and very few mathematics graduates leaving universities today know more than that many problems of applied mathematics lead to elliptic integrals, and that these therefore "cannot be solved." However, there are signs that the arid view of mathematics, which became dominant in the immediate postwar years, according to which the discipline's prime purpose is not the solving of problems but the cataloging of axiom systems, is now giving way to a more balanced appraisal and school pupils are once again being encouraged to develop their mathematical ingenuity in using mathematics, rather than to exhibit a precocious (and, one suspects, very partial) understanding of its foundations. In this changed climate, one is hopeful that elliptic functions will again appear in the repertoire of university lecturers, particularly if government pressure on the universities to place greater emphasis on the effectiveness of their teaching and the interests of the majority of their students continues to be maintained.

In this event, a difficulty which will arise is the lack of textbooks for recommendation to students for additional reading and as sources of exercises. During the last twenty years of his service as a university teacher, the author has often received enquiries from users of mathematics regarding books which he could recommend for self-study in this field. Upon checking the appropriate section of his and other university libraries, he has discovered the choice of suitable books to be severely restricted and, very significantly, those available to be in heavy demand. Although a number of elementary introductions to the theory were published in the last century and a very few early in this, these are now out of print and the only works easily available assume an extensive background knowledge of the theory of analytic functions, in the application of which the ordinary professional applied mathematician can no longer be expected to be well practiced. For most undergraduate students also, a textbook which treats elliptic functions as a footnote to the theory of analytic functions is not very satisfactory, since it entails delaying the introduction of the topic until very late in the course and so limits the number of illustrations which may be made of the utility of the functions for solving practical problems. It is preferable, therefore, to permit a student to acquire a working knowledge of these functions on the basis of elementary analysis involving little more than the convergence of infinite series, in the same way that the properties of the circular and hyperbolic functions of a complex variable are elucidated during the first year of a university course by reason of their manifold applications. It is true that the extremely powerful methods provided by analytic function theory permit very elegant proofs of some of the elliptic function relationships to be constructed, but to the beginner these arguments appear as clever tricks which only serve to deepen his mystification regarding the true provenance of the results they establish. An introductory book which can be recommended for self-study by the undergraduate student or ordinary working mathematician, therefore, will first define the functions by the more elementary processes of analysis, in a similar manner to the other well-known transcendental functions, and then proceed to derive their properties in a straightforward way, before perhaps turning, at a later stage, to a deeper analysis depending upon the general properties of analytic functions. The most elegant and economical account is rarely the most effective for didactic purposes. This, then, is the plan I have followed in attempting to fill the gap in the range of texts available to the university student or to practicing applied mathematicians and engineers who need to remedy a serious deficiency in their tool kit. The authors of a standard set of tables (*Smithsonian Elliptic Functions Tables* by G. W. and R. M. Spenceley) comment in their introduction: "In the popular field of undergraduate mathematics, a new elementary text on elliptic functions with a wide variety of problems and applications, supplemented by an adequate set of tables, would be a boon to teachers and students." I hope that this book satisfies this long-standing need which they, and others, have identified.

Chapter 1 introduces the theta functions by their Fourier expansions and establishes their principal properties and identities. These functions have relatively few applications and, analytically, their behavior is more complex than that of the elliptic functions. However, their definition involves no more than an appeal to the idea of convergence of a complex series and this will be familiar to all serious users of mathematical techniques. Further, the rate of convergence of a theta series is very rapid in all practical circumstances, so that the reader is immediately provided with a practicable method for computing these functions and, later, the elliptic functions which are simply related to them. All the tables at the end of this book have been computed on this basis using a small desk computer and the reader should experience no difficulty in extending this collection to suit his requirements, either by the addition of new tables or by the recomputation of the given tables to higher accuracy or for smaller steps of the arguments. Proving the theta function identities also offers the student useful practice in the manipulation of infinite series, especially the multiplication of absolutely convergent series and the rearrangement of terms in the product. Thus, we have followed the approach to elliptic functions pioneered by Jacobi in his lectures and which leads quite naturally to the definition of the Jacobi functions in Chapter 2. In many treatments, the Weierstrass $\mathscr{P}$-function is taken to be fundamental and the Jacobi functions are derived from this. But it is the Jacobi functions which are the more closely related to the circular and hyperbolic functions with which the reader will be familiar and it is these functions which are predominantly useful in applications to mechanics, electrodynamics, geometry, etc. Thus, in a text intended for reading by mathematical practitioners, it is essential that the Jacobi functions are introduced as soon as possible. Having defined these functions as ratios of theta functions, their double periodicity and the identities they satisfy follow immediately from corresponding results for the theta functions and are worked through in the remainder of Chapter 2. Chapter 3 is devoted to the inverse Jacobi functions; it presents the three standard forms of elliptic integrals described by Legendre (for real values of the integration variable only) and obtains the properties of the complete integrals as functions of the modulus k; Jacobi's epsilon and zeta functions are also brought in at this stage. The text next turns to consider applications of the theory as it has been developed thus far. Geometrical applications are studied in Chapter 4, including those to the ellipse, the ellipsoid, Cartesian ovals, and spherical trigonometry. Chapter 5 deals with physical applications and covers plane motion of a pendulum, orbits under various laws of attraction to a center (including Einstein's law), the rotation of a rigid body about a smooth pivot, current flow in a rectangular plate, and parallel plate capacitors.

Almost all the argument covered in the first five chapters can be understood by a reader whose mathematical equipment extends no further than the techniques covered in the first year of an undergraduate mathematics course or during ancillary mathematics lectures arranged for engineers and physicists. It will be found challenging, but the attractiveness of the diverse and unexpected relationships revealed and the interest of the problems which become amenable to solution by application of the theory should prove adequate compensation for the effort expended. The remaining four chapters are provided for those who now feel the urge to explore the subject further and, in particular, to apply in this very fertile field a knowledge of the general theory of analytic functions.

Chapter 6 commences the study of elliptic functions having any pair of assigned periods by first constructing the sigma functions from the theta functions and then proceeding to define Weierstrass's zeta function and $\mathscr{P}$-function. The inverse $\mathscr{P}$-function leads to a consideration of an elliptic integral in Weierstrass's standard form. Application of the Weierstrass functions to the solution of a number of geometrical and physical problems is made in Chapter 7; among the items discussed are general three-dimensional motion of a pendulum, the spinning top, and a projectile moving under constant gravity and air resistance proportional to the cube of its speed.

In Chapter 8, the theory of analytic functions is applied without restraint to investigate the general properties of doubly periodic functions. Further expansions of the Weierstrass and Jacobi functions are constructed and the representation of any elliptic function as a rational function of the sigma function or the $\mathscr{P}$-function is determined. The chapter closes with an investigation of the multivalued nature of the elliptic integrals, both complete and incomplete, when these are treated as contour integrals.

The final chapter is an introduction to the very extensive theory of modular transformations. This associates together in a comprehensive theory a number of particular transformations of the Jacobian modulus already encountered in earlier chapters.

At the end of the book, we have collected together a wide variety of relevant tables. The entries are generally to five significant figures, but are only intended to provide an indication of the numerical behavior of the quantities tabulated, without permitting linear interpolation for intermediate values of the argument. For accurate numerical evaluation, the reader will need to refer to one of the more extensive sets of tables listed in the bibliography or to recompute the tables (a useful and straightforward exercise).

A collection of over 200 exercises has been distributed between the chapters and it is hoped these will provide stimulating checks for the reader on his understanding of the main text and, in addition, a useful collection of subsidiary results.

In regard to my notation, I have generally conformed to that which will be most frequently encountered when other readily available texts are referred to. Thus, the four theta functions are denoted by $\theta_i(z|\tau)$ $(i = 1, 2, 3, 4)$ and represent the same functions considered by Whittaker and Watson in their *Course of Modern Analysis*. However, I have preferred to denote the basic primitive periods of Weierstrass's $\mathscr{P}$-function by $2\omega_1$ and $2\omega_3$ (Whittaker and Watson use $2\omega_1$ and $2\omega_2$), with $\omega_2 = -\omega_1 - \omega_3$, since in the important special case when ω_1 is real and ω_3 is purely imaginary, this leads to the inequalities $e_1 > e_2 > e_3$, where $e_\alpha = \mathscr{P}(\omega_\alpha)$. This notation was adopted in many of the earlier texts (e.g., *Applications of Elliptic Functions* by Greenhill). In relating the Weierstrass function to the theta functions, it is then necessary to take $\tau = \omega_3/\omega_1$. The sigma functions are denoted by $\sigma(u)$, $\sigma_1(u)$, $\sigma_2(u)$, $\sigma_3(u)$ and, except for an additional exponential factor, are replicas of the theta functions $\theta_1(z)$, $\theta_2(z)$, $\theta_3(z)$, $\theta_4(z)$ respectively (with $z = \pi u/2\omega_1$). The notation for the Jacobi functions is the one introduced by Gudermann and Glaisher; this is an international standard and, since alternative notations suggested by Neville and others have not been generally accepted, it was felt no purpose would be served in burdening the reader with an account of these.

I am now approaching the termination of a life, one of whose major enjoyments has been the study of mathematics. The three jewels whose effulgence has most dazzled me are Maxwell's theory of electromagnetism, Einstein's theory of relativity, and the theory of elliptic functions. I have now published textbooks on each of these topics, but the one from whose preparation and writing I have derived the greatest pleasure is the present work on elliptic functions. How enviable are Jacobi and Weierstrass to have been the creators of such a work of art! As a lesser mortal lays down his pen, he salutes them and hopes that his execution of their composition does not offend any who have ears to hear the music of the spheres.

D. F. LAWDEN

Contents

CHAPTER 1

Theta Functions

1.1. Theta Functions as Solutions of the Heat Conduction Equation

We shall introduce the theta functions by considering a specific heat conduction problem. The reader who is unfamiliar with the details of the following argument should return to this section in due course and, meanwhile, accept equations (1.1.8) and (1.1.12) as definitions of the theta functions θ_1 and θ_4 respectively.

Let θ be the temperature at time t at any point in a solid material whose conducting properties are uniform and isotropic. Then, if ρ is the material's density, s is its specific heat, and k its thermal conductivity, θ satisfies the partial differential equation

$$\kappa \nabla^2 \theta = \partial\theta/\partial t, \tag{1.1.1}$$

where $\kappa = k/s\rho$ is termed the *diffusivity*. In the special case where there is no variation of temperature in the x- and y-directions of a rectangular Cartesian frame $Oxyz$, the heat flow is everywhere parallel to the z-axis and the heat conduction equation reduces to the form

$$\kappa \frac{\partial^2 \theta}{\partial z^2} = \frac{\partial\theta}{\partial t} \tag{1.1.2}$$

and $\theta = \theta(z, t)$.

The specific problem we shall study is the flow of heat in an infinite slab of material, bounded by the planes $z = 0, \pi$, when the conditions over each boundary plane are kept uniform at every time t. The heat flow is then entirely in the z-direction and equation (1.1.2) is applicable.

First, suppose the boundary conditions are that the faces of the slab are maintained at zero temperature, i.e., $\theta = 0$ for $z = 0, \pi$ and all t. Initially, at $t = 0$, suppose $\theta = f(z)$ for $0 < z < \pi$. Then it is shown in texts devoted to the solution of the partial differential equations of physics (e.g., G. Stephenson, *An Introduction to Partial Differential Equations for Science Students*, 2nd ed., pp. 47–50, Longman, 1970) that the method of separation of variables leads to the solution

$$\theta(z, t) = \sum_{n=1}^{\infty} b_n e^{-n^2 \kappa t} \sin nz, \tag{1.1.3}$$

where b_n are Fourier coefficients determined by the equation

$$b_n = \frac{2}{\pi} \int_0^\pi f(z) \sin nz \, dz. \tag{1.1.4}$$

In the special case where $f(z) = \pi\delta(z - \tfrac{1}{2}\pi)$ (where $\delta(z)$ is Dirac's unit impulse function), the slab is initially at zero temperature everywhere, except in the neighborhood of the midplane $z = \tfrac{1}{2}\pi$, where the temperature is very high. To achieve this high temperature, it will be necessary to inject a quantity of heat h (joules) per unit area into this plane to raise its temperature from zero. h is given by

$$h = \rho s \pi \int_{(1/2)\pi - 0}^{(1/2)\pi + 0} \delta(z - \tfrac{1}{2}\pi) \, dz = \rho s \pi. \tag{1.1.5}$$

We now calculate that

$$b_n = 2 \int_0^\pi \delta(z - \tfrac{1}{2}\pi) \sin nz \, dz = 2 \sin (\tfrac{1}{2}n\pi). \tag{1.1.6}$$

Thus, heat diffusion over the slab is governed by the equation

$$\theta(z, t) = 2 \sum_{n=0}^{\infty} (-1)^n e^{-(2n+1)^2 \kappa t} \sin(2n + 1)z. \tag{1.1.7}$$

θ has been graphed against z for a set of increasing values of t in Fig. 1.1(a) and shows the effect of heat flowing outward from the hot central stratum across the boundary faces, eventually reducing the whole slab to zero temperature.

Writing $e^{-4\kappa t} = q$, the solution (1.1.7) assumes the form

$$\theta = \theta_1(z, q) = 2 \sum_{n=0}^{\infty} (-1)^n q^{(n+1/2)^2} \sin(2n + 1)z, \tag{1.1.8}$$

where the function $\theta_1(z, q)$ is defined by the infinite series. This is the first of the theta functions.

Another, related, theta function arises if we change the boundary conditions of our problem. Thus suppose the slab's faces are insulated so that no heat can leak across either. The corresponding boundary conditions are $\partial\theta/\partial z = 0$

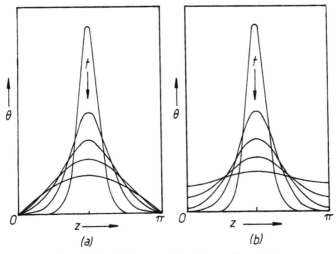

Figure 1.1. Diffusion of heat through a plate.

on $z = 0, \pi$, and separation of variables leads to the solution

$$\theta(z, t) = \tfrac{1}{2}a_0 + \sum_{n=1}^{\infty} a_n e^{-n^2\kappa t} \cos nz, \qquad (1.1.9)$$

where

$$a_n = \frac{2}{\pi} \int_0^\pi f(z) \cos nz \, dz. \qquad (1.1.10)$$

In the case $f(z) = \pi\delta(z - \tfrac{1}{2}\pi)$, we find that

$$a_n = 2 \cos \tfrac{1}{2}n\pi \qquad (1.1.11)$$

and the solution to the problem is therefore

$$\theta = \theta_4(z, q) = 1 + 2 \sum_{n=1}^{\infty} (-1)^n q^{n^2} \cos 2nz, \qquad (1.1.12)$$

where $q = e^{-4\kappa t}$ as before. This series defines the theta function $\theta_4(z, q)$. This time, heat again diffuses outward from the central stratum, but is trapped inside the slab so that ultimately the slab rests in equilibrium at a unit uniform temperature. Successive distributions of temperature over the slab are indicated in Fig. 1.1(b).

1.2. Definitions of the Four Theta Functions

The first theta function $\theta_1(z, q)$ is defined by the series (1.1.8) for all complex values of z and q such that $|q| < 1$. Replacing the sine function by its Euler representation by exponentials, we obtain an alternative definition by a doubly infinite series,

$$i\theta_1(z,q) = \sum_{n=-\infty}^{\infty} (-1)^n q^{(n+1/2)^2} e^{i(2n+1)z} = q^{1/4} \sum_{n=-\infty}^{\infty} (-1)^n q^{n(n+1)} e^{i(2n+1)z}.$$
$$(1.2.1)$$

To establish convergence, let u_n denote the nth term of this series. Then

$$|u_{n+1}/u_n| = |q^{2n+2} e^{2iz}| = |q|^{2n+2} e^{-2y}, \qquad (1.2.2)$$

if $z = x + iy$. As $n \to +\infty$, since $|q| < 1$, this ratio tends to zero and, by D'Alembert's test, therefore, the series converges at $+\infty$. As $n \to -\infty$, to apply this test we need to consider the ratio $|u_n/u_{n+1}|$. Clearly, the limit of this ratio is again zero and, hence, the series converges at $-\infty$ also.

An alternative notation is to write

$$q = e^{i\pi\tau}, \qquad (1.2.3)$$

where the imaginary part of τ must be positive to give $|q| < 1$. q is called the *nome* of the theta function and τ is its *parameter*. Thus, equation (1.2.1) can be written

$$i\theta_1(z|\tau) = \sum_{n=-\infty}^{\infty} (-1)^n \exp i\{(n+\tfrac{1}{2})^2 \pi\tau + (2n+1)z\}. \qquad (1.2.4)$$

(Note how the dependence on τ is indicated in $\theta_1(z|\tau)$.)

τ is regarded as the fundamental parameter from which q is derived by equation (1.2.3). Any power of q, say q^α, then follows from the equation

$$q^\alpha = e^{i\pi\tau\alpha} \qquad (1.2.5)$$

without ambiguity. The fractional power present in equation (1.2.1) is to be evaluated on this understanding.

Clearly, θ_1 is an odd function of z, which takes real values when z and q are real (i.e., τ is purely imaginary). At $z = 0$, $\theta_1 = 0$.

If $z = x + iy$ and $-Y \leqslant y \leqslant Y$, then

$$|(-1)^n q^{(n+1/2)^2} e^{i(2n+1)z}| = |q|^{(n+1/2)^2} e^{-(2n+1)y} \leqslant |q|^{(n+1/2)^2} e^{(2n+1)Y} \qquad (1.2.6)$$

for $n > 0$. But the series $\sum_{n=0}^{\infty} |q|^{(n+1/2)^2} e^{-(2n+1)Y}$ is convergent by D'Alembert's test and it follows that the series $\sum_{n=0}^{\infty} (-1)^n q^{(n+1/2)^2} e^{i(2n+1)z}$ is uniformly convergent in the strip $-Y \leqslant y \leqslant Y$. The inequality

$$|(-1)^n q^{(n+1/2)^2} e^{i(2n+1)z}| \leqslant |q|^{(n+1/2)^2} e^{-(2n+1)Y} \qquad (1.2.7)$$

is valid for $n < 0$ and establishes the uniform convergence of the series $\sum_{n=-1}^{-\infty} (-1)^n q^{(n+1/2)^2} e^{i(2n+1)z}$ in the same strip. Thus, the doubly infinite series for θ_1 is uniformly convergent in the strip $-Y \leqslant y \leqslant Y$. But, $e^{i(2n+1)z}$ is regular for all values of z and it follows that θ_1 is regular throughout the strip. Since Y is arbitrary, we have established that θ_1 is an integral function of z, i.e., is regular for all finite values of z.

In the early chapters of this book, we shall be primarily interested in the dependence of the theta functions on z. Thus, τ and q will usually be suppressed and the functions written simply as $\theta_r(z)$ $(r = 1, 2, 3, 4)$.

It is evident from the definition (1.1.8) that $\theta_1(z)$ is periodic with period 2π. Incrementing z by a quarter period, we define the theta function $\theta_2(z)$ thus:

$$\theta_2(z) = \theta_1(z + \tfrac{1}{2}\pi) = 2 \sum_{n=0}^{\infty} (-1)^n q^{(n+1/2)^2} \sin\{(2n+1)z + (n+\tfrac{1}{2})\pi\},$$

$$= 2 \sum_{n=0}^{\infty} q^{(n+1/2)^2} \cos(2n+1)z,$$

$$= \sum_{n=-\infty}^{\infty} q^{(n+1/2)^2} e^{i(2n+1)z}. \tag{1.2.8}$$

Evidently, $\theta_2(z)$ is an even integral function of z with period 2π.

$\theta_4(z)$ is defined by the series (1.1.12), or alternatively by the doubly infinite series

$$\theta_4(z) = \sum_{n=-\infty}^{\infty} (-1)^n q^{n^2} e^{2inz}. \tag{1.2.9}$$

It is an even, integral function of z with period π.

Incrementing z by $\tfrac{1}{2}\pi$, we obtain the final theta function θ_3 thus:

$$\theta_3(z) = \theta_4(z + \tfrac{1}{2}\pi) = 1 + 2 \sum_{n=1}^{\infty} q^{n^2} \cos 2nz,$$

$$= \sum_{n=-\infty}^{\infty} q^{n^2} e^{2inz}. \tag{1.2.10}$$

$\theta_3(z)$ is also an even, integral function of z with period π.

All the theta functions take real values for real values of z and q (see Table A, p. 270).

It is convenient to summarize our definitions as follows:

$$\theta_1 = -i \sum_{n=-\infty}^{\infty} (-1)^n q^{(n+1/2)^2} e^{i(2n+1)z} = 2 \sum_{n=0}^{\infty} (-1)^n q^{(n+1/2)^2} \sin(2n+1)z,$$
$$\tag{1.2.11}$$

$$\theta_2 = \sum_{n=-\infty}^{\infty} q^{(n+1/2)^2} e^{i(2n+1)z} = 2 \sum_{n=0}^{\infty} q^{(n+1/2)^2} \cos(2n+1)z, \tag{1.2.12}$$

$$\theta_3 = \sum_{n=-\infty}^{\infty} q^{n^2} e^{2inz} = 1 + 2 \sum_{n=1}^{\infty} q^{n^2} \cos 2nz, \tag{1.2.13}$$

$$\theta_4 = \sum_{n=-\infty}^{\infty} (-1)^n q^{n^2} e^{2inz} = 1 + 2 \sum_{n=1}^{\infty} (-1)^n q^{n^2} \cos 2nz. \tag{1.2.14}$$

1.3. Periodicity Properties

That θ_1, θ_2 are periodic with period 2π, and θ_3, θ_4 are periodic with period π, follows immediately from their Fourier expansions (1.2.11)–(1.2.14). However, these functions also have the property that, if their argument z is incremented by $\pi\tau$, they are unaffected except for multiplication by a simple factor. For example,

$$\theta_1(z + \pi\tau) = -i \sum_{n=-\infty}^{\infty} (-1)^n \exp\{(n + \tfrac{1}{2})^2 i\pi\tau + i(2n + 1)(z + \pi\tau)\}$$

$$= i \exp(-i\pi\tau - 2iz) \sum_{n=-\infty}^{\infty} (-1)^{n+1} \exp\{(n + \tfrac{3}{2})^2 i\pi\tau + i(2n + 3)z\}$$

$$= -(qe^{2iz})^{-1}\theta_1(z). \tag{1.3.1}$$

$\pi\tau$ is termed a quasi-period of $\theta_1(z)$ with periodicity factor $-(qe^{2iz})^{-1}$.

The reader is left to verify that $\pi\tau$ is also a quasi-period of the other theta functions, with the same periodicity factor or its negative. Further, $\pi + \pi\tau$ must also be a quasi-period with periodicity factor $\pm(qe^{2iz})^{-1}$. Writing $\lambda = qe^{2iz}$, the following set of identities may be verified:

$$\theta_1(z) = -\theta_1(z + \pi) = -\lambda\theta_1(z + \pi\tau) = \lambda\theta_1(z + \pi + \pi\tau), \tag{1.3.2}$$

$$\theta_2(z) = -\theta_2(z + \pi) = \lambda\theta_2(z + \pi\tau) = -\lambda\theta_2(z + \pi + \pi\tau), \tag{1.3.3}$$

$$\theta_3(z) = \theta_3(z + \pi) = \lambda\theta_3(z + \pi\tau) = \lambda\theta_3(z + \pi + \pi\tau), \tag{1.3.4}$$

$$\theta_4(z) = \theta_4(z + \pi) = -\lambda\theta_4(z + \pi\tau) = -\lambda\theta_4(z + \pi + \pi\tau). \tag{1.3.5}$$

It now follows that, if m, n are integers or zero, then $m\pi + n\pi\tau$ is a quasi-period of each of the theta functions, the appropriate periodicity factor being $\pm(q^{n^2}e^{2inz})^{-1}$ (see Exercise 1).

Incrementation of z by the half periods $\tfrac{1}{2}\pi, \tfrac{1}{2}\pi\tau$, and $\tfrac{1}{2}(\pi + \pi\tau)$ leads to the identities

$$\theta_1(z) = -\theta_2(z + \tfrac{1}{2}\pi) = -i\mu\theta_4(z + \tfrac{1}{2}\pi\tau) = -i\mu\theta_3(z + \tfrac{1}{2}\pi + \tfrac{1}{2}\pi\tau), \tag{1.3.6}$$

$$\theta_2(z) = \theta_1(z + \tfrac{1}{2}\pi) = \mu\theta_3(z + \tfrac{1}{2}\pi\tau) = \mu\theta_4(z + \tfrac{1}{2}\pi + \tfrac{1}{2}\pi\tau), \tag{1.3.7}$$

$$\theta_3(z) = \theta_4(z + \tfrac{1}{2}\pi) = \mu\theta_2(z + \tfrac{1}{2}\pi\tau) = \mu\theta_1(z + \tfrac{1}{2}\pi + \tfrac{1}{2}\pi\tau), \tag{1.3.8}$$

$$\theta_4(z) = \theta_3(z + \tfrac{1}{2}\pi) = -i\mu\theta_1(z + \tfrac{1}{2}\pi\tau) = i\mu\theta_2(z + \tfrac{1}{2}\pi + \tfrac{1}{2}\pi\tau), \tag{1.3.9}$$

where $\mu = q^{1/4}e^{iz}$.

Since $\theta_1(0) = 0$, it follows from equations (1.3.2) that $\theta_1(z)$ has zeros where $z = m\pi + n\pi\tau$. The zeros of the other theta functions then follow from equations (1.3.6); we summarize these data below:

$$\left.\begin{array}{ll} \theta_1(z) = 0 & \text{where } z = m\pi + n\pi\tau, \\ \theta_2(z) = 0 & \text{where } z = (m + \tfrac{1}{2})\pi + n\pi\tau, \\ \theta_3(z) = 0 & \text{where } z = (m + \tfrac{1}{2})\pi + (n + \tfrac{1}{2})\pi\tau, \\ \theta_4(z) = 0 & \text{where } z = m\pi + (n + \tfrac{1}{2})\pi\tau. \end{array}\right\} \tag{1.3.10}$$

1.4. Identities Involving Products of Theta Functions

A number of important identities relating theta functions can be obtained by multiplication of two of their series, followed by a rearrangement of the terms in the product series (which is permissible, since the series are absolutely convergent).

Thus, we have

$$\theta_1(x,q)\theta_1(y,q) = -\sum_m \sum_n (-1)^{m+n} q^{(m+1/2)^2+(n+1/2)^2} e^{i(2m+1)x+i(2n+1)y}, \quad (1.4.1)$$

the summations with respect to both integers m and n extending from $-\infty$ to $+\infty$.

We now change the integer variables of summation from (m,n) to (r,s) by the $(1-1)$ transformation equations

$$m+n=r, \qquad m-n=s. \tag{1.4.2}$$

If (m,n) are both even or both odd, (r,s) will both be even. If, however, (m,n) have opposite parity, (r,s) will both be odd. Hence, by permitting (r,s) to range over all pairs of even integers (positive, negative, and zero) and over all pairs of odd integers, every pair of integers (m,n) will be obtained exactly once.

Further, we can write

$$(m+\tfrac{1}{2})^2+(n+\tfrac{1}{2})^2 = \tfrac{1}{2}(r+1)^2+\tfrac{1}{2}s^2, \tag{1.4.3}$$

$$(2m+1)x+(2n+1)y = (r+1)(x+y)+s(x-y). \tag{1.4.4}$$

Thus

$$\theta_1(x,q)\theta_1(y,q) = -\sum (-1)^r q^{(1/2)(r+1)^2+(1/2)s^2} e^{i(r+1)(x+y)+is(x-y)}, \tag{1.4.5}$$

the summation being over all even pairs (r,s) and all odd pairs (r,s), as just explained. It follows that

$$\theta_1(x,q)\theta_1(y,q) = -\sum_r \sum_s q^{2(r+1/2)^2+2s^2} e^{i(2r+1)(x+y)+2is(x-y)}$$
$$+\sum_r \sum_s q^{2r^2+2(s+1/2)^2} e^{2ir(x+y)+i(2s+1)(x-y)}, \tag{1.4.6}$$

where the summations are now to be taken over all pairs of integers (r,s).

Each of the new double series can be expressed as a product of two single series, as follows:

$$\theta_1(x,q)\theta_1(y,q) = -\sum_r q^{2(r+1/2)^2} e^{i(2r+1)(x+y)} \sum_s q^{2s^2} e^{2is(x-y)}$$

$$+\sum_r q^{2r^2} e^{2ir(x+y)} \sum_s q^{2(s+1/2)^2} e^{i(2s+1)(x-y)}$$

$$= \theta_3(x+y,q^2)\theta_2(x-y,q^2)-\theta_2(x+y,q^2)\theta_3(x-y,q^2), \tag{1.4.7}$$

which is an identity between theta functions of the type we are seeking.

Further identities of the same type which can be established by this method are as follows:

$$\theta_1(x,q)\theta_2(y,q) = \theta_1(x+y,q^2)\theta_4(x-y,q^2) + \theta_4(x+y,q^2)\theta_1(x-y,q^2),$$
(1.4.8)

$$\theta_2(x,q)\theta_2(y,q) = \theta_2(x+y,q^2)\theta_3(x-y,q^2) + \theta_3(x+y,q^2)\theta_2(x-y,q^2),$$
(1.4.9)

$$\theta_3(x,q)\theta_3(y,q) = \theta_3(x+y,q^2)\theta_3(x-y,q^2) + \theta_2(x+y,q^2)\theta_2(x-y,q^2),$$
(1.4.10)

$$\theta_3(x,q)\theta_4(y,q) = \theta_4(x+y,q^2)\theta_4(x-y,q^2) - \theta_1(x+y,q^2)\theta_1(x-y,q^2),$$
(1.4.11)

$$\theta_4(x,q)\theta_4(y,q) = \theta_3(x+y,q^2)\theta_3(x-y,q^2) - \theta_2(x+y,q^2)\theta_2(x-y,q^2).$$
(1.4.12)

These will be left as exercises for the reader to practice the method (or, alternatively, to derive from (1.4.7) and (1.4.10) by incrementing x and/or y by $\frac{1}{2}\pi$).

We are now able to deduce a large number of identities involving products of four theta functions, all having the same nome.

For example, squaring and subtracting the identities (1.4.7) and (1.4.12), we first deduce that

$$\theta_4^2(x,q)\theta_4^2(y,q) - \theta_1^2(x,q)\theta_1^2(y,q)$$
$$= [\theta_3^2(x+y,q^2) - \theta_2^2(x+y,q^2)] \times [\theta_3^2(x-y,q^2) - \theta_2^2(x-y,q^2)].$$
(1.4.13)

Then, putting $y = 0$ in the identity (1.4.12) gives the result

$$\theta_3^2(x,q^2) - \theta_2^2(x,q^2) = \theta_4(x,q)\theta_4(0,q).$$
(1.4.14)

We now deduce from (1.4.13) that

$$\theta_4(x+y)\theta_4(x-y)\theta_4^2(0) = \theta_4^2(x)\theta_4^2(y) - \theta_1^2(x)\theta_1^2(y),$$
(1.4.15)

all theta functions in this identity having the same nome q.

Similar identities can now be derived by incrementing x and/or y by half periods $\frac{1}{2}\pi, \frac{1}{2}\pi\tau$, and $\frac{1}{2}\pi + \frac{1}{2}\pi\tau$ in the last identity. A complete list is given below:

$$\theta_1(x+y)\theta_1(x-y)\theta_4^2(0) = \theta_3^2(x)\theta_2^2(y) - \theta_2^2(x)\theta_3^2(y) = \theta_1^2(x)\theta_4^2(y) - \theta_4^2(x)\theta_1^2(y)$$
(1.4.16)

$$\theta_2(x+y)\theta_2(x-y)\theta_4^2(0) = \theta_4^2(x)\theta_2^2(y) - \theta_1^2(x)\theta_3^2(y) = \theta_2^2(x)\theta_4^2(y) - \theta_3^2(x)\theta_1^2(y)$$
(1.4.17)

$$\theta_3(x+y)\theta_3(x-y)\theta_4^2(0) = \theta_4^2(x)\theta_3^2(y) - \theta_1^2(x)\theta_2^2(y) = \theta_3^2(x)\theta_4^2(y) - \theta_2^2(x)\theta_1^2(y)$$
(1.4.18)

$$\theta_4(x+y)\theta_4(x-y)\theta_4^2(0) = \theta_3^2(x)\theta_3^2(y) - \theta_2^2(x)\theta_2^2(y) = \theta_4^2(x)\theta_4^2(y) - \theta_1^2(x)\theta_1^2(y)$$
$$(1.4.19)$$

By squaring and adding the identities (1.4.7) and (1.4.10), we find that

$$\theta_1^2(x,q)\theta_1^2(y,q) + \theta_3^2(x,q)\theta_3^2(y,q)$$
$$= [\theta_3^2(x+y,q^2) + \theta_2^2(x+y,q^2)] \times [\theta_3^2(x-y,q^2) + \theta_2^2(x-y,q^2)].$$
$$(1.4.20)$$

Putting $y = 0$ in (1.4.10) shows that

$$\theta_3^2(x,q^2) + \theta_2^2(x,q^2) = \theta_3(x,q)\theta_3(0,q). \qquad (1.4.21)$$

It follows that

$$\theta_3(x+y)\theta_3(x-y)\theta_3^2(0) = \theta_1^2(x)\theta_1^2(y) + \theta_3^2(x)\theta_3^2(y), \qquad (1.4.22)$$

all theta functions in this identity having nome q.

Incrementation of x and y by half periods in the last identity provides a complete set of identities of this type, viz.

$$\theta_1(x+y)\theta_1(x-y)\theta_3^2(0) = \theta_1^2(x)\theta_3^2(y) - \theta_3^2(x)\theta_1^2(y) = \theta_4^2(x)\theta_2^2(y) - \theta_2^2(x)\theta_4^2(y),$$
$$(1.4.23)$$

$$\theta_2(x+y)\theta_2(x-y)\theta_3^2(0) = \theta_2^2(x)\theta_3^2(y) - \theta_4^2(x)\theta_1^2(y) = \theta_3^2(x)\theta_2^2(y) - \theta_1^2(x)\theta_4^2(y),$$
$$(1.4.24)$$

$$\theta_3(x+y)\theta_3(x-y)\theta_3^2(0) = \theta_1^2(x)\theta_1^2(y) + \theta_3^2(x)\theta_3^2(y) = \theta_2^2(x)\theta_2^2(y) + \theta_4^2(x)\theta_4^2(y),$$
$$(1.4.25)$$

$$\theta_4(x+y)\theta_4(x-y)\theta_3^2(0) = \theta_1^2(x)\theta_2^2(y) + \theta_3^2(x)\theta_4^2(y) = \theta_2^2(x)\theta_1^2(y) + \theta_4^2(x)\theta_3^2(y).$$
$$(1.4.26)$$

Squaring and subtracting the identities (1.4.7) and (1.4.9), we are led to the result

$$\theta_2^2(x,q)\theta_2^2(y,q) - \theta_1^2(x,q)\theta_1^2(y,q)$$
$$= 4\theta_2(x+y,q^2)\theta_3(x+y,q^2)\theta_2(x-y,q^2)\theta_3(x-y,q^2). \quad (1.4.27)$$

Putting $y = 0$ in (1.4.9) yields

$$2\theta_2(x,q^2)\theta_3(x,q^2) = \theta_2(x,q)\theta_2(0,q). \qquad (1.4.28)$$

Thus, (1.4.27) reduces to

$$\theta_2(x+y)\theta_2(x-y)\theta_2^2(0) = \theta_2^2(x)\theta_2^2(y) - \theta_1^2(x)\theta_1^2(y). \qquad (1.4.29)$$

Incrementation by half periods in the last identity now generates the complete set of identities:

$$\theta_1(x+y)\theta_1(x-y)\theta_2^2(0) = \theta_1^2(x)\theta_2^2(y) - \theta_2^2(x)\theta_1^2(y) = \theta_4^2(x)\theta_3^2(y) - \theta_3^2(x)\theta_4^2(y),$$
$$(1.4.30)$$

$$\theta_2(x+y)\theta_2(x-y)\theta_2^2(0) = \theta_2^2(x)\theta_2^2(y) - \theta_1^2(x)\theta_1^2(y) = \theta_3^2(x)\theta_3^2(y) - \theta_4^2(x)\theta_4^2(y),$$
$$(1.4.31)$$

$$\theta_3(x+y)\theta_3(x-y)\theta_2^2(0) = \theta_3^2(x)\theta_2^2(y) + \theta_4^2(x)\theta_1^2(y) = \theta_2^2(x)\theta_3^2(y) + \theta_1^2(x)\theta_4^2(y),$$
$$(1.4.32)$$

$$\theta_4(x+y)\theta_4(x-y)\theta_2^2(0) = \theta_4^2(x)\theta_2^2(y) + \theta_3^2(x)\theta_1^2(y) = \theta_1^2(x)\theta_3^2(y) + \theta_2^2(x)\theta_4^2(y).$$
$$(1.4.33)$$

Yet further identities of similar, but distinct, types can be derived from the original set (1.4.7)–(1.4.12) by manipulating them in the following manner: Exchanging x and y in (1.4.11) and then multiplying by the identity (1.4.8), we calculate that

$$\theta_1(x,q)\theta_2(y,q)\theta_3(y,q)\theta_4(x,q)$$
$$= [\theta_1(x+y,q^2)\theta_4(x-y,q^2) + \theta_4(x+y,q^2)\theta_1(x-y,q^2)]$$
$$\times [\theta_4(x+y,q^2)\theta_4(x-y,q^2) + \theta_1(x+y,q^2)\theta_1(x-y,q^2)]. \quad (1.4.34)$$

Exchanging x and y in this identity and adding the result to it, we find that

$$\theta_1(x,q)\theta_2(y,q)\theta_3(y,q)\theta_4(x,q) + \theta_1(y,q)\theta_2(x,q)\theta_3(x,q)\theta_4(y,q)$$
$$= 2\theta_1(x+y,q^2)\theta_4(x+y,q^2)[\theta_4^2(x-y,q^2) + \theta_1^2(x-y,q^2)]. \quad (1.4.35)$$

Putting $y=0$ in (1.4.8) and $x=0$ in (1.4.11) yields the identities

$$\theta_1(x,q)\theta_2(0,q) = 2\theta_1(x,q^2)\theta_4(x,q^2), \quad (1.4.36)$$

$$\theta_4(y,q)\theta_3(0,q) = \theta_4^2(y,q^2) + \theta_1^2(y,q^2). \quad (1.4.37)$$

The identity (1.4.35) now reduces to the form

$$\theta_1(x+y)\theta_4(x-y)\theta_2(0)\theta_3(0) = \theta_1(x)\theta_4(x)\theta_2(y)\theta_3(y) + \theta_2(x)\theta_3(x)\theta_1(y)\theta_4(y). \quad (1.4.38)$$

Incrementing x by $\frac{1}{2}\pi$ transforms the last identity to the form

$$\theta_2(x+y)\theta_3(x-y)\theta_2(0)\theta_3(0) = \theta_2(x)\theta_3(x)\theta_2(y)\theta_3(y) - \theta_1(x)\theta_4(x)\theta_1(y)\theta_4(y). \quad (1.4.39)$$

If the exchange of variables x and y is omitted in the foregoing manipulation, the identities which result are

$$\theta_1(x+y)\theta_3(x-y)\theta_2(0)\theta_4(0) = \theta_1(x)\theta_3(x)\theta_2(y)\theta_4(y) + \theta_2(x)\theta_4(x)\theta_1(y)\theta_3(y), \quad (1.4.40)$$

$$\theta_2(x+y)\theta_4(x-y)\theta_2(0)\theta_4(0) = \theta_2(x)\theta_4(x)\theta_2(y)\theta_4(y) - \theta_1(x)\theta_3(x)\theta_1(y)\theta_3(y). \quad (1.4.41)$$

Multiplying the identities (1.4.10) and (1.4.12), we find

$$\theta_3(x,q)\theta_3(y,q)\theta_4(x,q)\theta_4(y,q)$$
$$= \theta_3^2(x+y,q^2)\theta_3^2(x-y,q^2) - \theta_2^2(x+y,q^2)\theta_2^2(x-y,q^2). \quad (1.4.42)$$

Putting $y = 0$ in (1.4.10) and (1.4.12) and solving, we calculate that

$$2\theta_3^2(x, q^2) = \theta_3(x, q)\theta_3(0, q) + \theta_4(x, q)\theta_4(0, q), \tag{1.4.43}$$

$$2\theta_2^2(x, q^2) = \theta_3(x, q)\theta_3(0, q) - \theta_4(x, q)\theta_4(0, q). \tag{1.4.44}$$

Substituting from these identities into (1.4.42), we then get

$$2\theta_3(x)\theta_3(y)\theta_4(x)\theta_4(y) = [\theta_3(x+y)\theta_4(x-y) + \theta_4(x+y)\theta_3(x-y)]\theta_3(0)\theta_4(0). \tag{1.4.45}$$

Incrementing both x and y by $\frac{1}{2}\pi\tau$ transforms this identity to the form

$$2\theta_2(x)\theta_2(y)\theta_1(x)\theta_1(y) = [\theta_4(x+y)\theta_3(x-y) - \theta_3(x+y)\theta_4(x-y)]\theta_3(0)\theta_4(0). \tag{1.4.46}$$

Subtracting (1.4.46) from (1.4.45), it is found that

$$\theta_3(x+y)\theta_4(x-y)\theta_3(0)\theta_4(0) = \theta_3(x)\theta_4(x)\theta_3(y)\theta_4(y) - \theta_1(x)\theta_2(x)\theta_1(y)\theta_2(y). \tag{1.4.47}$$

Incrementing x by $\frac{1}{2}\pi + \frac{1}{2}\pi\tau$ further transforms this to the form

$$\theta_1(x+y)\theta_2(x-y)\theta_3(0)\theta_4(0) = \theta_1(x)\theta_2(x)\theta_3(y)\theta_4(y) + \theta_3(x)\theta_4(x)\theta_1(y)\theta_2(y). \tag{1.4.48}$$

Putting $y = 0$ in the identities (1.4.16)–(1.4.19), (1.4.23)–(1.4.26), and (1.4.30)–(1.4.33), only four distinct relationships arise, viz.

$$\theta_3^2(x)\theta_3^2(0) = \theta_4^2(x)\theta_4^2(0) + \theta_2^2(x)\theta_2^2(0), \tag{1.4.49}$$

$$\theta_4^2(x)\theta_3^2(0) = \theta_1^2(x)\theta_2^2(0) + \theta_3^2(x)\theta_4^2(0), \tag{1.4.50}$$

$$\theta_4^2(x)\theta_2^2(0) = \theta_1^2(x)\theta_3^2(0) + \theta_2^2(x)\theta_4^2(0), \tag{1.4.51}$$

$$\theta_3^2(x)\theta_2^2(0) = \theta_1^2(x)\theta_4^2(0) + \theta_2^2(x)\theta_3^2(0). \tag{1.4.52}$$

We have here shown that the square of any theta function can be expressed linearly in terms of the squares of any two other theta functions.

Setting x to zero in (1.4.49) yields the fundamental result

$$\theta_3^4(0) = \theta_2^4(0) + \theta_4^4(0). \tag{1.4.53}$$

An immediate consequence of this result is that any three of the identities (1.4.49)–(1.4.52) are linearly dependent. For example, multiplying the first, second, and third identities by $\theta_4^2(0), \theta_3^2(0), \theta_2^2(0)$ respectively and then adding the first two and subtracting the third, it will be found that all terms cancel. It follows that these identities cannot be solved for the ratios $\theta_1(x):\theta_2(x):\theta_3(x):\theta_4(x)$; i.e., no theta function is a trivial multiple of any other.

1.5. The Identity $\theta_1'(0) = \theta_2(0)\theta_3(0)\theta_4(0)$

This important fundamental identity can be proved thus: Differentiating (1.4.8) with respect to x and then putting $x = y = 0$, since $\theta_1(0) = 0$ we have that

$$\theta_1'(0, q)\theta_2(0, q) = 2\theta_1'(0, q^2)\theta_4(0, q^2). \tag{1.5.1}$$

Next, putting $x = y = 0$ in the identities (1.4.9), (1.4.11), we show that

$$\theta_2^2(0, q) = 2\theta_2(0, q^2)\theta_3(0, q^2), \tag{1.5.2}$$

$$\theta_3(0, q)\theta_4(0, q) = \theta_4^2(0, q^2). \tag{1.5.3}$$

We now divide the result (1.5.1) by both the last equations to yield

$$\frac{\theta_1'(0, q)}{\theta_2(0, q)\theta_3(0, q)\theta_4(0, q)} = \frac{\theta_1'(0, q^2)}{\theta_2(0, q^2)\theta_3(0, q^2)\theta_4(0, q^2)}. \tag{1.5.4}$$

By repeated application of this result, we can show that

$$\frac{\theta_1'(0, q)}{\theta_2(0, q)\theta_3(0, q)\theta_4(0, q)} = \frac{\theta_1'(0, q^{2^n})}{\theta_2(0, q^{2^n})\theta_3(0, q^{2^n})\theta_4(0, q^{2^n})} \tag{1.5.5}$$

for all positive integers n.

Letting $n \to \infty$ in this identity, $q^{2^n} \to 0$ (since $|q| < 1$). Hence

$$\frac{\theta_1'(0, q)}{\theta_2(0, q)\theta_3(0, q)\theta_4(0, q)} = \lim_{q \to 0} \frac{\theta_1'(0, q)}{\theta_2(0, q)\theta_3(0, q)\theta_4(0, q)}. \tag{1.5.6}$$

Referring to equations (1.2.11)–(1.2.14), we find that

$$\theta_1'(0, q) = \sum (-1)^n (2n + 1) q^{(n + 1/2)^2} = 2q^{1/4} + O(q^{9/4}), \tag{1.5.7}$$

$$\theta_2(0, q) = \sum q^{(n + 1/2)^2} = 2q^{1/4} + O(q^{9/4}), \tag{1.5.8}$$

$$\theta_3(0, q) = \sum q^{n^2} = 1 + O(q), \tag{1.5.9}$$

$$\theta_4(0, q) = \sum (-1)^n q^{n^2} = 1 + O(q). \tag{1.5.10}$$

Thus, the limit in equation (1.5.6) is 1 and the identity

$$\theta_1'(0) = \theta_2(0)\theta_3(0)\theta_4(0) \tag{1.5.11}$$

has been established.

1.6. Theta Functions as Infinite Products

Consider the infinite product

$$F(t, q) = \prod_{n=1}^{\infty} (1 + q^{2n-1} t)(1 + q^{2n-1} t^{-1}), \tag{1.6.1}$$

where t is a nonzero, complex parameter and $|q| < 1$. Suppose t is confined to the bounded, closed domain $r \leqslant |t| \leqslant R$ and q to the bounded closed domain $|q| \leqslant \rho < 1$. Then we can demonstrate that the infinite product is absolutely and uniformly convergent by application of the M-test (see, e.g., E.T. Copson, *Theory of Functions of a Complex Variable*, pp. 104–5, Oxford University Press, 1935). Thus

$$|q^{2n-1}(t+t^{-1}) + q^{4n-2}| \leqslant |q|^{2n-1}\{|t| + |t|^{-1}\} + |q|^{4n-2}$$
$$\leqslant \rho^{2n-1}(R+1/r) + \rho^{4n-2} = M_n \qquad (1.6.2)$$

and the series $\sum M_n$ is convergent (being the sum of two geometric series with common ratios less than 1). Hence, the test shows that the product converges uniformly with respect to both t and q and therefore defines $F(t, q)$ as a regular function of t and q for all values of q with modulus less than 1 and all nonzero values of t.

Considered as a function of t, $F(t, q)$ has the following properties:

$$\text{(i)} \quad F(t^{-1}, q) = F(t, q), \qquad (1.6.3)$$

$$\text{(ii)} \quad qt F(q^2 t, q) = F(t, q). \qquad (1.6.4)$$

(i) is obvious from the definition (1.6.1). To prove (ii), observe that

$$F(q^2 t, q) = (1 + q^3 t)(1 + q^5 t) \cdots (1 + q^{-1} t^{-1})(1 + qt^{-1})(1 + q^3 t^{-1}) \cdots$$

$$= \frac{1 + q^{-1} t^{-1}}{1 + qt}(1 + qt)(1 + q^3 t) \cdots (1 + qt^{-1})(1 + q^3 t^{-1}) \cdots$$

$$= \frac{1}{qt} F(t, q), \qquad (1.6.5)$$

the rearrangement of factors in the infinite product being justified by its absolute convergence.

Since $F(t, q)$ is regular in t throughout any annular region $r < |t| < R$, it is expansible in a Laurent series about $t = 0$, thus:

$$F(t, q) = \sum_{n=-\infty}^{\infty} a_n t^n, \qquad (1.6.6)$$

the expansion being convergent for all values of t except $t = 0$. The property (1.6.3) implies that $a_n = a_{-n}$. The property (1.6.4) requires that

$$qt \Sigma a_n q^{2n} t^n = \Sigma a_n t^n. \qquad (1.6.7)$$

Equating coefficients of t^n ($n > 0$) in the two members of this identity, we have that

$$a_n = q^{2n-1} a_{n-1}. \qquad (1.6.8)$$

Repeated application of this result shows that

$$a_n = q^{2n-1} q^{2n-3} \cdots q^3 q a_0 = q^{n^2} a_0. \qquad (1.6.9)$$

We conclude that

$$F(t,q) = a_0 \Sigma q^{n^2} t^n. \tag{1.6.10}$$

Putting $t = -qe^{2iz}, qe^{2iz}, e^{2iz},$ and $-e^{2iz}$ in succession in the last identity and referring to equations (1.2.11)–(1.2.14) now leads immediately to the results

$$F(-qe^{2iz}, q) = ia_0 q^{-1/4} e^{-iz} \theta_1(z, q), \tag{1.6.11}$$

$$F(qe^{2iz}, q) = a_0 q^{-1/4} e^{-iz} \theta_2(z, q), \tag{1.6.12}$$

$$F(e^{2iz}, q) = a_0 \theta_3(z, q), \tag{1.6.13}$$

$$F(-e^{2iz}, q) = a_0 \theta_4(z, q), \tag{1.6.14}$$

valid for all values of z.

It remains to calculate a_0, which is independent of z and t but is a function of q. For $z = 0$ and real values of q, both $\theta_3(z, q)$ and $F(e^{2iz}, q)$ are positive; hence a_0 is real and positive.

Equation (1.6.11) can be rearranged into the form

$$a_0 q^{-1/4} \theta_1(z) = -ie^{iz} \prod_{n=1}^{\infty} (1 - q^{2n} e^{2iz})(1 - q^{2n-2} e^{-2iz}),$$

$$= -ie^{iz}(1 - e^{-2iz}) \prod_{n=1}^{\infty} (1 - q^{2n} e^{2iz})(1 - q^{2n} e^{-2iz}),$$

$$= 2 \sin z \prod_{n=1}^{\infty} (1 - 2q^{2n} \cos 2z + q^{4n}). \tag{1.6.15}$$

Differentiating with respect to z and then putting $z = 0$, we obtain

$$a_0 q^{-1/4} \theta_1'(0) = 2 \prod_{n=1}^{\infty} (1 - q^{2n})^2. \tag{1.6.16}$$

Equation (1.6.12) can be rearranged similarly into the form

$$a_0 q^{-1/4} \theta_2(z) = 2 \cos z \prod_{n=1}^{\infty} (1 + 2q^{2n} \cos 2z + q^{4n}). \tag{1.6.17}$$

Putting $z = 0$ yields

$$a_0 q^{-1/4} \theta_2(0) = 2 \prod_{n=1}^{\infty} (1 + q^{2n})^2. \tag{1.6.18}$$

Also, setting $z = 0$ in equations (1.6.13), (1.6.14), we conclude that

$$a_0 \theta_3(0) = \prod_{n=1}^{\infty} (1 + q^{2n-1})^2, \tag{1.6.19}$$

$$a_0 \theta_4(0) = \prod_{n=1}^{\infty} (1 - q^{2n-1})^2. \tag{1.6.20}$$

Substituting in the identity (1.5.11), we now find that

$$a_0^2 \prod_{n=1}^{\infty} (1 - q^{2n})^2 = \prod_{n=1}^{\infty} (1 + q^{2n})^2 (1 + q^{2n-1})^2 (1 - q^{2n-1})^2. \qquad (1.6.21)$$

Hence, since a_0 is positive for real q,

$$a_0 = + \prod_{n=1}^{\infty} (1 + q^{2n})(1 + q^{2n-1})(1 - q^{2n-1})/(1 - q^{2n}),$$

$$= \prod_{n=1}^{\infty} (1 + q^{2n})(1 + q^{2n-1})(1 - q^{2n})(1 - q^{2n-1})/(1 - q^{2n})^2,$$

$$= \prod_{n=1}^{\infty} (1 + q^{n})(1 - q^{n})/(1 - q^{2n})^2,$$

$$= 1 \bigg/ \prod_{n=1}^{\infty} (1 - q^{2n}). \qquad (1.6.22)$$

Equations (1.6.11)–(1.6.14) now yield expressions for the theta functions as infinite products, as follows:

$$\theta_1(z) = 2q^{1/4} \sin z \prod_{n=1}^{\infty} (1 - q^{2n})(1 - 2q^{2n} \cos 2z + q^{4n}), \qquad (1.6.23)$$

$$\theta_2(z) = 2q^{1/4} \cos z \prod_{n=1}^{\infty} (1 - q^{2n})(1 + 2q^{2n} \cos 2z + q^{4n}), \qquad (1.6.24)$$

$$\theta_3(z) = \prod_{n=1}^{\infty} (1 - q^{2n})(1 + 2q^{2n-1} \cos 2z + q^{4n-2}), \qquad (1.6.25)$$

$$\theta_4(z) = \prod_{n=1}^{\infty} (1 - q^{2n})(1 - 2q^{2n-1} \cos 2z + q^{4n-2}). \qquad (1.6.26)$$

By calculating the values of z for which the nth factor of each of these products vanishes, the zeros of the theta functions can be located; the reader may verify that these are in accord with the earlier results (1.3.10).

1.7. Jacobi's Transformation*

Consider a function $g(z)$ defined by a convergent series thus:

$$g(z) = \sum_{n=-\infty}^{\infty} f(z + 2n\pi). \qquad (1.7.1)$$

Clearly,

$$g(z + 2\pi) = g(z) \qquad (1.7.2)$$

---

* The reader who is unfamiliar with analytic function theory should pass over the details of this section until he is better prepared, noting meanwhile the transformations (1.7.12)–(1.7.15).

showing that $g(z)$ is periodic, with period 2π. We shall assume $f(z)$ to be regular throughout some strip $-Y \leqslant y \leqslant Y$, where $z = x + iy$, and the series to be uniformly convergent in this region. Then $g(z)$ is regular over the strip.

Hence, as proved in Appendix A, $g(z)$ can be expanded in a complex Fourier series thus:

$$g(z) = \sum_{r=-\infty}^{\infty} c_r e^{irz}, \tag{1.7.3}$$

where the coefficients c_r are given by

$$c_r = \frac{1}{2\pi} \int_0^{2\pi} g(t) e^{-irt} \, dt. \tag{1.7.4}$$

If we substitute for $g(t)$ from equation (1.7.1) into equation (1.7.4), we calculate that

$$c_r = \frac{1}{2\pi} \int_0^{2\pi} \sum_n f(t + 2n\pi) e^{-irt} \, dt = \frac{1}{2\pi} \sum_n \int_0^{2\pi} f(t + 2n\pi) e^{-irt} \, dt$$

$$= \frac{1}{2\pi} \sum_n \int_{2n\pi}^{2(n+1)\pi} f(u) e^{-ir(u - 2n\pi)} \, du = \frac{1}{2\pi} \sum_n \int_{2n\pi}^{2(n+1)\pi} f(u) e^{-iru} \, du$$

$$= \frac{1}{2\pi} \int_{-\infty}^{\infty} f(u) e^{-iru} \, du, \tag{1.7.5}$$

the inversion of the order of integration and summation being justified by the uniform convergence. The identity (1.7.3) is now seen to be equivalent to

$$\sum_{n=-\infty}^{\infty} f(z + 2n\pi) = \frac{1}{2\pi} \sum_{r=-\infty}^{\infty} e^{irz} \int_{-\infty}^{\infty} f(u) e^{-iru} \, du. \tag{1.7.6}$$

We shall now apply this identity to the series defining $\theta_3(z|\tau)$. This series has been shown to be uniformly convergent in the strip defined above, in section 1.2. First, we note that

$$\theta_3(z|\tau) = \sum_{n=-\infty}^{\infty} \exp(in^2\pi\tau + 2inz)$$

$$= \exp(z^2/i\pi\tau) \sum_{n=-\infty}^{\infty} \exp\left[\frac{i\tau}{4\pi}\left(\frac{2z}{\tau} + 2n\pi\right)^2\right]$$

$$= \exp(z^2/i\pi\tau) \sum_{n=-\infty}^{\infty} \exp\left[\frac{i\tau}{4\pi}(w + 2n\pi)^2\right], \tag{1.7.7}$$

where $w = 2z/\tau$. With $f(w) = \exp(i\tau w^2/4\pi)$, the identity (1.7.6) now shows that

$$\theta_3(z|\tau) = \frac{1}{2\pi} e^{z^2/i\pi\tau} \sum_{r=-\infty}^{\infty} e^{irw} \int_{-\infty}^{\infty} \exp\left(\frac{i\tau u^2}{4\pi} - iru\right) du. \tag{1.7.8}$$

But, using the integral calculated in Appendix B (with $a = (-i\tau/4\pi)^{1/2}$, $b = ir$), we find that

$$\int_{-\infty}^{\infty} \exp\left(\frac{i\tau u^2}{4\pi} - iru\right) du = 2\pi(-i\tau)^{-1/2} e^{-i\pi r^2/\tau}, \qquad (1.7.9)$$

where $(-i\tau)^{1/2}$ is to be taken with its argument in the interval $(-\frac{1}{4}\pi, \frac{1}{4}\pi)$ (which is possible, since the real part of $-i\tau$ is positive). It now follows that

$$\theta_3(z|\tau) = (-i\tau)^{-1/2} e^{z^2/i\pi\tau} \sum_{r=-\infty}^{\infty} \exp(-ir^2\pi\tau^{-1} + 2irz\tau^{-1})$$

$$= (-i\tau)^{-1/2} e^{z^2/i\pi\tau} \theta_3(\tau^{-1}z| - \tau^{-1}). \qquad (1.7.10)$$

Thus, writing

$$\tau' = -1/\tau, \qquad (1.7.11)$$

and replacing z by τz, the identity (1.7.10) is seen to be equivalent to the result

$$\theta_3(z|\tau') = (-i\tau)^{1/2} e^{i\tau z^2/\pi} \theta_3(\tau z|\tau). \qquad (1.7.12)$$

This is *Jacobi's transformation* for θ_3.

Incrementing z by $\frac{1}{2}\pi$ and referring to the identities (1.3.7), (1.3.9), it will be found that Jacobi's transformation of θ_3 leads to the further transformation equation

$$\theta_4(z|\tau') = (-i\tau)^{1/2} e^{i\tau z^2/\pi} \theta_2(\tau z|\tau). \qquad (1.7.13)$$

A transformation for θ_1 is next found by incrementing z by $\frac{1}{2}\pi\tau'$ in equation (1.7.13). The result is

$$\theta_1(z|\tau') = -i(-i\tau)^{1/2} e^{i\tau z^2/\pi} \theta_1(\tau z|\tau). \qquad (1.7.14)$$

The set of transformation equations is now completed by incrementing z by $\frac{1}{2}\pi$ in this last result to give

$$\theta_2(z|\tau') = (-i\tau)^{1/2} e^{i\tau z^2/\pi} \theta_4(\tau z|\tau). \qquad (1.7.15)$$

1.8. Landen's Transformation

This relates theta functions with parameter 2τ to theta functions with parameter τ, i.e., functions with nome q^2 to those with nome q.

Putting $y = x$ in the identities (1.4.8), (1.4.10)–(1.4.12), we first establish that

$$\theta_1(x|\tau)\theta_2(x|\tau) = \theta_1(2x|2\tau)\theta_4(0|2\tau), \qquad (1.8.1)$$

$$\theta_3^2(x|\tau) = \theta_3(2x|2\tau)\theta_3(0|2\tau) + \theta_2(2x|2\tau)\theta_2(0|2\tau), \qquad (1.8.2)$$

$$\theta_3(x|\tau)\theta_4(x|\tau) = \theta_4(2x|2\tau)\theta_4(0|2\tau), \qquad (1.8.3)$$

$$\theta_4^2(x|\tau) = \theta_3(2x|2\tau)\theta_3(0|2\tau) - \theta_2(2x|2\tau)\theta_2(0|2\tau). \qquad (1.8.4)$$

Then, putting $x = 0$ into equations (1.8.2)–(1.8.4) and solving, we find that

$$\theta_2^2(0|2\tau) = \tfrac{1}{2}[\theta_3^2(0|\tau) - \theta_4^2(0|\tau)], \tag{1.8.5}$$

$$\theta_3^2(0|2\tau) = \tfrac{1}{2}[\theta_3^2(0|\tau) + \theta_4^2(0|\tau)], \tag{1.8.6}$$

$$\theta_4^2(0|2\tau) = \theta_3(0|\tau)\theta_4(0|\tau). \tag{1.8.7}$$

Equations (1.8.1)–(1.8.4) can now be solved for theta functions with parameter 2τ in terms of functions with parameter τ, thus:

$$\theta_1(2x|2\tau) = \frac{\theta_1(x|\tau)\theta_2(x|\tau)}{\sqrt{[\theta_3(0|\tau)\theta_4(0|\tau)]}}, \tag{1.8.8}$$

$$\theta_2(2x|2\tau) = \frac{\theta_3^2(x|\tau) - \theta_4^2(x|\tau)}{\sqrt{[2\{\theta_3^2(0|\tau) - \theta_4^2(0|\tau)\}]}}, \tag{1.8.9}$$

$$\theta_3(2x|2\tau) = \frac{\theta_3^2(x|\tau) + \theta_4^2(x|\tau)}{\sqrt{[2\{\theta_3^2(0|\tau) + \theta_4^2(0|\tau)\}]}}, \tag{1.8.10}$$

$$\theta_4(2x|2\tau) = \frac{\theta_3(x|\tau)\theta_4(x|\tau)}{\sqrt{[\theta_3(0|\tau)\theta_4(0|\tau)]}}. \tag{1.8.11}$$

These are the *Landen transformation equations* for theta functions.

Incrementing x by $\tfrac{1}{2}\pi$ in the identities (1.8.8) and (1.8.11), we obtain alternative forms for the transformation equations (1.8.9), (1.8.10), viz.

$$\theta_2(2x|2\tau) = \frac{\theta_1(\tfrac{1}{4}\pi + x)\theta_1(\tfrac{1}{4}\pi - x)}{\sqrt{(\theta_3\theta_4)}} = \frac{\theta_2(\tfrac{1}{4}\pi + x)\theta_2(\tfrac{1}{4}\pi - x)}{\sqrt{(\theta_3\theta_4)}}, \tag{1.8.12}$$

$$\theta_3(2x|2\tau) = \frac{\theta_3(\tfrac{1}{4}\pi + x)\theta_3(\tfrac{1}{4}\pi - x)}{\sqrt{(\theta_3\theta_4)}} = \frac{\theta_4(\tfrac{1}{4}\pi + x)\theta_4(\tfrac{1}{4}\pi - x)}{\sqrt{(\theta_3\theta_4)}}, \tag{1.8.13}$$

where the theta functions on the right-hand side all have parameter τ.

Putting $x = 0$ in these identities, we find that

$$\theta_2(0|2\tau) = \theta_1^2(\tfrac{1}{4}\pi)/\sqrt{(\theta_3\theta_4)} = \theta_2^2(\tfrac{1}{4}\pi)/\sqrt{(\theta_3\theta_4)}, \tag{1.8.14}$$

$$\theta_3(0|2\tau) = \theta_3^2(\tfrac{1}{4}\pi)/\sqrt{(\theta_3\theta_4)} = \theta_4^2(\tfrac{1}{4}\pi)/\sqrt{(\theta_3\theta_4)}, \tag{1.8.15}$$

which are alternative forms of the results (1.8.5) and (1.8.6).

1.9. Derivatives of Ratios of the Theta Functions

Differentiating the identity (1.4.38) partially with respect to y yields the result

$$[\theta_1'(x + y)\theta_4(x - y) - \theta_1(x + y)\theta_4'(x - y)]\theta_2(0)\theta_3(0)$$
$$= \theta_1(x)\theta_4(x)[\theta_2'(y)\theta_3(y) + \theta_2(y)\theta_3'(y)] + \theta_2(x)\theta_3(x)[\theta_1'(y)\theta_4(y) + \theta_1(y)\theta_4'(y)]. \tag{1.9.1}$$

Then, putting $y = 0$ and rearranging, we get

$$\frac{d}{dx}(\theta_1/\theta_4) = \frac{\theta_4(x)\theta_1'(x) - \theta_1(x)\theta_4'(x)}{\theta_4^2(x)} = \frac{\theta_1'(0)\theta_4(0)}{\theta_2(0)\theta_3(0)} \cdot \frac{\theta_2(x)\theta_3(x)}{\theta_4^2(x)}, \tag{1.9.2}$$

recalling that $\theta_1(0) = \theta_2'(0) = \theta_3'(0) = \theta_4'(0) = 0$. Use of the identity (1.5.11) reduces this result to the form

$$\frac{d}{dx}(\theta_1/\theta_4) = \theta_4^2(0)\theta_2(x)\theta_3(x)/\theta_4^2(x). \tag{1.9.3}$$

The derivatives of the remaining ratios of the theta functions can be calculated in the same way from similar identities established in section 1.4, or by making these depend on the result (1.9.3) through the identities (1.4.49)–(1.4.52). Thus, writing (1.4.51) in the form

$$\left[\frac{\theta_1(x)}{\theta_4(x)}\right]^2 \theta_3^2(0) + \left[\frac{\theta_2(x)}{\theta_4(x)}\right]^2 \theta_4^2(0) = \theta_2^2(0), \tag{1.9.4}$$

and differentiating, we obtain

$$\left[\frac{\theta_1(x)}{\theta_4(x)}\right]\frac{d}{dx}\left(\frac{\theta_1}{\theta_4}\right)\theta_3^2(0) + \left[\frac{\theta_2(x)}{\theta_4(x)}\right]\frac{d}{dx}\left(\frac{\theta_2}{\theta_4}\right)\theta_4^2(0) = 0. \tag{1.9.5}$$

Substitution from (1.9.3) now proves that

$$\frac{d}{dx}(\theta_2/\theta_4) = -\theta_3^2(0)\theta_1(x)\theta_3(x)/\theta_4^2(x). \tag{1.9.6}$$

Using either method (or $\theta_1/\theta_3 = \theta_1/\theta_4 \div \theta_3/\theta_4$, etc.), the reader should establish the following identites:

$$\frac{d}{dx}(\theta_3/\theta_4) = -\theta_2^2(0)\theta_1(x)\theta_2(x)/\theta_4^2(x), \tag{1.9.7}$$

$$\frac{d}{dx}(\theta_1/\theta_3) = \theta_3^2(0)\theta_2(x)\theta_4(x)/\theta_3^2(x), \tag{1.9.8}$$

$$\frac{d}{dx}(\theta_2/\theta_3) = -\theta_4^2(0)\theta_1(x)\theta_4(x)/\theta_3^2(x), \tag{1.9.9}$$

$$\frac{d}{dx}(\theta_1/\theta_2) = \theta_2^2(0)\theta_3(x)\theta_4(x)/\theta_2^2(x). \tag{1.9.10}$$

Six other identities involving the reciprocal ratios now follow immediately. These are listed below:

$$\frac{d}{dx}(\theta_4/\theta_1) = -\theta_4^2(0)\theta_2(x)\theta_3(x)/\theta_1^2(x), \tag{1.9.11}$$

$$\frac{d}{dx}(\theta_4/\theta_2) = \theta_3^2(0)\theta_1(x)\theta_3(x)/\theta_2^2(x), \tag{1.9.12}$$

$$\frac{d}{dx}(\theta_4/\theta_3) = \theta_2^2(0)\theta_1(x)\theta_2(x)/\theta_3^2(x), \qquad (1.9.13)$$

$$\frac{d}{dx}(\theta_3/\theta_1) = -\theta_3^2(0)\theta_2(x)\theta_4(x)/\theta_1^2(x), \qquad (1.9.14)$$

$$\frac{d}{dx}(\theta_3/\theta_2) = \theta_4^2(0)\theta_1(x)\theta_4(x)/\theta_2^2(x), \qquad (1.9.15)$$

$$\frac{d}{dx}(\theta_2/\theta_1) = -\theta_2^2(0)\theta_3(x)\theta_4(x)/\theta_1^2(x). \qquad (1.9.16)$$

By a further application of the identities (1.4.49)–(1.4.52), the first-order differential equations satisfied by these ratios can be found. Thus,

$$\theta_4^2(0)\frac{\theta_2^2(x)}{\theta_4^2(x)} = \theta_2^2(0) - \theta_3^2(0)\frac{\theta_1^2(x)}{\theta_4^2(x)}, \qquad (1.9.17)$$

$$\theta_4^2(0)\frac{\theta_3^2(x)}{\theta_4^2(x)} = \theta_3^2(0) - \theta_2^2(0)\frac{\theta_1^2(x)}{\theta_4^2(x)}. \qquad (1.9.18)$$

Hence, substituting in the right-hand member of equation (1.9.3) and putting $\xi = \theta_1(x)/\theta_4(x)$, we find that

$$\left(\frac{d\xi}{dx}\right)^2 = [\theta_2^2(0) - \theta_3^2(0)\xi^2][\theta_3^2(0) - \theta_2^2(0)\xi^2]. \qquad (1.9.19)$$

This is the differential equation satisfied by θ_1/θ_4. The general solution must be $\theta_1(x + \alpha)/\theta_4(x + \alpha)$.

The calculation of the differential equations satisfied by the other ratios is left as an exercise for the reader (see Exercise 12).

EXERCISES

1. Show that the periodicity factors associated with the quasi-period $m\pi + n\pi\tau$ of the theta functions $\theta_r(z)$ $(r = 1, 2, 3, 4)$ are $(-1)^{m+n}\Omega$, $(-1)^m\Omega, \Omega, (-1)^n\Omega$ respectively, where $\Omega = (q^{n^2}e^{2inz})^{-1}$.

2. Prove that

   (i) $\theta_3(z|\tau) = \theta_3(2z|4\tau) + \theta_2(2z|4\tau),$

   (ii) $\theta_4(z|\tau) = \theta_3(2z|4\tau) - \theta_2(2z|4\tau).$

3. Prove that

   (i) $\theta_1(z, q) = 2\sum_{-\infty}^{\infty} q^{(1/4)(4n+1)^2}\sin(4n+1)z = -2\sum_{-\infty}^{\infty} q^{(1/4)(4n-1)^2}\sin(4n-1)z,$

   (ii) $\theta_2(z, q) = 2\sum_{-\infty}^{\infty} q^{(1/4)(4n+1)^2}\cos(4n+1)z = 2\sum_{-\infty}^{\infty} q^{(1/4)(4n-1)^2}\cos(4n-1)z.$

4. Prove that

$$\theta_1(x-y)\theta_1(x+y)\theta_1(z-w)\theta_1(z+w) + \theta_1(z-x)\theta_1(z+x)\theta_1(y-w)\theta_1(y+w)$$
$$+ \theta_1(y-z)\theta_1(y+z)\theta_1(x-w)\theta_1(x+w) = 0.$$

5. If $2s = x + y + z + w$, prove that

$$\theta_2(s-x)\theta_2(s-y)\theta_2(s-z)\theta_2(s-w) + \theta_3(s-x)\theta_3(s-y)\theta_3(s-z)\theta_3(s-w)$$
$$= \theta_2(x)\theta_2(y)\theta_2(z)\theta_2(w) + \theta_3(x)\theta_3(y)\theta_3(z)\theta_3(w),$$
$$\theta_2(s-x)\theta_2(s-y)\theta_2(s-z)\theta_2(s-w) - \theta_3(s-x)\theta_3(s-y)\theta_3(s-z)\theta_3(s-w)$$
$$= \theta_1(x)\theta_1(y)\theta_1(z)\theta_1(w) - \theta_4(x)\theta_4(y)\theta_4(z)\theta_4(w).$$

Deduce that

$$2\theta_2(s-x)\theta_2(s-y)\theta_2(s-z)\theta_2(s-w) = \theta_1(x)\theta_1(y)\theta_1(z)\theta_1(w) + \theta_2(x)\theta_2(y)\theta_2(z)\theta_2(w)$$
$$+ \theta_3(x)\theta_3(y)\theta_3(z)\theta_3(w) - \theta_4(x)\theta_4(y)\theta_4(z)\theta_4(w)$$
$$2\theta_3(s-x)\theta_3(s-y)\theta_3(s-z)\theta_3(s-w) = -\theta_1(x)\theta_1(y)\theta_1(z)\theta_1(w) + \theta_2(x)\theta_2(y)\theta_2(z)\theta_2(w)$$
$$+ \theta_3(x)\theta_3(y)\theta_3(z)\theta_3(w) + \theta_4(x)\theta_4(y)\theta_4(z)\theta_4(w).$$

Incrementing x, y, z and w, all by $\frac{1}{2}\pi$, establish the identities

$$2\theta_1(s-x)\theta_1(s-y)\theta_1(s-z)\theta_1(s-w) = \theta_1(x)\theta_1(y)\theta_1(z)\theta_1(w) + \theta_2(x)\theta_2(y)\theta_2(z)\theta_2(w)$$
$$- \theta_3(x)\theta_3(y)\theta_3(z)\theta_3(w) + \theta_4(x)\theta_4(y)\theta_4(z)\theta_4(w)$$
$$2\theta_4(s-x)\theta_4(s-y)\theta_4(s-z)\theta_4(s-w) = \theta_1(x)\theta_1(y)\theta_1(z)\theta_1(w) - \theta_2(x)\theta_2(y)\theta_2(z)\theta_2(w)$$
$$+ \theta_3(x)\theta_3(y)\theta_3(z)\theta_3(w) + \theta_4(x)\theta_4(y)\theta_4(z)\theta_4(w).$$

6. Prove that

$$\theta_1^4(x) + \theta_3^4(x) = \theta_2^4(x) + \theta_4^4(x).$$

7. Incrementing both x and y in the identity (1.4.7) by $\frac{1}{2}\pi + \frac{1}{2}\pi\tau$, obtain the identity (1.4.10).

8. Prove that

$$\frac{\theta'(z)}{\theta(z)} - \frac{\theta'(z+\pi\tau)}{\theta(z+\pi\tau)} = 2i, \qquad \frac{1}{4}\pi i \frac{\partial^2 \theta}{\partial z^2} + \frac{\partial \theta}{\partial \tau} = 0,$$

for all the theta functions.

9. Prove that $\theta_1'(0) = 2q^{1/4}\prod_{n=1}^{\infty}(1-q^{2n})^3$.

10. Obtain the duplication formulae:

$$\theta_1(2x)\theta_2(0)\theta_3(0)\theta_4(0) = 2\theta_1(x)\theta_2(x)\theta_3(x)\theta_4(x),$$
$$\theta_2(2x)\theta_2(0)\theta_4^2(0) = \theta_2^2(x)\theta_4^2(x) - \theta_1^2(x)\theta_3^2(x),$$
$$\theta_3(2x)\theta_3(0)\theta_4^2(0) = \theta_3^2(x)\theta_4^2(x) - \theta_1^2(x)\theta_2^2(x),$$
$$\theta_4(2x)\theta_4^3(0) = \theta_3^4(x) - \theta_2^4(x) = \theta_4^4(x) - \theta_1^4(x).$$

Deduce that $\theta_2(\frac{1}{3}\pi)\theta_3(\frac{1}{3}\pi)\theta_4(\frac{1}{3}\pi) = \frac{1}{2}\theta_2(0)\theta_3(0)\theta_4(0).$

11. Prove the identity

$$\left[\prod_{n=1}^{\infty}(1-q^{2n-1})\right]^8 + 16q\left[\prod_{n=1}^{\infty}(1+q^{2n})\right]^8 = \left[\prod_{n=1}^{\infty}(1+q^{2n-1})\right]^8.$$

12. If $\eta = \theta_4(x)/\theta_1(x)$, $\zeta = \theta_1(x)/\theta_2(x)$, prove that

$$\left(\frac{d\eta}{dx}\right)^2 = [\theta_3^2(0) - \eta^2\theta_2^2(0)][\theta_2^2(0) - \eta^2\theta_3^2(0)],$$

$$\left(\frac{d\zeta}{dx}\right)^2 = [\theta_3^2(0) + \zeta^2\theta_4^2(0)][\theta_4^2(0) + \zeta^2\theta_3^2(0)].$$

13. If $\theta_1(z) = \phi(z)\sin z$, express $\phi(z)$ as an infinite product and by logarithmic differentiation prove that

$$\phi''(0)/\phi(0) = 8 \sum_{n=1}^{\infty} \frac{q^{2n}}{(1-q^{2n})^2}.$$

Hence prove that

$$\theta_1'''(0)/\theta_1'(0) = 24 \sum_{n=1}^{\infty} \frac{q^{2n}}{(1-q^{2n})^2} - 1.$$

14. By logarithmic differentiation of the products for $\theta_2(z), \theta_3(z), \theta_4(z)$, obtain the results

(i) $\theta_2''(0)/\theta_2(0) = -1 - 8 \sum_{n=1}^{\infty} \frac{q^{2n}}{(1+q^{2n})^2},$

(ii) $\theta_3''(0)/\theta_3(0) = -8 \sum_{n=1}^{\infty} \frac{q^{2n-1}}{(1+q^{2n-1})^2},$

(iii) $\theta_4''(0)/\theta_4(0) = 8 \sum_{n=1}^{\infty} \frac{q^{2n-1}}{(1-q^{2n-1})^2}.$

15. From the results of the last two exercises, establish that

$$\theta_1'''(0)/\theta_1'(0) = \theta_2''(0)/\theta_2(0) + \theta_3''(0)/\theta_3(0) + \theta_4''(0)/\theta_4(0).$$

16. If the argument of all the theta functions is zero, prove that

(i) $\dfrac{\theta_4''}{\theta_4} - \dfrac{\theta_3''}{\theta_3} = \theta_2^4,$

(ii) $\dfrac{\theta_4''}{\theta_4} - \dfrac{\theta_2''}{\theta_2} = \theta_3^4,$

(iii) $\dfrac{\theta_3''}{\theta_3} - \dfrac{\theta_2''}{\theta_2} = \theta_4^4.$

17. By logarithmic differentiation of the identities (1.6.23)–(1.6.26), prove that

$$\frac{\theta_1'(z)}{\theta_1(z)} = \cot z + 4\sin 2z \sum_{n=1}^{\infty} \frac{q^{2n}}{1 - 2q^{2n}\cos 2z + q^{4n}}$$

$$\frac{\theta_2'(z)}{\theta_2(z)} = -\tan z - 4\sin 2z \sum_{n=1}^{\infty} \frac{q^{2n}}{1 + 2q^{2n}\cos 2z + q^{4n}}$$

$$\frac{\theta_3'(z)}{\theta_3(z)} = -4\sin 2z \sum_{n=1}^{\infty} \frac{q^{2n-1}}{1+2q^{2n-1}\cos 2z + q^{4n-2}}$$

$$\frac{\theta_4'(z)}{\theta_4(z)} = 4\sin 2z \sum_{n=1}^{\infty} \frac{q^{2n-1}}{1-2q^{2n-1}\cos 2z + q^{4n-2}}.$$

18. If $f(\sigma) = \sum_{n=0}^{\infty}(2n+1)^2 e^{-n(n+1)\sigma}$, prove that $\sigma^{3/2}f(\sigma) \to \pi^{1/2}$ as $\sigma \to 0$. (Hint: Relate $f(\sigma)$ to θ_2 and use Jacobi's transformation.)

19. An infinite heat-conducting slab occupies the region between the planes $x = \pm a$. At $t = 0$, the temperature u at a point (x, y, z) within the slab is given by $u = T|x|$. If the surfaces of the slab are insulated, show that the temperature at any later time t is given by

$$u = \tfrac{1}{2}aT\left[1 - \frac{1}{\pi^2}\int_0^{\alpha} q^{-1}\theta_2(z,q)\,dq \right]$$

where $z = \pi x/a$ and $\alpha = \exp(-4\kappa\pi^2 t/a^2)$. (Assume u satisfies the heat conduction equation (1.1.2) and that $\partial u/\partial x$ vanishes at the slab's surfaces.)

20. Prove that

$$\sum_{n=-\infty}^{\infty} \exp(-n^2 t) = \sqrt{(\pi/t)} \sum_{n=-\infty}^{\infty} \exp(-n^2\pi^2/t).$$

Taking Laplace transforms, derive the identity

$$\coth x = \frac{1}{x} + \sum_{n=1}^{\infty} \frac{2x}{x^2 + n^2\pi^2}.$$

21. Obtain the fourth-order Landen transformation equations

$$\theta_1(4x|4\tau) = A\theta_1(x)\theta_1(\tfrac{1}{4}\pi - x)\theta_1(\tfrac{1}{4}\pi + x)\theta_2(x),$$
$$\theta_2(4x|4\tau) = A\theta_2(\tfrac{1}{8}\pi - x)\theta_2(\tfrac{1}{8}\pi + x)\theta_2(\tfrac{3}{8}\pi - x)\theta_2(\tfrac{3}{8}\pi + x),$$
$$\theta_3(4x|4\tau) = A\theta_3(\tfrac{1}{8}\pi - x)\theta_3(\tfrac{1}{8}\pi + x)\theta_3(\tfrac{3}{8}\pi - x)\theta_3(\tfrac{3}{8}\pi + x),$$
$$\theta_4(4x|4\tau) = A\theta_4(x)\theta_4(\tfrac{1}{4}\pi - x)\theta_4(\tfrac{1}{4}\pi + x)\theta_3(x),$$

where $A = 1/(\theta_3\theta_4\theta_3(\tfrac{1}{4}\pi))$.

CHAPTER 2

Jacobi's Elliptic Functions

2.1. Definitions of Jacobi's Elliptic Functions

The elliptic functions $\operatorname{sn} u, \operatorname{cn} u$, and $\operatorname{dn} u$ are defined as ratios of theta functions as below:

$$\operatorname{sn} u = \frac{\theta_3(0)}{\theta_2(0)} \cdot \frac{\theta_1(z)}{\theta_4(z)}, \tag{2.1.1}$$

$$\operatorname{cn} u = \frac{\theta_4(0)}{\theta_2(0)} \cdot \frac{\theta_2(z)}{\theta_4(z)}, \tag{2.1.2}$$

$$\operatorname{dn} u = \frac{\theta_4(0)}{\theta_3(0)} \cdot \frac{\theta_3(z)}{\theta_4(z)}, \tag{2.1.3}$$

where $z = u/\theta_3^2(0)$. $\operatorname{sn} u$ is read as "es en yew" or as "san yew"; $\operatorname{cn} u$ and $\operatorname{dn} u$ can similarly be read letter by letter or as "can u" and "dan u" respectively.

If the parameter τ is purely imaginary (q real), the elliptic functions are all real for real values of u. The numerical multipliers have been chosen so that $\operatorname{cn} 0 = \operatorname{dn} 0 = 1$.

Squaring and adding equations (2.1.1), (2.1.2), and referring to the identity (1.4.51), we obtain the identity

$$\operatorname{sn}^2 u + \operatorname{cn}^2 u = 1. \tag{2.1.4}$$

This is the counterpart for the elliptic sine and cosine of a familiar trigonometric identity and explains the presence of the numerical factor $\theta_3(0)/\theta_2(0)$ in the definition of $\operatorname{sn} u$. Similarly, the identities (1.4.50) and (1.4.49) are found to be equivalent to the relationships

$$\operatorname{dn}^2 u + \{\theta_2^4(0)/\theta_3^4(0)\} \operatorname{sn}^2 u = 1, \tag{2.1.5}$$

$$\mathrm{dn}^2 u - \{\theta_2^4(0)/\theta_3^4(0)\}\, \mathrm{cn}^2 u = \theta_4^4(0)/\theta_3^4(0). \tag{2.1.6}$$

Defining parameters k and k' by the equations

$$k = \theta_2^2(0)/\theta_3^2(0), \qquad k' = \theta_4^2(0)/\theta_3^2(0), \tag{2.1.7}$$

the identity (1.4.53) shows that

$$k^2 + k'^2 = 1. \tag{2.1.8}$$

k is termed the *modulus* of the elliptic functions and k' is the *complementary modulus*. The identities (2.1.4)–(2.1.6) can now be rewritten in the forms

$$\mathrm{sn}^2 u + \mathrm{cn}^2 u = 1, \tag{2.1.9}$$

$$\mathrm{dn}^2 u + k^2\, \mathrm{sn}^2 u = 1, \tag{2.1.10}$$

$$\mathrm{dn}^2 u - k^2\, \mathrm{cn}^2 u = k'^2. \tag{2.1.11}$$

Like the identities (1.4.49)–(1.4.51) from which they are derived, these equations are not linearly independent (as is easily verified).

k and k' are functions of the nome q; using equations (1.2.12)–(1.2.14) and (1.6.24)–(1.6.26), we find that

$$k = 4q^{1/2}\left(\frac{1 + q^2 + q^6 + q^{12} + \cdots}{1 + 2q + 2q^4 + 2q^9 + \cdots}\right)^2 = 4q^{1/2}\prod_{n=1}^{\infty}\left(\frac{1 + q^{2n}}{1 + q^{2n-1}}\right)^4, \tag{2.1.12}$$

$$k' = \left(\frac{1 - 2q + 2q^4 - 2q^9 + \cdots}{1 + 2q + 2q^4 + 2q^9 + \cdots}\right)^2 = \prod_{n=1}^{\infty}\left(\frac{1 - q^{2n-1}}{1 + q^{2n-1}}\right)^4. \tag{2.1.13}$$

If q is real and $0 \leqslant q < 1$, then $0 \leqslant k < 1, 0 < k' \leqslant 1$. When working with elliptic functions, the modulus and complementary modulus are regarded as given quantities and then q is to be derived from the last pair of equations. As q increases from 0 to 1, each factor of the product (2.1.13) steadily decreases from 1 to 0 and it follows that k' decreases monotonically from 1 to 0. Thus, given values of k and k' in the interval $(0, 1)$, there will be a unique value of q in the same interval satisfying the equations (2.1.12) and (2.1.13). More generally, it will be shown (equation (2.2.3)) that $q = \exp(-\pi K'/K)$ and that K, K' can be defined as analytic functions of k (section 8.12). Hence, for any complex value of k, at least one value of q can be found to satisfy equation (2.1.12); thus, elliptic functions to any modulus can be represented by theta functions. q and τ are found to be multivalued functions of k (see section 2.3) and it would appear at first sight, therefore, that the Jacobi functions, as defined above, would also be multivalued functions of k. This is not the case, however, as will be established in section 2.5. Values of q corresponding to values of $k^2 = 0(0.1)1$ are set out in Table B (p. 278).

When it is required to state the modulus explicitly, the elliptic functions of Jacobi will be written $\mathrm{sn}(u, k)$, $\mathrm{cn}(u, k)$, and $\mathrm{dn}(u, k)$.

Equation (2.1.12) shows that $k \to 0$ as $q \to 0$ (and, therefore, $k' \to 1$). But, for small values of q, it follows from equations (1.2.11)–(1.2.14) that

$$\theta_1(z) = 2q^{1/4}\sin z + O(q^{9/4}), \tag{2.1.14}$$

$$\theta_2(z) = 2q^{1/4}\cos z + O(q^{9/4}), \tag{2.1.15}$$

$$\theta_3(z) = 1 + O(q), \tag{2.1.16}$$

$$\theta_4(z) = 1 + O(q). \tag{2.1.17}$$

From equations (2.1.1)–(2.1.3), we therefore deduce that as $k \to 0$

$$\operatorname{sn}(u, k) \to \sin u, \tag{2.1.18}$$

$$\operatorname{cn}(u, k) \to \cos u, \tag{2.1.19}$$

$$\operatorname{dn}(u, k) \to 1. \tag{2.1.20}$$

This explains the notation which has been adopted for sn and cn.

Nine other elliptic functions are defined by taking reciprocals and quotients; the notation will be clear from the definitions below:

$$
\left.
\begin{aligned}
&\operatorname{ns} u = 1/\operatorname{sn} u, \quad\quad \operatorname{nc} u = 1/\operatorname{cn} u, \quad\quad \operatorname{nd} u = 1/\operatorname{dn} u, \\[2mm]
&\operatorname{sc} u = \frac{\operatorname{sn} u}{\operatorname{cn} u}, \quad\quad \operatorname{cd} u = \frac{\operatorname{cn} u}{\operatorname{dn} u}, \quad\quad \operatorname{ds} u = \frac{\operatorname{dn} u}{\operatorname{sn} u}, \\[2mm]
&\operatorname{cs} u = \frac{\operatorname{cn} u}{\operatorname{sn} u}, \quad\quad \operatorname{dc} u = \frac{\operatorname{dn} u}{\operatorname{cn} u}, \quad\quad \operatorname{sd} u = \frac{\operatorname{sn} u}{\operatorname{dn} u}.
\end{aligned}
\right\} \tag{2.1.21}
$$

ns u is pronounced "en es yew" or "nas yew," etc. Clearly, $\operatorname{sc}(u, k) \to \tan u$ as $k \to 0$ and the alternative notation $\operatorname{sn} u/\operatorname{cn} u = \operatorname{tn} u$ is sometimes used, therefore.

Since the theta functions are integral functions, the elliptic functions $\operatorname{sn} u$, $\operatorname{cn} u$, and $\operatorname{dn} u$ are regular for all values of the complex variable u, except those for which $\theta_4(z)$ vanishes, viz.

$$u = \theta_3^2(0)[m\pi + (n + \tfrac{1}{2})\pi\tau]. \tag{2.1.22}$$

At all points of this lattice in the u-plane, the elliptic functions $\operatorname{sn} u$, $\operatorname{cn} u$, and $\operatorname{dn} u$ have simple poles, whose residues will be calculated later (section 2.8).

The zeros of the elliptic functions are determined by the zeros of the theta functions $\theta_1, \theta_2, \theta_3$. Referring to (1.3.10), we note that

$$
\left.
\begin{aligned}
&\operatorname{sn} u = 0 \quad \text{where } u = \theta_3^2(0)(m\pi + n\pi\tau), \\
&\operatorname{cn} u = 0 \quad \text{where } u = \theta_3^2(0)\{(m + \tfrac{1}{2})\pi + n\pi\tau\}, \\
&\operatorname{dn} u = 0 \quad \text{where } u = \theta_3^2(0)\{(m + \tfrac{1}{2})\pi + (n + \tfrac{1}{2})\pi\tau\}.
\end{aligned}
\right\} \tag{2.1.23}
$$

2.2. Double Periodicity of the Elliptic Functions

It has been shown in section 1.3 that the theta functions each have one period and one quasi-period. However, the periodicity factors associated with the quasi-periods are found to cancel in equations (2.1.1)–(2.1.3), with the result that the elliptic functions have two distinct periods.

Thus, using the identities (1.3.2)–(1.3.5), we see that

$$\operatorname{sn}(u + 2\pi\theta_3^2(0)) = \operatorname{sn} u, \tag{2.2.1}$$

$$\operatorname{sn}(u + \pi\tau\theta_3^2(0)) = \operatorname{sn} u. \tag{2.2.2}$$

$\operatorname{sn} u$ accordingly has two periods $2\pi\theta_3^2(0)$ and $\pi\tau\theta_3^2(0)$ whose ratio $\frac{1}{2}\tau$ must be complex (with positive imaginary part). If τ is purely imaginary (q real), the first period is real and the second is purely imaginary. In all circumstances, we shall write

$$K = \tfrac{1}{2}\pi\theta_3^2(0), \qquad iK' = \tfrac{1}{2}\pi\tau\theta_3^2(0) = \tau K, \tag{2.2.3}$$

so that, when τ is purely imaginary, K and K' are both real and positive. The periods of $\operatorname{sn} u$ are then seen to be $4K$ and $2iK'$, i.e.,

$$\operatorname{sn}(u + 4K) = \operatorname{sn}(u + 2iK') = \operatorname{sn} u. \tag{2.2.4}$$

The reader should now verify that

$$\operatorname{cn}(u + 4K) = \operatorname{cn}(u + 2K + 2iK') = \operatorname{cn} u. \tag{2.2.5}$$

Thus, $\operatorname{cn} u$ has periods $4K$ and $2K + 2iK'$. Also

$$\operatorname{dn}(u + 2K) = \operatorname{dn}(u + 4iK') = \operatorname{dn} u, \tag{2.2.6}$$

showing that $\operatorname{dn} u$ has periods $2K, 4iK'$.

The second of equations (2.2.3) indicates that the ratio of the quantities iK' and K can be chosen arbitrarily, subject to the condition that the imaginary part of this ratio must be positive. Any doubly periodic functions, the ratio of whose periods is complex and which is regular save for poles, is termed an *elliptic function*.

Equations (2.2.3) define K, K' as functions of q, thus:

$$K = \tfrac{1}{2}\pi \prod_{n=1}^{\infty} (1 - q^{2n})^2 (1 + q^{2n-1})^4 = \tfrac{1}{2}\pi(1 + 2q + 2q^4 + \cdots)^2, \tag{2.2.7}$$

$$K' = -\tfrac{1}{2}\ln q \prod_{n=1}^{\infty} (1 - q^{2n})^2 (1 + q^{2n-1})^4 = -\tfrac{1}{2}\ln q(1 + 2q + 2q^4 + \cdots)^2. \tag{2.2.8}$$

Since q is a multivalued function of k, K and K' will be multivalued functions of the modulus also. The significance of this feature will be explained in the next section and will be more fully investigated in section 8.12. As $q \to 0$, $K \to \tfrac{1}{2}\pi$, and $K' \to +\infty$, which is to be expected, since the circular functions $\sin u$ and $\cos u$ have only one period, 2π. K has been tabulated against k^2 in Table B (p. 278).

Using the new notation, the poles of the Jacobian elliptic functions (equation (2.1.22)) can be reexpressed as being at the points

$$u = 2mK + i(2n + 1)K'. \tag{2.2.9}$$

The zeros (see (2.1.23)) are as follows:

$$\left.\begin{array}{ll} \text{sn}\, u = 0 & \text{where } u = 2mK + 2inK', \\ \text{cn}\, u = 0 & \text{where } u = (2m+1)K + 2inK', \\ \text{dn}\, u = 0 & \text{where } u = (2m+1)K + i(2n+1)K'. \end{array}\right\} \qquad (2.2.10)$$

The factorization of the theta functions shows that these zeros are all simple.
    Sets of identities derived from equations (1.3.2)–(1.3.5) and (1.3.6)–(1.3.9) can now be established.
    For incrementation by $2K, 2iK'$, and $2(K+iK')$, we find

$$\text{sn}\, u = -\,\text{sn}(u + 2K) = \text{sn}(u + 2iK') = -\,\text{sn}(u + 2K + 2iK'), \qquad (2.2.11)$$

$$\text{cn}\, u = -\,\text{cn}(u + 2K) = -\,\text{cn}(u + 2iK') = \text{cn}(u + 2K + 2iK'), \qquad (2.2.12)$$

$$\text{dn}\, u = \text{dn}(u + 2K) = -\,\text{dn}(u + 2iK') = -\,\text{dn}(u + 2K + 2iK'). \qquad (2.2.13)$$

Putting $u = 0$ in these identities, we deduce the results

$$\text{sn}\, 2K = \text{sn}\, 2iK' = \text{sn}(2K + 2iK') = 0, \qquad (2.2.14)$$

$$-\,\text{cn}\, 2K = -\,\text{cn}\, 2iK' = \text{cn}(2K + 2iK') = 1, \qquad (2.2.15)$$

$$\text{dn}\, 2K = -\,\text{dn}\, 2iK' = -\,\text{dn}(2K + 2iK') = 1. \qquad (2.2.16)$$

For incrementation by K, K', and $K + iK'$, we can show that

$$\text{sn}(u + K) = \text{cd}\, u, \qquad \text{sn}(u + iK') = \frac{1}{k}\,\text{ns}\, u, \qquad \text{sn}(u + K + iK') = \frac{1}{k}\,\text{dc}\, u,$$
$$(2.2.17)$$

$$\text{cn}(u + K) = -\,k'\,\text{sd}\, u, \qquad \text{cn}(u + iK') = \frac{1}{ik}\,\text{ds}\, u, \qquad \text{cn}(u + K + iK') = \frac{k'}{ik}\,\text{nc}\, u,$$
$$(2.2.18)$$

$$\text{dn}(u + K) = k'\,\text{nd}\, u, \qquad \text{dn}(u + iK') = -\,i\,\text{cs}\, u, \qquad \text{dn}(u + K + iK') = ik'\,\text{sc}\, u.$$
$$(2.2.19)$$

Putting $u = 0$ in (2.2.17)–(2.2.19), we find that

$$\begin{array}{llll} \text{sn}\, K = 1, & \text{sn}\, iK' = \infty, & \text{sn}(K + iK') = 1/k, & (2.2.20) \\ \text{cn}\, K = 0, & \text{cn}\, iK' = \infty, & \text{cn}(K + iK') = k'/ik, & (2.2.21) \\ \text{dn}\, K = k', & \text{dn}\, iK' = \infty, & \text{dn}(K + iK') = 0. & (2.2.22) \end{array}$$

All these results should be verified by the reader as exercises.
    For real values of u, the graph of $\text{sn}\, u$ is similar to that of $\sin u$, except that zeros occur at $u = 0, 2K, 4K$, etc. However, it is only necessary to graph and tabulate the function over the quarter-period $(0, K)$, since

$$\text{sn}(2K - u) = \text{sn}\, u, \qquad \text{sn}(2K + u) = -\,\text{sn}\, u, \qquad (2.2.23)$$

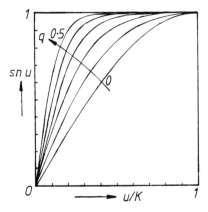

Figure 2.1. Graphs of sn$(u|q)$.

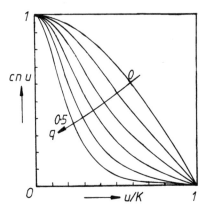

Figure 2.2. Graphs of cn$(u|q)$.

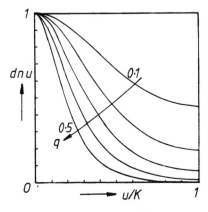

Figure 2.3. Graphs of dn$(u|q)$.

by equations (2.2.11). Graphs for nome $q = 0(0.1)0.5$ plotting u/K as abscissa are shown in Fig. 2.1.

The graph of cn u is similar to that of $\cos u$; its quarter-period graph has been plotted in Fig. 2.2 for $q = 0(0.1)0.5$.

In the case of dn u, $2K$ is a period and, since $\mathrm{dn}(2K - u) = \mathrm{dn}\, u$, we need only plot its graph over the half-period $(0, K)$. This has been done for $q = 0(0.1)0.5$ in Fig. 2.3.

Values of sn x, cn x, and dn x for real x and $k = 0.1(0.1)0.9$ are provided in Table C (p. 280).

2.3. Primitive Periods and Period Parallelograms

The periods of an elliptic function $f(u)$ are often denoted by $2\omega_1, 2\omega_3$. (Note: ω_2 is defined to be such that $\omega_1 + \omega_2 + \omega_3 = 0$; this notation is convenient for application to the Weierstrass function in Chapter 6.) Then

$$f(u + 2\omega_1 + 2\omega_3) = f(u + 2\omega_1) = f(u), \qquad (2.3.1)$$

showing that the sum of two periods is also a period. By adding a period to itself a number of times, we also conclude that an integral multiple of a period is a period. Also, note that

$$f(u) = f(u - 2\omega_1 + 2\omega_1) = f(u - 2\omega_1), \qquad (2.3.2)$$

so that $-2\omega_1$ (and $-2\omega_3$) is a period. Together, these results imply that $2m\omega_1 + 2n\omega_3$ is a period, where m, n are positive or negative integers or zero.

The ratio ω_3/ω_1 is denoted by τ and we adopt the convention that the order in which the two periods are taken is such that this ratio has a positive imaginary part; this is in accord with the assumption made in section 1.2. q is then given by equation (1.2.3) as before.

In the particular case of sn u, we have $2\omega_1 = 4K$, $2\omega_3 = 2iK'$, and $\omega_3/\omega_1 = iK'/2K = \frac{1}{2}\tau$ according to equation (2.2.3); the notations appear to be inconsistent. However, note that cn u has a period $4iK' = 2(2K + 2iK') - 4K$ and, hence, that the three Jacobian functions have common periods $4K, 4iK'$. If, therefore, $2\omega_1, 2\omega_3$ are taken to denote these common periods, the discrepancy disappears. Thus, K and iK' are often referred to as the *quarter-periods* of the set of Jacobian functions (K' by itself is sometimes loosely termed a quarter-period).

$2\omega_1$ and $2\omega_3$ will be assumed to be *fundamental periods*; i.e., no submultiple of either is a period of the elliptic function. Then, as remarked earlier, $2m\omega_1 + 2n\omega_3$ is a period. However, it is not necessarily the case that any period of the function can be expressed in this form. Thus, $4K, 4iK'$ are fundamental periods of cn u, but its period $2K + 2iK'$ cannot be expressed as a sum of multiples of these fundamental periods. Nevertheless, it was proved by Jacobi (see section 8.1.) that any elliptic function (except a constant) does always possess a pair of *primitive periods* having the property that any other period is a sum of multiples of these. This means that no elliptic function (except a constant) can have more than two independent periods. Primitive periods for $cn\,u$ are $4K$ are $2K + 2iK'$, the fundamental period $4iK'$ being expressible thus:

$$4iK' = 2(2K + 2iK') - 4K. \tag{2.3.3}$$

In future, unless explicitly stated otherwise, we shall take $2\omega_1, 2\omega_3$ to be primitive periods. Then, suppose we construct new periods $2\omega_1', 2\omega_3'$ by the equations

$$\omega_1' = a\omega_1 + b\omega_3, \qquad \omega_3' = c\omega_1 + d\omega_3, \tag{2.3.4}$$

where a, b, c, d are integers such that $ad - bc = 1$. Then the new periods are also primitive for, solving for ω_1 and ω_3, we obtain

$$\omega_1 = d\omega_1' - b\omega_3', \qquad \omega_3 = -c\omega_1' + a\omega_3', \tag{2.3.5}$$

showing that any period expressible as a sum of multiples of $2\omega_1, 2\omega_3$ is also expressible as a sum of multiples of $2\omega_1', 2\omega_3'$. Further, if $\omega_3/\omega_1 = \alpha + i\beta$ $(\beta > 0)$, then

$$\mathscr{I}(\omega_3'/\omega_1') = \mathscr{I}\frac{c + d(\alpha + i\beta)}{a + b(\alpha + i\beta)} = \frac{\beta}{(a + b\alpha)^2 + b^2\beta^2}, \tag{2.3.6}$$

showing that the imaginary part of ω_3'/ω_1' is also positive. Thus, we can always construct any number of pairs of primitive periods when one pair is known. For example, with $a = 1$, $b = -1$, $c = 0$, $d = 1$, we calculate that

$$2\omega_1' = 4K - (2K + 2iK') = 2K - 2iK', \qquad 2\omega_3' = 2K + 2iK' \tag{2.3.7}$$

are primitive periods for $cn\,u$.

Consider the points in the complex plane representing the periods $\Omega_{m,n} = 2m\omega_1 + 2n\omega_3$ of $f(u)$ for all integral values of m and n. These constitute a *lattice*. A parallelogram whose vertices are the points $\Omega_{m,n}, \Omega_{m+1,n}, \Omega_{m+1,n+1}, \Omega_{m,n+1}$ is called a *period parallelogram*; clearly, the u-plane separates into a mosaic of period parallelograms as indicated by the unbroken lines in Fig. 2.4. Since ω_3/ω_1 has positive imaginary part, its argument lies between 0 and π; this implies that the complex vector representing ω_1 must be rotated anticlockwise through an angle less than $180°$ to align it with the vector ω_3; thus, the vertices of the period parallelogram referred to earlier have been listed in the anticlockwise sense around the figure.

Let P be a point in, or on, the "base parallelogram" having vertices $\Omega_{0,0}, \Omega_{1,0}, \Omega_{1,1}, \Omega_{0,1}$ and let Q be a point similarly situated with respect to any other parallelogram with vertices $\Omega_{m,n}$, etc. Then, if u is the affix of $P, u + \Omega_{m,n}$ is the affix of Q and $f(u)$ takes the same value at the two points. P and Q are said to be *congruent points* and all points congruent to P are congruent to one another. Evidently, if we know the value of $f(u)$ at all points of a period parallelogram, its value at the congruent points of any other

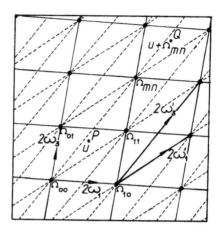

Figure 2.4. Period parallelograms.

period parallelogram will be known; thus, an elliptic function is completely determined by its values over a period parallelogram. To be precise, we need know only the values taken over the interior of a parallelogram and a pair of adjacent sides.

Instead of a period parallelogram, it is often convenient to work with the parallelogram obtained by translating a period parallelogram without rotation so that its vertices are the points $\Omega_{m,n} + u, \Omega_{m+1,n} + u, \Omega_{m+1,n+1} + u,$ $\Omega_{m,n+1} + u$. Such a parallelogram will be called a *cell*. A knowledge of $f(u)$ over a cell is also sufficient to specify the elliptic function. By proper choice of u, the infinities of $f(u)$ (which we show in Chapter 8 to be invariably present) can always be banished from the perimeter and be made to lie inside the cell; this facilitates much complex analysis.

Suppose we transform from primitive periods $2\omega_1, 2\omega_3$ to new primitive periods $2\omega_1', 2\omega_3'$ by the equations (2.3.4). The period lattice is unaffected, but the mosaic of period parallelograms undergoes a change as shown by the broken lines in Fig. 2.4. However, the new period parallelograms have the same area as the old. For, if $\omega_1 = \xi_1 + i\eta_1, \omega_3 = \xi_3 + i\eta_3$, the area of the parallelogram determined by these vectors is $\xi_1\eta_3 - \xi_3\eta_1 = \mathcal{I}(\omega_1^*\omega_3)$. (Note: Asterisks will, in future, denote complex conjugates). Thus, the parallelogram whose adjacent sides are the vectors ω_1', ω_3' has area

$$\begin{aligned}
\mathcal{I}(\omega_1'^*\omega_3') &= \mathcal{I}(a\omega_1^* + b\omega_3^*)(c\omega_1 + d\omega_3) \\
&= \mathcal{I}(ad\omega_1^*\omega_3 + bc\omega_1\omega_3^*) \\
&= \mathcal{I}(ad - bc)\omega_1^*\omega_3 \\
&= \mathcal{I}(\omega_1^*\omega_3),
\end{aligned} \qquad (2.3.8)$$

which proves the areas are equal. The mosaic of period parallelograms for an elliptic function is not a unique structure.

Since $\tau = \omega_3/\omega_1$, the parameter τ associated with an elliptic function is

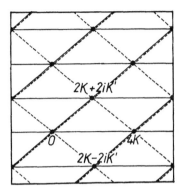

Figure 2.5. Period parallelograms for $\mathrm{cn}\,u$.

multivalued, any two values τ and τ' being related by a transformation

$$\tau' = \omega_3'/\omega_1' = \frac{c\omega_1 + d\omega_3}{a\omega_1 + b\omega_3} = \frac{c + d\tau}{a + b\tau}, \qquad (2.3.9)$$

where a, b, c, d are integers such that $ad - bc = 1$. In particular, the Jacobi functions $\mathrm{sn}(u, k)$, $\mathrm{cn}(u, k)$, and $\mathrm{dn}(u, k)$ with a prescribed value of the modulus k are associated with an infinity of possible values of τ (or q), each of which yields a possible pair of quarter-periods K and iK'. This explains the circumstance that K and K' are multivalued functions of k.

As an example, consider $\mathrm{cn}\,u$ with k real and $0 < k < 1$. We can take K and K' to be real and positive and, taking the primitive periods to be $2\omega_1 = 4K$, $2\omega_3 = 2K + 2iK'$, we generate the mosaic of period parallelograms indicated by the continuous lines in Fig. 2.5. The associated value of the parameter is $(2K + 2iK')/4K = \frac{1}{2}(1 + \tau)$ (τ here denotes the ratio of the quarter-periods, iK'/K). If, instead, we use the primitive periods $2\omega_1' = 2K - 2iK'$, $2\omega_3' = 2K + 2iK'$, the mosaic generated is that defined by the broken lines in Fig. 2.5. The parameter value is then $(2K + 2iK')/(2K - 2iK') = (1 + \tau)/(1 - \tau)$.

2.4. Addition Theorems

Dividing the identity (1.4.38) by (1.4.19) and then putting $x = u/\theta_3^2(0)$, $y = v/\theta_3^2(0)$ we find, using the definitions (2.1.1)–(2.1.3), that

$$\mathrm{sn}(u + v) = \frac{\mathrm{sn}\,u\,\mathrm{cn}\,v\,\mathrm{dn}\,v + \mathrm{sn}\,v\,\mathrm{cn}\,u\,\mathrm{dn}\,u}{1 - k^2\,\mathrm{sn}^2u\,\mathrm{sn}^2v}. \qquad (2.4.1)$$

This is the addition theorem for $\mathrm{sn}\,u$.

Similarly, dividing the identities (1.4.41), (1.4.47) by (1.4.19), we obtain the addition theorems for cn u and dn u, viz.

$$cn(u+v) = \frac{cn\,u\,cn\,v - sn\,u\,sn\,v\,dn\,u\,dn\,v}{1 - k^2\,sn^2u\,sn^2v}, \tag{2.4.2}$$

$$dn(u+v) = \frac{dn\,u\,dn\,v - k^2\,sn\,u\,sn\,v\,cn\,u\,cn\,v}{1 - k^2\,sn^2u\,sn^2v}. \tag{2.4.3}$$

Putting $u = v$, the following duplication formulae result:

$$sn\,2u = \frac{2\,sn\,u\,cn\,u\,dn\,u}{1 - k^2\,sn^4u}, \tag{2.4.4}$$

$$cn\,2u = \frac{2\,cn^2u}{1 - k^2\,sn^4u} - 1 = 1 - \frac{2\,sn^2u\,dn^2u}{1 - k^2\,sn^4u}, \tag{2.4.5}$$

$$dn\,2u = \frac{2\,dn^2u}{1 - k^2\,sn^4u} - 1 = 1 - \frac{2k^2\,sn^2u\,cn^2u}{1 - k^2\,sn^4u}. \tag{2.4.6}$$

From these results the reader may verify that the formulae listed below follow:

$$sn\,u = \sqrt{\left[\frac{1 - cn\,2u}{1 + dn\,2u}\right]}, \tag{2.4.7}$$

$$cn\,u = \sqrt{\left[\frac{cn\,2u + dn\,2u}{1 + dn\,2u}\right]}, \tag{2.4.8}$$

$$dn\,u = \sqrt{\left[\frac{cn\,2u + dn\,2u}{1 + cn\,2u}\right]}. \tag{2.4.9}$$

Putting $u = \frac{1}{2}K$ in the last three identities and referring to the results (2.2.20)–(2.2.22), we find that

$$sn\tfrac{1}{2}K = 1/\sqrt{(1+k')}, \qquad cn\tfrac{1}{2}K = \sqrt{k'}/\sqrt{(1+k')}, \qquad dn\tfrac{1}{2}K = \sqrt{k'}. \tag{2.4.10}$$

Replacing v by $-v$ in equation (2.4.1) and then adding the result to this equation, we obtain the identity

$$sn(u+v) + sn(u-v) = \frac{2\,sn\,u\,cn\,v\,dn\,v}{1 - k^2\,sn^2u\,sn^2v}. \tag{2.4.11}$$

The following identities can be established from (2.4.2) and (2.4.3) by a similar process:

$$cn(u+v) + cn(u-v) = \frac{2\,cn\,u\,cn\,v}{1 - k^2\,sn^2u\,sn^2v}, \tag{2.4.12}$$

$$dn(u+v) + dn(u-v) = \frac{2\,dn\,u\,dn\,v}{1 - k^2\,sn^2u\,sn^2v}. \tag{2.4.13}$$

If, instead of adding identities, we subtract, it is proved that

$$\operatorname{cn}(u-v) - \operatorname{cn}(u+v) = \frac{2\operatorname{sn}u\operatorname{sn}v\operatorname{dn}u\operatorname{dn}v}{1 - k^2\operatorname{sn}^2u\operatorname{sn}^2v}, \qquad (2.4.14)$$

$$\operatorname{dn}(u-v) - \operatorname{dn}(u+v) = \frac{2k^2\operatorname{sn}u\operatorname{sn}v\operatorname{cn}u\operatorname{cn}v}{1 - k^2\operatorname{sn}^2u\operatorname{sn}^2v}. \qquad (2.4.15)$$

The corresponding identity for sn does not differ essentially from (2.4.11).

If we divide the identity (1.4.16) by (1.4.19), the following identity emerges:

$$\operatorname{sn}(u+v)\operatorname{sn}(u-v) = \frac{\operatorname{sn}^2u - \operatorname{sn}^2v}{1 - k^2\operatorname{sn}^2u\operatorname{sn}^2v}. \qquad (2.4.16)$$

Similarly, dividing (1.4.17) by (1.4.19), we get

$$\operatorname{cn}(u+v)\operatorname{cn}(u-v) = \frac{\operatorname{cn}^2v - \operatorname{sn}^2u\operatorname{dn}^2v}{1 - k^2\operatorname{sn}^2u\operatorname{sn}^2v} = \frac{\operatorname{dn}^2u\operatorname{dn}^2v - k'^2}{k^2(1 - k^2\operatorname{sn}^2u\operatorname{sn}^2v)}, \qquad (2.4.17)$$

and (1.4.18) by (1.4.19) gives

$$\operatorname{dn}(u+v)\operatorname{dn}(u-v) = \frac{\operatorname{dn}^2v - k^2\operatorname{sn}^2u\operatorname{cn}^2v}{1 - k^2\operatorname{sn}^2u\operatorname{sn}^2v} = \frac{k'^2 + k^2\operatorname{cn}^2u\operatorname{cn}^2v}{1 - k^2\operatorname{sn}^2u\operatorname{sn}^2v}. \qquad (2.4.18)$$

Mixed product formulae can also be established. Thus, dividing (1.4.39), (1.4.40), and (1.4.48) each by (1.4.19), we show that

$$\operatorname{cn}(u+v)\operatorname{dn}(u-v) = \frac{\operatorname{cn}u\operatorname{cn}v\operatorname{dn}u\operatorname{dn}v - k'^2\operatorname{sn}u\operatorname{sn}v}{1 - k^2\operatorname{sn}^2u\operatorname{sn}^2v}, \qquad (2.4.19)$$

$$\operatorname{sn}(u+v)\operatorname{dn}(u-v) = \frac{\operatorname{sn}u\operatorname{dn}u\operatorname{cn}v + \operatorname{sn}v\operatorname{dn}v\operatorname{cn}u}{1 - k^2\operatorname{sn}^2u\operatorname{sn}^2v}, \qquad (2.4.20)$$

$$\operatorname{sn}(u+v)\operatorname{cn}(u-v) = \frac{\operatorname{sn}u\operatorname{cn}u\operatorname{dn}v + \operatorname{sn}v\operatorname{cn}v\operatorname{dn}u}{1 - k^2\operatorname{sn}^2u\operatorname{sn}^2v}. \qquad (2.4.21)$$

Other useful product formulae can be derived from earlier identities. Thus, dividing (2.4.14) by (2.4.13) and (2.4.15) by (2.4.12), we obtain the results

$$\operatorname{sn}u\operatorname{sn}v = \frac{\operatorname{cn}(u-v) - \operatorname{cn}(u+v)}{\operatorname{dn}(u-v) + \operatorname{dn}(u+v)} = \frac{1}{k^2}\frac{\operatorname{dn}(u-v) - \operatorname{dn}(u+v)}{\operatorname{cn}(u-v) + \operatorname{cn}(u+v)}. \qquad (2.4.22)$$

Also, from (2.4.19), it follows that

$$\operatorname{cn}(u+v)\operatorname{dn}(u-v) + \operatorname{cn}(u-v)\operatorname{dn}(u+v) = \frac{2\operatorname{cn}u\operatorname{cn}v\operatorname{dn}u\operatorname{dn}v}{1 - k^2\operatorname{sn}^2u\operatorname{sn}^2v}, \qquad (2.4.23)$$

$$\operatorname{cn}(u-v)\operatorname{dn}(u+v) - \operatorname{cn}(u+v)\operatorname{dn}(u-v) = \frac{2k'^2\operatorname{sn}u\operatorname{sn}v}{1 - k^2\operatorname{sn}^2u\operatorname{sn}^2v}. \qquad (2.4.24)$$

Then, dividing (2.4.23) by (2.4.12) and (2.4.13) and dividing (2.4.24) by (2.4.14) and (2.4.15), we find that

$$\operatorname{cn} u \operatorname{cn} v = \frac{\operatorname{cn}(u+v)\operatorname{dn}(u-v) + \operatorname{cn}(u-v)\operatorname{dn}(u+v)}{\operatorname{dn}(u+v) + \operatorname{dn}(u-v)} \tag{2.4.25}$$

$$= \frac{k'^2}{k^2} \cdot \frac{\operatorname{dn}(u-v) - \operatorname{dn}(u+v)}{\operatorname{cn}(u-v)\operatorname{dn}(u+v) - \operatorname{cn}(u+v)\operatorname{dn}(u-v)}, \tag{2.4.26}$$

$$\operatorname{dn} u \operatorname{dn} v = \frac{\operatorname{cn}(u-v)\operatorname{dn}(u+v) + \operatorname{cn}(u+v)\operatorname{dn}(u-v)}{\operatorname{cn}(u+v) + \operatorname{cn}(u-v)} \tag{2.4.27}$$

$$= k'^2 \frac{\operatorname{cn}(u-v) - \operatorname{cn}(u+v)}{\operatorname{cn}(u-v)\operatorname{dn}(u+v) - \operatorname{cn}(u+v)\operatorname{dn}(u-v)}. \tag{2.4.28}$$

Putting $k = 0$ in all the identities of this section, they either become trivial or reduce to well-known trigonometric formulae.

2.5. Derivatives of Jacobi's Elliptic Functions

Differentiating the equation (2.1.1) with respect to u making use of the identity (1.9.3), we find that

$$\frac{d}{du}\operatorname{sn} u = \frac{\theta_4^2(0)}{\theta_2(0)\theta_3(0)} \cdot \frac{\theta_2(z)\theta_3(z)}{\theta_4^2(z)}. \tag{2.5.1}$$

Thus, $\operatorname{sn}'(0) = 1$; this explains the presence of the factor $1/\theta_3^2(0)$ in the relation $z = u/\theta_3^2(0)$. Making use of the equations (2.1.2) and (2.1.3), equation (2.5.1) reduces to the form

$$\frac{d}{du}\operatorname{sn} u = \operatorname{cn} u \operatorname{dn} u. \tag{2.5.2}$$

Similarly, by differentiating equations (2.1.2) and (2.1.3) and making use of the identities (1.9.6), (1.9.7), and (2.1.7), we establish the results

$$\frac{d}{du}\operatorname{cn} u = -\operatorname{sn} u \operatorname{dn} u, \tag{2.5.3}$$

$$\frac{d}{du}\operatorname{dn} u = -k^2 \operatorname{sn} u \operatorname{cn} u. \tag{2.5.4}$$

Given the value of k and the initial conditions $\operatorname{sn} 0 = 0$, $\operatorname{cn} 0 = \operatorname{dn} 0 = 1$, the equations (2.5.2)–(2.5.4) determine the three Jacobi functions uniquely.*

---

* It follows from results established in section 2.8 that the singularities of the right-hand members of equations (2.5.2)–(2.5.4) are double poles with zero residues. The path along which these equations are integrated does not affect the result, therefore, and their integrals are single-valued.

This establishes that these functions are single-valued functions of k (indeed, of k^2, i.e., $sn(u, -k) = sn(u, k)$, etc.).

It should now be observed that the denominators in the identities (2.4.1)–(2.4.3) can be expressed in the forms (i) $sn\,u\,sn'v + sn\,v\,sn'u$, (ii) $cn\,u\,cn\,v - cn'u\,cn'v$, (iii) $dn\,u\,dn\,v - k^{-2}\,dn'u\,dn'v$. This invites comparison with the trigonometric identities for $sin(u + v)$ and $cos(u + v)$.

The following further derivatives can now be verified:

$$\frac{d}{du}ns\,u = -cs\,u\,ds\,u, \tag{2.5.5}$$

$$\frac{d}{du}nc\,u = sc\,u\,dc\,u, \tag{2.5.6}$$

$$\frac{d}{du}nd\,u = k^2\,sd\,u\,cd\,u, \tag{2.5.7}$$

$$\frac{d}{du}sc\,u = dc\,u\,nc\,u, \tag{2.5.8}$$

$$\frac{d}{du}cs\,u = -ds\,u\,ns\,u, \tag{2.5.9}$$

$$\frac{d}{du}sd\,u = cd\,u\,nd\,u, \tag{2.5.10}$$

$$\frac{d}{du}ds\,u = -cs\,u\,ns\,u, \tag{2.5.11}$$

$$\frac{d}{du}cd\,u = -k'^2\,sd\,u\,nd\,u, \tag{2.5.12}$$

$$\frac{d}{du}dc\,u = k'^2\,sc\,u\,nc\,u. \tag{2.5.13}$$

Higher-order derivatives are now calculable by repeated application of these results and then, using Maclaurin's theorem, power series expansions can be found for those functions which are regular in the neighborhood of $u = 0$. The reader may verify the terms given below:

$$sn\,u = u - \frac{1}{3!}(1 + k^2)u^3 + \frac{1}{5!}(1 + 14k^2 + k^4)u^5 - \cdots, \tag{2.5.14}$$

$$cn\,u = 1 - \frac{1}{2!}u^2 + \frac{1}{4!}(1 + 4k^2)u^4 - \cdots, \tag{2.5.15}$$

$$dn\,u = 1 - \frac{1}{2!}k^2u^2 + \frac{1}{4!}(4k^2 + k^4)u^4 - \cdots. \tag{2.5.16}$$

2.6. Elliptic Functions with Imaginary Argument

In this section, we shall investigate the effect on the elliptic functions of applying Jacobi's transformation (section 1.7) to the component theta functions.

Writing $\tau' = -1/\tau$, we first observe that the identities (1.7.12), (1.7.15) imply that

$$k(\tau') = \frac{\theta_2^2(0|\tau')}{\theta_3^2(0|\tau')} = \frac{\theta_4^2(0|\tau)}{\theta_3^2(0|\tau)} = k'(\tau) \tag{2.6.1}$$

(we have used equations (2.1.7)). That is, the modulus for the transformed parameter τ' is the complementary modulus for the original parameter τ. Also, since the transformation is a reciprocal one, it must follow that

$$k'(\tau') = k(\tau) \tag{2.6.2}$$

(verify from equations (2.1.7)).

Further use of the identity (1.7.12) shows that

$$\theta_3^2(0|\tau') = (-i\tau)\theta_3^2(0|\tau), \tag{2.6.3}$$

implying by equations (2.2.3) that

$$K(\tau') = (-i\tau)K(\tau) = K'(\tau). \tag{2.6.4}$$

But equation (2.6.1) shows that the modulus corresponding to τ' is k'; hence $K(\tau') = K(k')$ and the result (2.6.4) may be written

$$K(k') = K'(k). \tag{2.6.5}$$

Equation (2.6.4) is valid without restriction. However, since K and K' are multivalued functions of the modulus k, equation (2.6.5) is only acceptable provided corresponding values of the two sides of the equation are taken; which values correspond is decided by the fundamental equation (2.6.4). Using equation (2.6.5), values of K in Table B now also provide values of K'.

We can now transform equation (2.1.1) to give

$$\text{sn}(u|\tau') = \frac{\theta_3(0|\tau')}{\theta_2(0|\tau')} \cdot \frac{\theta_1(z'|\tau')}{\theta_4(z'|\tau')} = \frac{\theta_3(0|\tau)}{\theta_4(0|\tau)} \cdot \frac{(-i)\theta_1(\tau z'|\tau)}{\theta_2(\tau z'|\tau)}, \tag{2.6.6}$$

where $z' = u/\theta_3^2(0|\tau')$. Thus,

$$\tau z' = \tau u/\theta_3^2(0|\tau') = iu/\theta_3^2(0|\tau) = iz. \tag{2.6.7}$$

Equation (2.6.6) can now be written in the form

$$\text{sn}(u|\tau') = -i\frac{\theta_3(0|\tau)}{\theta_4(0|\tau)} \cdot \frac{\theta_1(iz|\tau)}{\theta_2(iz|\tau)} = -i\,\text{sc}(iu|\tau), \tag{2.6.8}$$

or

$$\text{sn}(u, k') = -i\,\text{sc}(iu, k), \tag{2.6.9}$$

showing how to transform sn to complementary modulus. (Since Jacobi's functions are single valued with respect to the modulus, this identity is valid without restriction.)

Similarly, transforming equations (2.1.2) and (2.1.3), we find that

$$cn(u, k') = nc(iu, k), \tag{2.6.10}$$

$$dn(u, k') = dc(iu, k). \tag{2.6.11}$$

The remaining nine Jacobi functions can now be transformed to complementary modulus immediately. However, these transformations are most often used to change from an imaginary argument iu to a real argument u and will accordingly be collected together in the form most suited to this purpose below:

$$\left.\begin{array}{lll}
sn(iu, k) = i\,sc(u, k'), & cn(iu, k) = nc(u, k'), & dn(iu, k) = dc(u, k'), \\
ns(iu, k) = -i\,cs(u, k'), & nc(iu, k) = cn(u, k'), & nd(iu, k) = cd(u, k'), \\
cd(iu, k) = nd(u, k'), & ds(iu, k) = -i\,ds(u, k'), & sc(iu, k) = i\,sn(u, k'), \\
dc(iu, k) = dn(u, k'), & sd(iu, k) = i\,sd(u, k'), & cs(iu, k) = -i\,ns(u, k').
\end{array}\right\} \tag{2.6.12}$$

Putting $u = \frac{1}{2}K'(k)$ in the transformation equations for $sn(iu, k)$, $cn(iu, k)$, and $dn(iu, k)$, and using the results (2.4.10), we now calculate that

$$sn(\tfrac{1}{2}iK', k) = i\,sc\{\tfrac{1}{2}K'(k), k'\} = i\,sc\{\tfrac{1}{2}K(k'), k'\} = i/\sqrt{k}, \tag{2.6.13}$$

$$cn(\tfrac{1}{2}iK', k) = nc\{\tfrac{1}{2}K'(k), k'\} = nc\{\tfrac{1}{2}K(k'), k'\} = \sqrt{(1 + k)}/\sqrt{k}, \tag{2.6.14}$$

$$dn(\tfrac{1}{2}iK', k) = dc\{\tfrac{1}{2}K'(k), k'\} = dc\{\tfrac{1}{2}K(k'), k'\} = \sqrt{(1 + k)}. \tag{2.6.15}$$

Letting $k \to 0$ (so that $k' \to 1$) in equation (2.6.9), it follows from the results (2.1.18)–(2.1.20) that

$$\lim_{k \to 1} sn(u, k) = \lim_{k' \to 1} sn(u, k') = \lim_{k \to 0} -i\,sc(iu, k) = -i\tan(iu) = \tanh u. \tag{2.6.16}$$

Applying the same procedure to equations (2.6.12), we deduce the results

$$\left.\begin{array}{l}
\lim_{k \to 1} cn(u, k) = \lim_{k \to 1} dn(u, k) = \operatorname{sech} u, \\[1ex]
\lim_{k \to 1} cs(u, k) = \lim_{k \to 1} ds(u, k) = \operatorname{cosech} u, \\[1ex]
\lim_{k \to 1} sc(u, k) = \lim_{k \to 1} sd(u, k) = \sinh u, \\[1ex]
\lim_{k \to 1} nc(u, k) = \lim_{k \to 1} nd(u, k) = \cosh u, \\[1ex]
\lim_{k \to 1} cd(u, k) = \lim_{k \to 1} dc(u, k) = 1, \\[1ex]
\lim_{k \to 1} ns(u, k) = \coth u.
\end{array}\right\} \tag{2.6.17}$$

The approach of sn u, cn u, and dn u toward their hyperbolic function limits as k and $q \to 1$ is clearly indicated in Figs. 2.1–2.3. Elliptic functions may be regarded as bridging the gap between circular and hyperbolic functions.

2.7. Integrals of the Elliptic Functions

All twelve Jacobian elliptic functions can be integrated without difficulty. Making use of the derivatives calculated at (2.5.2)–(2.5.13), the reader will be able to verify the following indefinite integrals by differentiation:

$$\int \mathrm{sn}\, u \, du = \frac{1}{k} \ln(\mathrm{dn}\, u - k \, \mathrm{cn}\, u), \tag{2.7.1}$$

$$\int \mathrm{cn}\, u \, du = \frac{1}{k} \sin^{-1}(k \, \mathrm{sn}\, u), \tag{2.7.2}$$

$$\int \mathrm{dn}\, u \, du = \sin^{-1}(\mathrm{sn}\, u), \tag{2.7.3}$$

$$\int \mathrm{cs}\, u \, du = -\ln(\mathrm{ns}\, u + \mathrm{ds}\, u), \tag{2.7.4}$$

$$\int \mathrm{ds}\, u \, du = \ln(\mathrm{ns}\, u - \mathrm{cs}\, u), \tag{2.7.5}$$

$$\int \mathrm{dc}\, u \, du = \ln(\mathrm{nc}\, u + \mathrm{sc}\, u). \tag{2.7.6}$$

It then follows from the result (2.7.1) that

$$\int \mathrm{sn}(u + K) \, du = \frac{1}{k} \ln\{\mathrm{dn}(u + K) - k \, \mathrm{cn}(u + K)\} \tag{2.7.7}$$

and, hence, by application of the quarter-period formulae (2.2.17)–(2.2.19), we get

$$\int \mathrm{cd}\, u \, du = \frac{1}{k} \ln(\mathrm{nd}\, u + k \, \mathrm{sd}\, u), \tag{2.7.8}$$

having neglected an additive constant. This result, also, is easily verified by differentiation.

Similarly, by replacing u by $u + K$ in the formulae (2.7.2)–(2.7.6), we calculate that

$$\int \mathrm{sd}\, u \, du = -\frac{1}{kk'} \sin^{-1}(k \, \mathrm{cd}\, u), \tag{2.7.9}$$

$$\int \text{nd} \, u \, du = \frac{1}{k'} \sin^{-1}(\text{cd} \, u), \tag{2.7.10}$$

$$\int \text{sc} \, u \, du = \frac{1}{k'} \ln(\text{dc} \, u + k' \, \text{nc} \, u), \tag{2.7.11}$$

$$\int \text{nc} \, u \, du = \frac{1}{k'} \ln(\text{dc} \, u + k' \, \text{sc} \, u), \tag{2.7.12}$$

$$\int \text{ns} \, u \, du = -\ln(\text{ds} \, u + \text{cs} \, u). \tag{2.7.13}$$

The last five results, also, are easily checked by differentiation.

Jacobi defined a function am u by means of the equation

$$\text{am} \, u = \int_0^u \text{dn} \, v \, dv. \tag{2.7.14}$$

It then follows that

$$\frac{d}{du} \text{am} \, u = \text{dn} \, u \tag{2.7.15}$$

and, using the result (2.7.3), that

$$\text{sn} \, u = \sin(\text{am} \, u). \tag{2.7.16}$$

Also,

$$\text{cn} \, u = \sqrt{(1 - \text{sn}^2 u)} = \cos(\text{am} \, u), \tag{2.7.17}$$

the positive sign being clearly appropriate. Note that am $K = \frac{1}{2}\pi$, am $2K = \pi$, etc.

2.8. Poles of sn u, cn u, and dn u

It has been remarked that these elliptic functions all have singularities at the points $u = 2mK + i(2n + 1)K'$ (equation (2.2.9)).

Thus, inside the period parallelogram for sn u having vertices $-K$, $3K$, $3K + 2iK'$, $-K + 2iK'$, there are two singularities at $u = iK'$ and $u = 2K + iK'$. All other singularities are at points congruent to these where, due to the double periodicity, sn u behaves in like manner.

Now, using equations (2.2.17) and (2.5.14),

$$\text{sn}(iK' + u) = \frac{1}{k \, \text{sn} \, u} = \frac{1}{ku} [1 - \tfrac{1}{6}(1 + k^2)u^2 + \cdots]^{-1}$$

$$= \frac{1}{ku} + \frac{1}{6k}(1 + k^2)u + \cdots, \tag{2.8.1}$$

showing that sn u has a simple pole at $u = iK'$ with residue $1/k$.

Also, referring to equations (2.2.11), we calculate that

$$\text{sn}(2K + iK' + u) = -\text{sn}(iK' + u) = -\frac{1}{ku} + \cdots. \tag{2.8.2}$$

We conclude that $\text{sn}\,u$ has a simple pole at $u = 2K + iK'$ with residue $-1/k$.
$\text{cn}\,u$ also has the pair of singularities $u = iK'$, $2K + iK'$ inside its period parallelogram with vertices $-2K$, $2K$, $4K + 2iK'$, $2iK'$. In the vicinity of iK',

$$\text{cn}(iK' + u) = \frac{1}{ik}\frac{\text{dn}\,u}{\text{sn}\,u} = \frac{1}{iku}(1 - \tfrac{1}{2}k^2u^2 + \cdots)[1 - \tfrac{1}{6}(1 + k^2)u^2 + \cdots]^{-1}$$

$$= \frac{1}{iku} + \frac{1}{6ik}(1 - 2k^2)u + \cdots, \tag{2.8.3}$$

showing that $\text{cn}\,u$ has a simple pole with residue $-i/k$ at this point. Since $\text{cn}(2K + iK' + u) = -\text{cn}(iK' + u)$, the residue at the other simple pole is i/k.

In the case of $\text{dn}\,u$, a period parallelogram having vertices $-K$, K, $K + 4iK'$, $-K + 4iK'$ encloses the singularites $u = iK'$, $3iK'$; all other singularites are congruent to these. Near iK',

$$\text{dn}(iK' + u) = \frac{1}{i}\frac{\text{cn}\,u}{\text{sn}\,u} = \frac{1}{iu}(1 - \tfrac{1}{2}u^2 + \cdots)[1 - \tfrac{1}{6}(1 + k^2)u^2 + \cdots]^{-1}$$

$$= \frac{1}{iu} - \frac{1}{6i}(2 - k^2)u + \cdots. \tag{2.8.4}$$

Thus $\text{dn}\,u$ has a simple pole at iK' with residue $-i$. Now $\text{dn}(3iK' + u) = -\text{dn}(iK' + u)$ and the simple pole at $3iK'$ accordingly has residue i.

We summarize these findings in the table below:

Function	Singularities	Residues	
$\text{sn}\,u$	$iK', 2K + iK'$	$1/k, -1/k$	(2.8.5)
$\text{cn}\,u$	$iK', 2K + iK'$	$-i/k, i/k$	
$\text{dn}\,u$	$iK', 3iK'$	$-i, i$	

Multiplication of series shows that

$$\text{sn}(iK' + u)\,\text{cn}(iK' + u) = -\frac{i}{k^2u^2} - \frac{i}{6k^2}(2 - k^2) + \cdots;$$

i.e., $\text{sn}\,u\,\text{cn}\,u$ has a pole at $u = iK'$ with zero residue. The statement in the note at the bottom of p. 36 is now easily verified.

EXERCISES

1. Show that

$$\prod_{n=1}^{\infty}(1 - q^{2n-1}) = \prod_{n=1}^{\infty}\left(\frac{1 - q^n}{1 - q^{2n}}\right) = \prod_{n=1}^{\infty}(1 + q^n)^{-1}.$$

2. Obtain the following results:

(i) $\displaystyle\prod_{n=1}^{\infty} (1 - q^{2n-1})^6 = 2q^{1/4}k'k^{-1/2}$,

(ii) $\displaystyle\prod_{n=1}^{\infty} (1 + q^{2n-1})^6 = 2q^{1/4}(kk')^{-1/2}$,

(iii) $\displaystyle\prod_{n=1}^{\infty} (1 - q^{2n})^6 = 2\pi^{-3}q^{-1/2}kk'K^3$,

(iv) $\displaystyle\prod_{n=1}^{\infty} (1 + q^{2n})^6 = \tfrac{1}{4}q^{-1/2}kk'^{-1/2}$,

(v) $\displaystyle\prod_{n=1}^{\infty} (1 - q^n)^6 = 4\pi^{-3}q^{-1/4}k^{1/2}k'^2K^3$,

(vi) $\displaystyle\prod_{n=1}^{\infty} (1 + q^n)^6 = \tfrac{1}{2}q^{-1/4}k^{1/2}k'^{-1}$.

(Hint: Use the result of Exercise 1.)

3. Obtain the following alternative forms of the addition theorems:

$$\operatorname{sn}(u+v) = \frac{s_1^2 - s_2^2}{s_1 c_2 d_2 - s_2 c_1 d_1} = \frac{s_1 c_1 d_2 + s_2 c_2 d_1}{c_1 c_2 + s_1 d_1 s_2 d_2} = \frac{s_1 d_1 c_2 + s_2 d_2 c_1}{d_1 d_2 + k^2 s_1 c_1 s_2 c_2},$$

$$\operatorname{cn}(u+v) = \frac{s_1 c_1 d_2 - s_2 c_2 d_1}{s_1 c_2 d_2 - s_2 c_1 d_1} = \frac{1 - s_1^2 - s_2^2 + k^2 s_1^2 s_2^2}{c_1 c_2 + s_1 d_1 s_2 d_2} = \frac{c_1 d_1 c_2 d_2 - k'^2 s_1 s_2}{d_1 d_2 + k^2 s_1 c_1 s_2 c_2},$$

$$\operatorname{dn}(u+v) = \frac{s_1 c_2 d_1 - s_2 c_1 d_2}{s_1 c_2 d_2 - s_2 c_1 d_1} = \frac{c_1 c_2 d_1 d_2 + k'^2 s_1 s_2}{c_1 c_2 + s_1 d_1 s_2 d_2} = \frac{1 - k^2 s_1^2 - k^2 s_2^2 + k^2 s_1^2 s_2^2}{d_1 d_2 + k^2 s_1 c_1 s_2 c_2},$$

where $s_1 = \operatorname{sn} u$, $s_2 = \operatorname{sn} v$, etc. (Hint: Replace u by $u + K$ and $u + iK'$ in the original forms.)

4. If $s = \operatorname{sn} u$, $S = \operatorname{sn} 2u$, etc., obtain the identities

$$s^2 = \frac{1-D}{k^2(1+C)} = \frac{D - k^2 C - k'^2}{k^2(D-C)} = \frac{D-C}{k'^2 + D - k^2 C},$$

$$c^2 = \frac{D + k^2 C - k'^2}{k^2(1+C)} = \frac{k'^2(1-D)}{k^2(D-C)} = \frac{k'^2(1+C)}{k'^2 + D - k^2 C},$$

$$d^2 = \frac{k'^2 + D + k^2 C}{1+D} = \frac{k'^2(1-C)}{D-C} = \frac{k'^2(1+D)}{k'^2 + D - k^2 C}.$$

5. Prove that $\operatorname{cs} u \,\operatorname{cs}(K - u) = k'$.

6. Obtain the identities

$$\operatorname{ns} u - \operatorname{cs} u = \operatorname{sn} \tfrac{1}{2}u \,\operatorname{dc} \tfrac{1}{2}u,$$
$$\operatorname{ns} u + \operatorname{ds} u = \operatorname{ds} \tfrac{1}{2}u \,\operatorname{nc} \tfrac{1}{2}u,$$
$$\operatorname{ds} u + \operatorname{cs} u = \operatorname{cn} \tfrac{1}{2}u \,\operatorname{ds} \tfrac{1}{2}u.$$

7. Show that, if $0 < k < 1$ and $k = \sin \alpha$,

$$\text{sn} \tfrac{1}{2}(K + iK') = e^{(1/4)\pi i - (1/2)i\alpha} \sqrt{(\text{cosec } \alpha)},$$
$$\text{cn} \tfrac{1}{2}(K + iK') = e^{-(1/4)\pi i} \sqrt{(\cot \alpha)},$$
$$\text{dn} \tfrac{1}{2}(K + iK') = e^{-(1/2)i\alpha} \sqrt{(\cos \alpha)}.$$

8. Express sn u, cn u, and dn u as infinite products, thus:

$$\text{sn } u = 2q^{1/4}k^{-1/2} \sin x \prod_{n=1}^{\infty} \left[\frac{1 - 2q^{2n} \cos 2x + q^{4n}}{1 - 2q^{2n-1} \cos 2x + q^{4n-2}} \right],$$

$$\text{cn } u = 2q^{1/4}k'^{1/2}k^{-1/2} \cos x \prod_{n=1}^{\infty} \left[\frac{1 + 2q^{2n} \cos 2x + q^{4n}}{1 - 2q^{2n-1} \cos 2x + q^{4n-2}} \right],$$

$$\text{dn } u = k'^{1/2} \prod_{n=1}^{\infty} \left[\frac{1 + 2q^{2n-1} \cos 2x + q^{4n-2}}{1 - 2q^{2n-1} \cos 2x + q^{4n-2}} \right],$$

where $x = \pi u/2K$. If $0 < k < 1$ and $k = \sin \alpha$, deduce that

$$e^{-(1/2)i\alpha} = \prod_{n=0}^{\infty} \left[\frac{1 - (-1)^n i q^{n+1/2}}{1 + (-1)^n i q^{n+1/2}} \right].$$

(Hint: Use a result from Exercise 7 above.) Taking logarithms, obtain the identity

$$\tfrac{1}{4}\alpha = \sum_{n=0}^{\infty} (-1)^n \tan^{-1}(q^{n+1/2}).$$

9. Show that

$$1 + \text{cn}(u + v)\,\text{cn}(u - v) = \frac{\text{cn}^2 u + \text{cn}^2 v}{1 - k^2 \, \text{sn}^2 u \, \text{sn}^2 v},$$

$$1 + \text{dn}(u + v)\,\text{dn}(u - v) = \frac{\text{dn}^2 u + \text{dn}^2 v}{1 - k^2 \, \text{sn}^2 u \, \text{sn}^2 v}.$$

10. By use of the identities sn$(u + K) = $ cd u etc., derive the following identities from equations (2.1.9)–(2.1.11):

$$\text{cd}^2 u + k'^2 \, \text{sd}^2 u = 1, \qquad k'^2 \, \text{nd}^2 u + k^2 \, \text{cd}^2 u = 1, \qquad \text{nd}^2 u - k^2 \, \text{sd}^2 u = 1.$$

Obtain these identities directly from equations (2.1.9)–(2.1.11).

11. Show that

$$\text{sn}(u + \tfrac{1}{2}K) = (1 + k')^{-1/2} \frac{k' \, \text{sn } u + \text{cn } u \, \text{dn } u}{1 - (1 - k') \text{sn}^2 u},$$

$$\text{sn}(u + \tfrac{1}{2}iK') = k^{-1/2} \frac{(1 + k) \text{sn } u + i \, \text{cn } u \, \text{dn } u}{1 + k \, \text{sn}^2 u}.$$

12. Writing $s = $ sn u, $c = $ cn u, $d = $ dn u, and

$$A = 3s - 4(1 + k^2)s^3 + 6k^2 s^5 - k^4 s^9,$$
$$B = c[1 - 4s^2 + 6k^2 s^4 - 4k^4 s^6 + k^4 s^8],$$
$$C = d[1 - 4k^2 s^2 + 6k^2 s^4 - 4k^2 s^6 + k^4 s^8],$$
$$D = 1 - 6k^2 s^4 + 4k^2(1 + k^2)s^6 - 3k^4 s^8,$$

prove that

$$\text{sn } 3u = A/D, \qquad \text{cn } 3u = B/D, \qquad \text{dn } 3u = C/D.$$

13. If $x + y + z + w = 0$, prove that

$$k'^2 - k^2 k'^2 \text{ sn } x \text{ sn } y \text{ sn } z \text{ sn } w + k^2 \text{ cn } x \text{ cn } y \text{ cn } z \text{ cn } w - \text{dn } x \text{ dn } y \text{ dn } z \text{ dn } w = 0.$$

(Hint: Make use of the identity given in Exercise 5 in Chapter 1.)

14. Prove the following identities:

(i) $\text{dn } u \text{ nd } v = \dfrac{\text{ds}(u+v) + \text{ds}(u-v)}{\text{ns}(u+v) + \text{ns}(u-v)} = \dfrac{\text{ns}(u+v) - \text{ns}(u-v)}{\text{ds}(u+v) - \text{ds}(u-v)}$,

(ii) $\text{sd } u \text{ cn } v = \dfrac{\text{sn}(u+v) + \text{sn}(u-v)}{\text{dn}(u+v) + \text{dn}(u-v)} = \dfrac{1}{k^2} \dfrac{\text{dn}(u-v) - \text{dn}(u+v)}{\text{sn}(u+v) - \text{sn}(u-v)}$,

(iii) $\text{sc } u \text{ dn } v = \dfrac{\text{sn}(u+v) + \text{sn}(u-v)}{\text{cn}(u+v) + \text{cn}(u-v)} = \dfrac{\text{cn}(u-v) - \text{cn}(u+v)}{\text{sn}(u+v) - \text{sn}(u-v)}$,

(iv) $\text{sn } u \text{ cd } v = \dfrac{\text{sd}(u+v) + \text{sd}(u-v)}{\text{nd}(u+v) + \text{nd}(u-v)} = \dfrac{1}{k^2} \dfrac{\text{nd}(u+v) - \text{nd}(u-v)}{\text{sd}(u+v) - \text{sd}(u-v)}$,

(v) $\text{sn } u \text{ dc } v = \dfrac{\text{sc}(u+v) + \text{sc}(u-v)}{\text{nc}(u+v) + \text{nc}(u-v)} = \dfrac{\text{nc}(u+v) - \text{nc}(u-v)}{\text{sc}(u+v) - \text{sc}(u-v)}$,

(vi) $\text{dc } u \text{ cd } v = \dfrac{\text{ds}(u+v) + \text{ds}(u-v)}{\text{cs}(u+v) + \text{cs}(u-v)} = \dfrac{\text{cs}(u+v) - \text{cs}(u-v)}{\text{ds}(u+v) - \text{ds}(u-v)}$,

(vii) $\text{sc } u \text{ nd } v = \dfrac{\text{sd}(u+v) + \text{sd}(u-v)}{\text{cd}(u+v) + \text{cd}(u-v)} = \dfrac{1}{k'^2} \dfrac{\text{cd}(u-v) - \text{cd}(u+v)}{\text{sd}(u+v) - \text{sd}(u-v)}$,

(viii) $\text{ds } u \text{ cn } v = \dfrac{\text{dc}(u+v) + \text{dc}(u-v)}{\text{sc}(u+v) + \text{sc}(u-v)} = k'^2 \dfrac{\text{sc}(u+v) - \text{sc}(u-v)}{\text{dc}(u+v) - \text{dc}(u-v)}$,

(ix) $\text{cn } u \text{ nc } v = \dfrac{\text{cs}(u+v) + \text{cs}(u-v)}{\text{ns}(u+v) + \text{ns}(u-v)} = \dfrac{\text{ns}(u+v) - \text{ns}(u-v)}{\text{cs}(u+v) - \text{cs}(u-v)}$.

15. If $k' = \frac{1}{4}(a^{-1} - a)^2$, where $0 < a < 1$, prove that

$$\text{sn}^2 \tfrac{1}{4}K = \dfrac{4a^3}{(1 + a^2)(1 + 2a - a^2)}.$$

16. By differentiating the identities (1.4.16)–(1.4.19) twice with respect to y and then putting $y = 0$, obtain the results:

$$\dfrac{d}{dz}\left[\dfrac{\theta_1'(z)}{\theta_1(z)}\right] = \dfrac{\theta_4''(0)}{\theta_4(0)} - \theta_3^4(0)\,\text{ns}^2\{z\theta_3^2(0)\},$$

$$\dfrac{d}{dz}\left[\dfrac{\theta_2'(z)}{\theta_2(z)}\right] = \dfrac{\theta_4''(0)}{\theta_4(0)} - \theta_3^4(0)\,\text{dc}^2\{z\theta_3^2(0)\},$$

$$\dfrac{d}{dz}\left[\dfrac{\theta_3'(z)}{\theta_3(z)}\right] = \dfrac{\theta_4''(0)}{\theta_4(0)} - \theta_2^4(0)\,\text{cd}^2\{z\theta_3^2(0)\},$$

$$\frac{d}{dz}\left[\frac{\theta_4'(z)}{\theta_4(z)}\right] = \frac{\theta_4''(0)}{\theta_4(0)} - \theta_2^4(0)\,\mathrm{sn}^2\{z\theta_3^2(0)\}.$$

Deduce that

$$\frac{\theta_1'(z)}{\theta_1(z)} = \frac{\theta_4''(0)}{\theta_4(0)}(z - \tfrac{1}{2}\pi) + \frac{2K}{\pi}\int_{2Kz/\pi}^K \mathrm{ns}^2 u\, du$$

and three other similar results.

17. Obtain the identities:

(i) $\mathrm{ns}(u+v) + \mathrm{ns}(u-v) = \dfrac{2\,\mathrm{sn}\,u\,\mathrm{cn}\,v\,\mathrm{dn}\,v}{\mathrm{sn}^2 u - \mathrm{sn}^2 v}$

(ii) $\mathrm{nc}(u+v) + \mathrm{nc}(u-v) = \dfrac{2\,\mathrm{cn}\,u\,\mathrm{cn}\,v}{1 - \mathrm{sn}^2 u - \mathrm{sn}^2 v + k^2\,\mathrm{sn}^2 u\,\mathrm{sn}^2 v}$

(iii) $\mathrm{nc}(u+v) - \mathrm{nc}(u-v) = \dfrac{2\,\mathrm{sn}\,u\,\mathrm{dn}\,u\,\mathrm{sn}\,v\,\mathrm{dn}\,v}{1 - \mathrm{sn}^2 u - \mathrm{sn}^2 v + k^2\,\mathrm{sn}^2 u\,\mathrm{sn}^2 v}$

(iv) $\mathrm{nd}(u+v) + \mathrm{nd}(u-v) = \dfrac{2\,\mathrm{dn}\,u\,\mathrm{dn}\,v}{1 - k^2\,\mathrm{sn}^2 u - k^2\,\mathrm{sn}^2 v + k^2\,\mathrm{sn}^2 u\,\mathrm{sn}^2 v}$

(v) $\mathrm{nd}(u+v) - \mathrm{nd}(u-v) = \dfrac{2k^2\,\mathrm{sn}\,u\,\mathrm{cn}\,u\,\mathrm{sn}\,v\,\mathrm{cn}\,v}{1 - k^2\,\mathrm{sn}^2 u - k^2\,\mathrm{sn}^2 v + k^2\,\mathrm{sn}^2 u\,\mathrm{sn}^2 v}$

18. Show that

(i) $1 - k^2\,\mathrm{sn}^2(u+v)\,\mathrm{sn}^2(u-v) = \dfrac{(1 - k^2\,\mathrm{sn}^4 u)(1 - k^2\,\mathrm{sn}^4 v)}{(1 - k^2\,\mathrm{sn}^2 u\,\mathrm{sn}^2 v)^2}$

(ii) $\dfrac{\mathrm{cn}^2\frac{1}{2}(u+K)\,\mathrm{dn}^2\frac{1}{2}(u+K)}{\mathrm{sn}^2\frac{1}{2}(u+K)} = k'^2\dfrac{1 - \mathrm{sn}\,u}{1 + \mathrm{sn}\,u}$

19. Prove that

(i) $k^2(\mathrm{dn}\,u + k'\,\mathrm{sn}\,u)^2 = (1 - k'\,\mathrm{dn}\,u + k^2\,\mathrm{sn}\,u)(1 + k'\,\mathrm{dn}\,u - k^2\,\mathrm{sn}\,u)$

(ii) $(\mathrm{dn}\,u + k')^2 = (1 + k'\,\mathrm{dn}\,u + k^2\,\mathrm{sn}\,u)(1 + k'\,\mathrm{dn}\,u - k^2\,\mathrm{sn}\,u)$

and deduce that

$$k^2\,\mathrm{sn}^4\tfrac{1}{2}(u+K) = \frac{1 - k'\,\mathrm{dn}\,u + k^2\,\mathrm{sn}\,u}{1 + k'\,\mathrm{dn}\,u + k^2\,\mathrm{sn}\,u}.$$

20. Establish the identities:

(i) $\{1 \pm \mathrm{sn}(u+v)\}\{1 \pm \mathrm{sn}(u-v)\} = (\mathrm{cn}\,v \pm \mathrm{sn}\,u\,\mathrm{dn}\,v)^2/D$

(ii) $\{1 \pm \mathrm{sn}(u+v)\}\{1 \mp \mathrm{sn}(u-v)\} = (\mathrm{cn}\,u \pm \mathrm{sn}\,v\,\mathrm{dn}\,u)^2/D$

(iii) $\{1 \pm \mathrm{cn}(u+v)\}\{1 \pm \mathrm{cn}(u-v) = (\mathrm{cn}\,u \pm \mathrm{cn}\,v)^2/D$

(iv) $\{1 \pm \mathrm{cn}(u+v)\}\{1 \mp \mathrm{cn}(u-v)\} = (\mathrm{sn}\,u\,\mathrm{dn}\,v \pm \mathrm{sn}\,v\,\mathrm{dn}\,u)^2/D$

(v) $\{1 \pm \mathrm{dn}(u+v)\}\{1 \pm \mathrm{dn}(u-v)\} = (\mathrm{dn}\,u \pm \mathrm{dn}\,v)^2/D$

(vi) $\{1 \mp \mathrm{dn}(u+v)\}\{1 \pm \mathrm{dn}(u-v)\} = k^2(\mathrm{sn}\,u\,\mathrm{cn}\,v \pm \mathrm{sn}\,v\,\mathrm{cn}\,u)^2/D$

(vii) $\{1 \pm k\,\text{sn}(u+v)\}\{1 \pm k\,\text{sn}(u-v)\} = (\text{dn}\,v \pm k\,\text{sn}\,u\,\text{cn}\,v)^2/D$

(viii) $\{1 \pm k\,\text{sn}(u+v)\}\{1 \mp k\,\text{sn}(u-v)\} = (\text{dn}\,u \pm k\,\text{sn}\,v\,\text{cn}\,u)^2/D$

where $D = 1 - k^2\,\text{sn}^2 u\,\text{sn}^2 v$.

21. Prove the following identities:

(i) $\text{sn}(u+v)\,\text{cn}(u-v) + \text{sn}(u-v)\,\text{cn}(u+v) = \dfrac{2\,\text{sn}\,u\,\text{cn}\,u\,\text{dn}\,v}{1 - k^2\,\text{sn}^2 u\,\text{sn}^2 v}$

(ii) $\text{cn}(u+v)\,\text{cn}(u-v) + \text{sn}(u+v)\,\text{sn}(u-v) = \dfrac{\text{cn}^2 v - \text{sn}^2 v\,\text{dn}^2 u}{1 - k^2\,\text{sn}^2 u\,\text{sn}^2 v}$

(iii) $\text{dn}(u+v)\,\text{dn}(u-v) - k^2\,\text{cn}(u+v)\,\text{cn}(u-v) = k'^2\dfrac{1 + k^2\,\text{sn}^2 u\,\text{sn}^2 v}{1 - k^2\,\text{sn}^2 u\,\text{sn}^2 v}$

(iv) $\text{dn}(u+v)\,\text{dn}(u-v) - \text{cn}(u+v)\,\text{cn}(u-v) = k'^2\dfrac{\text{sn}^2 u + \text{sn}^2 v}{1 - k^2\,\text{sn}^2 u\,\text{sn}^2 v}$

(v) $\text{sn}(u+v)\,\text{dn}(u-v) + \text{sn}(u-v)\,\text{dn}(u+v) = \dfrac{2\,\text{sn}\,u\,\text{dn}\,u\,\text{cn}\,v}{1 - k^2\,\text{sn}^2 u\,\text{sn}^2 v}$

22. Prove that $\text{sn}(K/3) = 1 - \text{cn}(K/3) = \dfrac{1}{1 + \text{dn}(K/3)}$.

(Hint: Use the identities (2.4.4) and (2.4.5).)

23. If $k^2 = 27/32$, show that $\text{sn}(K/3) = 2/3$, $\text{cn}(2K/3) = 1/3$, $\text{dn}(2K/3) = 1/2$.

24. Establish the following identities:

(i) $1 + k^2\,\text{sn}(u+v)\,\text{sn}(u-v) = (\text{dn}^2 v + k^2\,\text{sn}^2 u\,\text{cn}^2 v)/D$

(ii) $1 + \text{sn}(u+v)\,\text{sn}(u-v) = (\text{cn}^2 v + \text{sn}^2 u\,\text{dn}^2 v)/D$

(iii) $1 - \text{cn}(u+v)\,\text{cn}(u-v) = (\text{sn}^2 u\,\text{dn}^2 v + \text{sn}^2 v\,\text{dn}^2 u)/D$

(iv) $1 - \text{dn}(u+v)\,\text{dn}(u-v) = k^2(\text{sn}^2 u\,\text{cn}^2 v + \text{sn}^2 v\,\text{cn}^2 u)/D$

where $D = 1 - k^2\,\text{sn}^2 u\,\text{sn}^2 v$.

25. Show that

$$\frac{\text{cn}(u+v) - \text{dn}(u+v)}{\text{sn}(u+v)} = \frac{\text{cn}\,u\,\text{dn}\,v - \text{cn}\,v\,\text{dn}\,u}{\text{sn}\,u - \text{sn}\,v}.$$

Deduce that, if $u + v + w + x = 2K$, then

$(\text{cn}\,u\,\text{dn}\,v - \text{cn}\,v\,\text{dn}\,u)(\text{cn}\,w\,\text{dn}\,x - \text{cn}\,x\,\text{dn}\,w) = k'^2(\text{sn}\,u - \text{sn}\,v)(\text{sn}\,w - \text{sn}\,x).$

26. Prove that

$$\int \frac{\text{cn}\,u}{1 + \text{sn}\,u}\,du = \frac{1}{k'}\ln\left(\frac{\text{dn}\,u + k'\,\text{sn}\,u}{\text{dn}\,u + k'}\right).$$

(Hint: Multiply integrand by $(1 - s)/(1 - s)$.) Show, similarly, that

$$\int \frac{\text{dn}\,u}{1 + \text{sn}\,u}\,du = -\frac{\text{cn}\,u}{1 + \text{sn}\,u}.$$

$$\int \frac{\operatorname{sn} u}{1 + \operatorname{cn} u}\, du = \ln\left(\frac{1 + \operatorname{dn} u}{\operatorname{cn} u + \operatorname{dn} u}\right)$$

$$\int \frac{\operatorname{dn} u}{1 + \operatorname{cn} u}\, du = \frac{\operatorname{sn} u}{1 + \operatorname{cn} u}$$

$$\int \frac{\operatorname{sn} u}{1 + \operatorname{dn} u}\, du = \frac{1}{k^2} \ln\left(\frac{1 + \operatorname{cn} u}{\operatorname{cn} u + \operatorname{dn} u}\right)$$

$$\int \frac{\operatorname{cn} u}{1 + \operatorname{dn} u}\, du = \frac{\operatorname{sn} u}{1 + \operatorname{dn} u}$$

$$\int \frac{du}{\operatorname{cn} u + \operatorname{dn} u} = \frac{\operatorname{sn} u}{\operatorname{cn} u + \operatorname{dn} u}.$$

27. Prove that

$$\left(\frac{\operatorname{sn} u \operatorname{dn} u}{\operatorname{cn} u}\right)^2 = \frac{1 - \operatorname{cn} 2u}{1 + \operatorname{cn} 2u}.$$

28. Prove that

$$\operatorname{sc}(iK') = i, \qquad \operatorname{dc}(iK') = k, \qquad \operatorname{ds}(iK') = -ik.$$

29. Show that

$$\operatorname{sn}(u + iv) = \frac{\operatorname{sn} u \operatorname{dn} v + i \operatorname{cn} u \operatorname{dn} u \operatorname{sn} v \operatorname{cn} v}{1 - \operatorname{dn}^2 u \operatorname{sn}^2 v},$$

where functions of v are taken to modulus k'. Hence show that, if $0 < k < 1$ and $\operatorname{sn} w$ is real, then $w = (2n + 1)K + iv$ or $w = u + inK'$ (n an integer, u and v real). Deduce that $\operatorname{sn} w$ takes all real values.

30. Show that, if $\operatorname{cn} w$ is real, then $w = 2nK + iv$ or $w = u + 2inK'$ and deduce that $\operatorname{cn} w$ takes all real values.

31. Show that, if $\operatorname{dn} w$ is real, then $w = nK + iv$ or $w = u + 2inK'$ and deduce that $\operatorname{dn} w$ takes all real values.

32. Establish the identities

(i) $\operatorname{cn}(u + v) = \operatorname{cn} u \operatorname{cn} v - \operatorname{sn} u \operatorname{sn} v \operatorname{dn}(u + v)$,

(ii) $\operatorname{dn} u \operatorname{sn}(u + v) = \operatorname{cn} u \operatorname{sn} v + \operatorname{sn} u \operatorname{cn} v \operatorname{dn}(u + v)$,

(iii) $\operatorname{cn} u \operatorname{sn}(u + v) = \operatorname{dn} u \operatorname{sn} v + \operatorname{sn} u \operatorname{dn} v \operatorname{cn}(u + v)$,

(iv) $\operatorname{cn} u \operatorname{cn} v \operatorname{dn}(u + v) = k'^2 \operatorname{sn} u \operatorname{sn} v + \operatorname{dn} u \operatorname{dn} v \operatorname{cn}(u + v)$,

(v) $\operatorname{cn} u \operatorname{cn}(u + v) = \operatorname{cn} v - \operatorname{sn} u \operatorname{dn} v \operatorname{sn}(u + v)$,

(vi) $\operatorname{dn} u \operatorname{dn}(u + v) = \operatorname{dn} v - k^2 \operatorname{sn} u \operatorname{cn} v \operatorname{sn}(u + v)$.

33. Show that, if $\operatorname{sn}(K/3) = s$, then $k^2 = (2s - 1)/\{s^3(2 - s)\}$.

34. Prove $\displaystyle\int \operatorname{ns} u \operatorname{nd} u\, du = -\ln(\operatorname{ns} u + \operatorname{cs} u) - \frac{k}{k'} \tan^{-1}\left(\frac{k}{k'} \operatorname{cn} u\right).$

35. Taking the functions in identities (2.4.11)–(2.4.13) to have modulus k' and making

use of the results (2.6.12), obtain the identities

(i) $\mathrm{sc}(u+v)+\mathrm{sc}(u-v)=2k^2\,\mathrm{sn}\,u\,\mathrm{cn}\,u\,\mathrm{dn}\,v/D$,

(ii) $\mathrm{nc}(u+v)+\mathrm{nc}(u-v)=2k^2\,\mathrm{cn}\,u\,\mathrm{cn}\,v/D$,

(iii) $\mathrm{dc}(u+v)+\mathrm{dc}(u-v)=2k^2\,\mathrm{cn}\,u\,\mathrm{dn}\,u\,\mathrm{cn}\,v\,\mathrm{dn}\,v/D$,

where $D=\mathrm{dn}^2u\,\mathrm{dn}^2v-k'^2$.

36. Prove that, as $u\to0$,

$$\mathrm{sn}\,u=\frac{\sin\{u\sqrt{(1+k^2)}\}}{\sqrt{(1+k^2)}}+O(u^5),$$

$$\mathrm{cn}\,u=\cos u+O(u^4),$$

$$\mathrm{dn}\,u=\cos ku+O(u^4).$$

37. Prove that

(i) $y=\mathrm{sn}\,x$ is a solution of the equation

$$\frac{d^2y}{dx^2}+(1+k^2)y-2k^2y^3=0.$$

(ii) $y=\mathrm{cn}\,x$ is a solution of the equation

$$\frac{d^2y}{dx^2}+(1-2k^2)y+2k^2y^3=0.$$

(iii) $y=\mathrm{dn}\,x$ is a solution of the equation

$$\frac{d^2y}{dx^2}+(k^2-2)y+2y^3=0.$$

38. If $0<u<K$, establish the inequalities

$$\frac{1}{\mathrm{cn}\,u}>\frac{u}{\mathrm{sn}\,u}>1>\mathrm{dn}\,u>\mathrm{cn}\,u.$$

39. Show that

$$\int\frac{\mathrm{cn}\,u}{1-k\,\mathrm{sn}\,u}\,du=(1+k\,\mathrm{sn}\,u)/(k\,\mathrm{dn}\,u).$$

CHAPTER 3

Elliptic Integrals

3.1. Elliptic Integral of the First Kind

Throughout this chapter, the argument u and modulus k of all elliptic functions will be assumed real and, further, we suppose $0 < k < 1$, unless stated otherwise.

We start by calculating the derivative of the inverse function $\mathrm{sn}^{-1}(x, k)$. Like $\sin^{-1} x$, this function is multivalued, since an arbitrary integral multiple of $4K$ may be added or subtracted (also, since $\mathrm{sn}\,(2K - u) = \mathrm{sn}\,u$, there are other real values; further, the complex period of sn provides other complex values). In this chapter, however, we shall make the value definite by requiring all inverse functions to be taken in the range $(0, K)$. Putting $u = \mathrm{sn}^{-1}(x, k)$, then $x = \mathrm{sn}\,u$ and, hence,

$$\mathrm{d}x/\mathrm{d}u = \mathrm{cn}\,u\,\mathrm{dn}\,u = \sqrt{\{(1 - x^2)(1 - k^2 x^2)\}}, \tag{3.1.1}$$

having used the identities (2.1.9) and (2.1.10). Integration of this equation over the range $(0, x)$ $(0 \leqslant x \leqslant 1)$ for x and the corresponding range $(0, u)$ $(0 \leqslant u \leqslant K)$ for u, yields the result

$$\mathrm{sn}^{-1}(x, k) = u = \int_0^x \{(1 - t^2)(1 - k^2 t^2)\}^{-1/2}\,\mathrm{d}t \qquad (0 \leqslant x \leqslant 1). \tag{3.1.2}$$

This integral is called an *elliptic integral of the first kind* and we note that it can be evaluated by reference to tables of the function $\mathrm{sn}(u, k)$ (used inversely). Observe that the expression under the root sign is biquadratic in x—in section 3.3 we shall provide a general definition of an elliptic integral which features the square root of a polynomial of the fourth degree.

Since $\text{sn}^{-1}(1) = K$, a special case of equation (3.1.2) is

$$K = \int_0^1 \{(1 - t^2)(1 - k^2 t^2)\}^{-1/2} \, dt. \tag{3.1.3}$$

This is the *complete elliptic integral of the first kind* and it will be studied in detail in section 3.8.

Since $K'(k) = K(k')$, we also have

$$K'(k) = \int_0^1 \{(1 - t^2)(1 - k'^2 t^2)\}^{-1/2} \, dt. \tag{3.1.4}$$

Changing the variable by $s = \sqrt{(1 - k'^2 t^2)}/k$ (or $t = \sqrt{(1 - k^2 s^2)}/k'$) we obtain the alternative formula

$$K'(k) = \int_1^{1/k} \{(s^2 - 1)(1 - k^2 s^2)\}^{-1/2} \, ds. \tag{3.1.5}$$

By changing the variable in the integral (3.1.2) using the substitution $t = s/b$, we calculate that

$$\text{sn}^{-1}(x, k) = a \int_0^{bx} \{(a^2 - s^2)(b^2 - s^2)\}^{-1/2} \, ds, \tag{3.1.6}$$

where $0 < b < b/k = a$. It now follows that

$$\int_0^x \{(a^2 - t^2)(b^2 - t^2)\}^{-1/2} \, dt = \frac{1}{a} \text{sn}^{-1}\left(\frac{x}{b}, \frac{b}{a}\right), \tag{3.1.7}$$

where $0 \leqslant x \leqslant b < a$; this is a generalization of equation (3.1.2).

Changing the variable by $t = \sin\theta$, the integral (3.1.2) is reduced to Legendre's form, viz.

$$F(\phi, k) = \text{sn}^{-1}(\sin\phi, k) = \int_0^\phi (1 - k^2 \sin^2\theta)^{-1/2} \, d\theta, \tag{3.1.8}$$

where $x = \sin\phi$. Then, expanding the integrand in ascending powers of k^2 and integrating term by term, we find that

$$\text{sn}^{-1}(\sin\phi, k) = \phi + \tfrac{1}{4}k^2(\phi - \sin\phi\cos\phi) + \cdots, \tag{3.1.9}$$

which is equivalent to

$$u = \text{sn}^{-1}(x, k) = \sin^{-1}x + \tfrac{1}{4}k^2(\sin^{-1}x - x\sqrt{(1 - x^2)}) + \cdots. \tag{3.1.10}$$

This series can now be inverted to expand $\text{sn}\, u$ in powers of k^2, thus:

$$\begin{aligned}
\text{sn}(u, k) = x &= \sin[u + \tfrac{1}{4}k^2(x\sqrt{(1 - x^2)} - \sin^{-1}x) + O(k^4)] \\
&= \sin u + \tfrac{1}{4}k^2(x\sqrt{(1 - x^2)} - \sin^{-1}x)\cos u + O(k^4) \\
&= \sin u + \tfrac{1}{4}k^2(\sin u \cos u - u)\cos u + O(k^4), \tag{3.1.11}
\end{aligned}$$

after substituting $x = \sin u + O(k^2)$ in the term $O(k^2)$. Higher-order terms can be found explicitly by repetition of this procedure.

Values of $F(\phi, k)$ will be found in Table D (p. 285).

3.2. Further Elliptic Integrals Reducible to Canonical Form

By differentiating the remaining eleven inverse elliptic functions using the results (2.5.3)–(2.5.13), it is easy to express these functions as elliptic integrals whose forms are similar to the one given in equation (3.1.2). However, since all these inverse functions can be transformed into the form of an inverse sn, it must be possible, by some change of variable, to reduce the corresponding elliptic integrals to the canonical form (3.1.2).

For example, put $u = \mathrm{cn}^{-1} x$, so that $x = \mathrm{cn}\, u$. Then

$$dx/du = -\,\mathrm{sn}\, u \,\mathrm{dn}\, u = -\sqrt{\{(1 - x^2)(k'^2 + k^2 x^2)\}}. \qquad (3.2.1)$$

Integration over the range $(0, u)$ for u and $(1, x)$ for x now leads to the result

$$\mathrm{cn}^{-1}(x, k) = \int_x^1 \frac{dt}{\sqrt{\{(1 - t^2)(k'^2 + k^2 t^2)\}}} \qquad (0 \leqslant x \leqslant 1). \qquad (3.2.2)$$

This can be generalized by making the substitutions $t = s/b$, $k = b/\sqrt{(a^2 + b^2)}$. After some manipulation, we arrive at the formula

$$\frac{1}{\sqrt{(a^2 + b^2)}} \mathrm{cn}^{-1}\left[\frac{x}{b}, \frac{b}{\sqrt{(a^2 + b^2)}}\right]$$

$$= \int_x^b \frac{dt}{\sqrt{\{(a^2 + t^2)(b^2 - t^2)\}}} \qquad (0 \leqslant x \leqslant b). \qquad (3.2.3)$$

But $\mathrm{sn}\, u = (1 - \mathrm{cn}^2 u)^{1/2}$, so that $\mathrm{cn}^{-1} x = \mathrm{sn}^{-1}(1 - x^2)^{1/2}$. It follows that the last result can be expressed in the form

$$\frac{1}{\sqrt{(a^2 + b^2)}} \mathrm{sn}^{-1}\left[\frac{\sqrt{(b^2 - x^2)}}{b}, \frac{b}{\sqrt{(a^2 + b^2)}}\right] = \int_x^b \frac{dt}{\sqrt{\{(a^2 + t^2)(b^2 - t^2)\}}} \qquad (3.2.4)$$

Comparison with (3.1.2) now suggests that our present elliptic integral can be reduced to canonical form by the substitution $s = \sqrt{(b^2 - t^2)}/b$. Making this substitution, we find that

$$\int_x^b \frac{dt}{\sqrt{\{(a^2 + t^2)(b^2 - t^2)\}}} = \frac{1}{\sqrt{(a^2 + b^2)}} \int_0^{\sqrt{(1 - x^2/b^2)}} \frac{ds}{\sqrt{\{(1 - s^2)(1 - k^2 s^2)\}}},$$

$$(3.2.5)$$

where $k = b/\sqrt{(a^2 + b^2)}$. The result (3.2.4) can now be obtained by application of the formula (3.1.2).

As an exercise, the reader may now construct the elliptic integrals which can be evaluated in terms of the remaining ten inverse functions and also find, in each case, the change of variable which reduces the integral to canonical form. We list these results below:

$$\int_x^b \{(a^2 - t^2)(b^2 - t^2)\}^{-1/2}\, dt$$

$$= \frac{1}{a} cd^{-1}\left[\frac{x}{b}, \frac{b}{a}\right], \qquad 0 \leqslant x \leqslant b < a, \quad s^2 = \frac{a^2(b^2 - t^2)}{b^2(a^2 - t^2)}, \qquad (3.2.6)$$

$$\int_0^x \{(a^2 + t^2)(b^2 - t^2)\}^{-1/2}\, dt$$

$$= \frac{1}{\sqrt{(a^2 + b^2)}} sd^{-1}\left[\frac{\sqrt{(a^2 + b^2)}x}{ab}, \frac{b}{\sqrt{(a^2 + b^2)}}\right], \qquad 0 \leqslant x \leqslant b,$$

$$s^2 = \frac{(a^2 + b^2)t^2}{b^2(a^2 + t^2)}, \qquad (3.2.7)$$

$$\int_a^x \{(t^2 - a^2)(t^2 - b^2)\}^{-1/2}\, dt$$

$$= \frac{1}{a} dc^{-1}\left(\frac{x}{a}, \frac{b}{a}\right), \qquad b < a \leqslant x, \quad s^2 = \frac{t^2 - a^2}{t^2 - b^2}, \qquad (3.2.8)$$

$$\int_x^\infty \{(t^2 - a^2)(t^2 - b^2)\}^{-1/2}\, dt$$

$$= \frac{1}{a} ns^{-1}\left(\frac{x}{a}, \frac{b}{a}\right), \qquad b < a \leqslant x, \quad s = a/t, \qquad (3.2.9)$$

$$\int_b^x \{(a^2 - t^2)(t^2 - b^2)\}^{-1/2}\, dt$$

$$= \frac{1}{a} nd^{-1}\left[\frac{x}{b}, \frac{\sqrt{(a^2 - b^2)}}{a}\right], \qquad b \leqslant x \leqslant a, \quad s^2 = \frac{a^2(t^2 - b^2)}{(a^2 - b^2)t^2}, \qquad (3.2.10)$$

$$\int_x^a \{(a^2 - t^2)(t^2 - b^2)\}^{-1/2}\, dt$$

$$= \frac{1}{a} dn^{-1}\left[\frac{x}{a}, \frac{\sqrt{(a^2 - b^2)}}{a}\right], \qquad b \leqslant x \leqslant a, \quad s^2 = \frac{a^2 - t^2}{a^2 - b^2}, \qquad (3.2.11)$$

$$\int_a^x \{(t^2 - a^2)(t^2 + b^2)\}^{-1/2}\, dt$$

$$= \frac{1}{\sqrt{(a^2 + b^2)}} nc^{-1}\left[\frac{x}{a}, \frac{b}{\sqrt{(a^2 + b^2)}}\right], \qquad a \leqslant x, \quad s^2 = 1 - a^2/t^2, \qquad (3.2.12)$$

$$\int_x^\infty \{(t^2 - a^2)(t^2 + b^2)\}^{-1/2} \, dt$$

$$= \frac{1}{\sqrt{(a^2 + b^2)}} ds^{-1}\left[\frac{x}{\sqrt{(a^2 + b^2)}}, \frac{b}{\sqrt{(a^2 + b^2)}}\right], \quad a \leqslant x, \quad s^2 = \frac{a^2 + b^2}{t^2 + b^2},$$

$$\text{(3.2.13)}$$

$$\int_0^x \{(t^2 + a^2)(t^2 + b^2)\}^{-1/2} \, dt$$

$$= \frac{1}{a} sc^{-1}\left[\frac{x}{b}, \frac{\sqrt{(a^2 - b^2)}}{a}\right], \quad 0 < b < a, \quad 0 \leqslant x, \quad s^2 = \frac{t^2}{t^2 + a^2}, \quad \text{(3.2.14)}$$

$$\int_x^\infty \{(t^2 + a^2)(t^2 + b^2)\}^{-1/2} \, dt$$

$$= \frac{1}{a} cs^{-1}\left[\frac{x}{a}, \frac{\sqrt{(a^2 - b^2)}}{a}\right], \quad 0 < b < a, \quad 0 \leqslant x, \quad s^2 = \frac{a^2}{t^2 + a^2}. \quad \text{(3.2.15)}$$

The above listed formulae can be linked in complementary pairs. For example, if $u = cd^{-1}x$, then $x = cd\, u = sn(K - u)$ (equations (2.2.17)) and it follows that $sn^{-1}x = K - u = K - cd^{-1}x$, i.e.,

$$sn^{-1}x + cd^{-1}x = K(k). \quad \text{(3.2.16)}$$

As a consequence, if we add the results (3.1.7) and (3.2.6), we get

$$\int_0^b \{(a^2 - t^2)(b^2 - t^2)\}^{-1/2} \, dt = \frac{1}{a} K(b/a), \quad \text{(3.2.17)}$$

which is derivable from equation (3.1.3) by putting $t = s/b, k = b/a$.

Similarly, from the first of the identities (2.2.18), we can show that

$$cn^{-1}x + sd^{-1}(x/k') = K(k), \quad \text{(3.2.18)}$$

thus linking the formulae (3.2.3), (3.2.7), whose sum yields

$$\int_0^b \{(a^2 + t^2)(b^2 - t^2)\}^{-1/2} \, dt = \frac{1}{\sqrt{(a^2 + b^2)}} K\left[\frac{b}{\sqrt{(a^2 + b^2)}}\right]. \quad \text{(3.2.19)}$$

This result can also be shown to be equivalent to the complete integral formula (3.1.3) by making the substitutions $t^2 = \{(a^2 + b^2)s^2\}/\{b^2(a^2 + s^2)\}, k = b/\sqrt{(a^2 + b^2)}$ in the latter result.

The reader should establish four other identities linking the elliptic integrals as follows:

$$dc^{-1}x + ns^{-1}x = K(k), \quad \text{(3.2.20)}$$

$$nd^{-1}x + dn^{-1}(k'x) = K(k), \quad \text{(3.2.21)}$$

$$\mathrm{nc}^{-1}x + \mathrm{ds}^{-1}(k'x) = K(k), \tag{3.2.22}$$

$$\mathrm{sc}^{-1}x + \mathrm{cs}^{-1}(k'x) = K(k). \tag{3.2.23}$$

(3.2.20) links (3.2.8) and (3.2.9); (3.2.21) links (3.2.10) and (3.2.11); (3.2.22) links (3.2.12) and (3.2.13); and (3.2.23) links (3.2.14) and (3.2.15). Four new integrals for K arise as a result of these linkages.

3.3. General Elliptic Integrals

If $R(x, y)$ is a rational function of x and y, and y^2 is a cubic or quartic polynomial in x whose zeros are all different, then the integral

$$\int R(x, y)\,dx \tag{3.3.1}$$

is termed a *general elliptic integral*. If y^2 had a pair of identical zeros α, then $(x - \alpha)^2$ might be removed as a factor and $y = (x - \alpha)w$, where w is the square root of a quadratic; the integrand would then be of the form $R_0(x, w)$, where R_0 is rational in x and w, and the integral could be evaluated in terms of elementary functions.

Writing $R(x, y) = P(x, y)/Q(x, y)$, where P and Q are polynomials in x and y, we shall suppose the coefficients in these polynomials to be real. Since $Q(x, y)Q(x, -y)$ is an even function of y, it must be a polynomial in y^2 and, hence, can be expressed as a polynomial $L(x)$ in x alone. Consider the polynomial $yP(x, y)Q(x, -y)$; replacing powers of y^2 everywhere by powers of the cubic or quartic polynomial in x, we obtain the result

$$yP(x, y)Q(x, -y) = M(x) + yN(x), \tag{3.3.2}$$

where $M(x), N(x)$ are polynomials in x alone. Thus,

$$R(x, y) = \frac{yP(x, y)Q(x, -y)}{yQ(x, y)Q(x, y)} = \frac{M(x) + yN(x)}{yL(x)}$$

$$= y^{-1}R_1(x) + R_2(x), \tag{3.3.3}$$

where $R_1(x), R_2(x)$ are rational functions of x alone. Since $R_2(x)$ can be integrated by elementary methods, it now remains to calculate the integral

$$\int y^{-1}R_1(x)\,dx. \tag{3.3.4}$$

If y^2 is a quartic, we can always separate it into a pair of quadratic factors S_1, S_2 having real coefficients, thus:

$$y^2 = S_1 S_2 = (a_1 x^2 + 2b_1 x + c_1)(a_2 x^2 + 2b_2 x + c_2). \tag{3.3.5}$$

If y^2 is a cubic, then we take $a_2 = 0$. Consider the quadratic expression

$S_1 + \lambda S_2$; this has coincident zeros and is a perfect square whenever

$$D(\lambda) \equiv (a_1 + \lambda a_2)(c_1 + \lambda c_2) - (b_1 + \lambda b_2)^2 = 0. \tag{3.3.6}$$

There are four cases to consider: (i) Suppose at least two of the zeros of y^2 are complex. If we take these to be the zeros of S_1, then $a_1 c_1 - b_1^2 > 0$ and $D > 0$ when $\lambda = 0$. When $\lambda = -a_1/a_2$, then $D < 0$ provided $a_1/a_2 \neq b_1/b_2$ (the case where $a_1/a_2 = b_1/b_2$ is treated as case (iv) later). (If $a_2 = 0$, then $D(\lambda)$ is clearly negative for sufficiently large λ.) Hence $D = 0$ has real, distinct, roots. (ii) If y^2 is a quartic and all its zeros are real, let them be $\alpha > \beta > \gamma > \delta$ and choose S_1, S_2 so that (α, β) are zeros of S_1 and (γ, δ) are zeros of S_2. The condition that $D = 0$ should have real distinct roots is found to be

$$(a_1 c_2 - a_2 c_1)^2 + 4(a_1 b_2 - a_2 b_1)(c_1 b_2 - c_2 b_1) > 0. \tag{3.3.7}$$

Substituting $2b_1 = -a_1(\alpha + \beta), c_1 = a_1 \alpha \beta, 2b_2 = -a_2(\gamma + \delta), c_2 = a_2 \gamma \delta$, this condition reduces to the form

$$a_1^2 a_2^2 (\alpha - \gamma)(\beta - \gamma)(\alpha - \delta)(\beta - \delta) > 0, \tag{3.3.8}$$

which is clearly satisfied. (iii) If y^2 is a cubic and all its zeros are real, let them be $\alpha > \beta > \gamma$. Taking (α, β) to be the zeros of S_1 and γ the zero of S_2, we substitute $2b_1 = -a_1(\alpha + \beta), c_1 = a_1 \alpha \beta, a_2 = 0, c_2 = -2b_2 \gamma$ in the condition (3.3.7) to give

$$4a_1^2 b_2^2 (\alpha - \gamma)(\beta - \gamma) > 0, \tag{3.3.9}$$

which again is clearly satisfied. Hence, in cases (i)–(iii), $D = 0$ has real, distinct, roots.

In these cases, let λ_1, λ_2 be the roots of $D(\lambda) = 0$. Then

$$\left.\begin{array}{l} S_1 + \lambda_1 S_2 = (a_1 + \lambda_1 a_2)(x - p)^2, \\ S_1 + \lambda_2 S_2 = (a_1 + \lambda_2 a_2)(x - q)^2. \end{array}\right\} \tag{3.3.10}$$

Solving for S_1 and S_2, we show that these quadratics can be expressed in the form

$$S_1 = A_1(x - p)^2 + B_1(x - q)^2, \qquad S_2 = A_2(x - p)^2 + B_2(x - q)^2, \tag{3.3.11}$$

where A_1, B_1, A_2, B_2, p, q are all real numbers.

(iv) There remains the case where $a_1/a_2 = b_1/b_2$. Then, clearly,

$$S_1 = a_1(x - p)^2 + B_1, \qquad S_2 = a_2(x - p)^2 + B_2. \tag{3.3.12}$$

If S_1 and S_2 have been expressed in the forms (3.3.11), we change the variable in the integral (3.3.4) by the transformation

$$t = (x - p)/(x - q). \tag{3.3.13}$$

Then

$$\frac{dx}{y} = \frac{dx}{\sqrt{(S_1 S_2)}} = \pm \frac{(p - q)^{-1} dt}{\sqrt{\{(A_1 t^2 + B_1)(A_2 t^2 + B_2)\}}}. \tag{3.3.14}$$

If, however, S_1 and S_2 are expressible as at (3.3.12), the change of variable $t = x - p$ achieves this form more simply.

For numerical computational purposes, it is more convenient to derive the transformation (3.3.13) directly, thus:

If $(ax^4 + bx^3 + cx^2 + dx + e)$ denotes the quartic y^2, we seek a transformation

$$x = \frac{t + \lambda}{\mu t + 1} \qquad (3.3.15)$$

which brings it to the form

$$(At^4 + Bt^2 + C)/(\mu t + 1)^4. \qquad (3.3.16)$$

After substituting from equation (3.3.15) into the quartic, we find that the conditions for the vanishing of the terms in t^3 and t are respectively as follows:

$$\left.\begin{array}{l} 4a\lambda + b(3\lambda\mu + 1) + 2c\mu(\lambda\mu + 1) + d\mu^2(\lambda\mu + 3) + 4e\mu^3 = 0, \\ 4a\lambda^3 + b\lambda^2(\lambda\mu + 3) + 2c\lambda(\lambda\mu + 1) + d(3\lambda\mu + 1) + 4e\mu = 0. \end{array}\right\} \qquad (3.3.17)$$

Eliminating a between these equations and taking out a factor $(\lambda\mu - 1)$ (note that the transformation (3.3.15) degenerates if $\lambda\mu = 1$), we arrive at the equation

$$2b\lambda^2 + 2c\lambda(1 + \lambda\mu) + d(1 + 4\lambda\mu + \lambda^2\mu^2) + 4e\mu(1 + \lambda\mu) = 0. \quad (3.3.18)$$

But, the second of the conditions (3.3.17) is equivalent to

$$\mu = -\frac{4a\lambda^3 + 3b\lambda^2 + 2c\lambda + d}{b\lambda^3 + 2c\lambda^2 + 3d\lambda + 4e}. \qquad (3.3.19)$$

Substituting for μ from this last result into the condition (3.3.18), after some rearrangement we arrive at the following sextic equation for λ:

$$P\lambda^6 + Q\lambda^5 + R\lambda^4 + S\lambda^3 + T\lambda^2 + U\lambda + V = 0, \qquad (3.3.20)$$

where

$$\left.\begin{array}{l} P = b^3 + 8a^2d - 4abc, \\ Q = 2(16a^2e - 4ac^2 + b^2c + 2abd), \\ R = 5(b^2d + 8abe - 4acd), \\ S = 20(b^2e - ad^2), \\ T = -5(bd^2 + 8ade - 4bce), \\ U = -2(16ae^2 - 4c^2e + cd^2 + 2bde), \\ V = -d^3 - 8be^2 + 4cde. \end{array}\right\} \qquad (3.3.21)$$

Our earlier analysis establishes that there is a real transformation of the type (3.3.15) effecting the required reduction and the sextic (3.3.20) must, accordingly, possess a real root (and hence two real roots). Such a root is easily calculated using an iterative process programmed on a computer (see

Appendix C for a BASIC program which can be run on a BBC Acorn computer) and μ then follows from equation (3.3.19).

If μ is replaced by $1/\mu$ in the conditions (3.3.17), it may be checked that they are then symmetric in λ and μ. We conclude that $1/\mu$ must also be a root of the sextic (3.3.20). These two distinct real roots lead to alternative transformations by which the reduction can be effected (of course, if $\mu = 0$, the root $1/\mu$ fails to provide a transformation). There may be further real roots of the sextic, generating yet more transformations.

It remains to show that one at least of these transformations reduces the quartic to a form (3.3.16) in which $(At^4 + Bt^2 + C)$ can be separated into real quadratic factors $A(t^2 + \alpha)(t^2 + \beta)$, so that the standard forms (3.2.6)–(3.2.15) become applicable.

Suppose that y^2 has been reduced to the form (3.3.16), with $B^2 < 4AC$, so that separation into real factors is not possible. Then we have reduced the integral (3.3.4) to the same form, but with $y^2 = ax^4 + cx^2 + e$, where $c^2 < 4ae$. Now apply the transformation (3.3.15) once again. After putting $b = d = 0$ in the sextic (3.3.20), we find this reduces to

$$2(16a^2e - 4ac^2)\lambda^5 - 2(16ae^2 - 4c^2e)\lambda = 0, \tag{3.3.22}$$

which has the root $\lambda = (e/a)^{1/4}$; since $ae > \frac{1}{4}c^2 \geqslant 0$, this root is real. Substituting this value of λ in equation (3.3.19), we find $\mu = -(a/e)^{1/4} = -1/\lambda$. Thus, the transformation is

$$x = \lambda \frac{\lambda + t}{\lambda - t}. \tag{3.3.23}$$

It can now be checked that this transformation reduces the quartic thus:

$$ax^4 + cx^2 + e = (e/a)^2(\lambda - t)^{-4}(At^4 + 2Bt^2 + C), \tag{3.3.24}$$

where

$$A = 2a + c(a/e)^{1/2}, \qquad B = 6(ae)^{1/2} - c, \qquad C = 2e + c(e/a)^{1/2}. \tag{3.3.25}$$

Since $B^2 - AC = 16(ae)^{1/2}\{2(ae)^{1/2} - c\} > 0$, the quartic now separates into quadratic factors as required.

In general, the resultant of the two transformations of the type (3.3.15) which have been performed will not be a transformation of this type. However, the resultant transformation can be expressed as

$$x = \frac{pt + q}{rt + s} = \frac{\dfrac{p}{s}t + \dfrac{q}{s}}{\dfrac{r}{p}\left(\dfrac{p}{s}t\right) + 1}, \tag{3.3.26}$$

which is a transformation of type (3.3.15) from x to pt/s. If, therefore, we perform the transformation

$$x = \frac{t + q/s}{(r/p)t + 1} \tag{3.3.27}$$

on the original quartic, it will then separate into the quadratic factors already obtained, but with t replaced by st/p. However, the factors retain the form $A(t^2 + \alpha)(t^2 + \beta)$ after this replacement and thus we have proved that a transformation of the type (3.3.15) exists which will bring the quartic to the form (3.3.16) with a biquadratic expression factorizable into real factors.

The integral (3.3.4) has now been reduced to the form

$$\int R_3(t)\{(A_1 t^2 + B_1)(A_2 t^2 + B_2)\}^{-1/2}\, dt, \tag{3.3.28}$$

where R_3 is rational. If $R_3(t) = P(t)/Q(t)$, here P and Q are polynomials, we can separate R_3 into even and odd components thus:

$$R_3(t) = \frac{P(t)Q(-t)}{Q(t)Q(-t)} = \frac{M(t^2) + tN(t^2)}{L(t^2)} = R_4(t^2) + tR_5(t^2), \tag{3.3.29}$$

where L, M, N are polynomials and R_4, R_5 are rational. The integral (3.3.28) now separates thus:

$$\int R_4(t^2)X^{-1/2}\, dt + \int tR_5(t^2)X^{-1/2}\, dt, \tag{3.3.30}$$

where $X = \{(A_1 t^2 + B_1)(A_2 t^2 + B_2)\}$. The second of these integrals can be reduced by means of the transformation $t^2 = u$ to one which can be evaluated in terms of elementary functions. Applying the method of partial fractions to $R_4(t^2)$, this rational function can be expressed as a sum of terms of the types t^{2m} and $(t^2 + \gamma)^{-n}$, where m is a positive or negative integer or zero and n is a positive integer (γ may be complex). Thus, the first of the integrals (3.3.30) separates into integrals of the types

$$I_{2m} = \int t^{2m}\{(A_1 t^2 + B_1)(A_2 t^2 + B_2)\}^{-1/2}\, dt, \tag{3.3.31}$$

$$J_n = \int (t^2 + \gamma)^{-n}\{(A_1 t^2 + B_1)(A_2 t^2 + B_2)\}^{-1/2}\, dt. \tag{3.3.32}$$

Reduction formulae can now be found for these integrals thus: Writing $W = \{(A_1 t^2 + B_1)(A_2 t^2 + B_2)\}^{1/2}$, we find that

$$\frac{d}{dt}(t^{2m-1}W) = (2m - 1)t^{2m-2}(A_1 t^2 + B_1)(A_2 t^2 + B_2)W^{-1}$$
$$+ t^{2m}(2A_1 A_2 t^2 + A_1 B_2 + A_2 B_1)W^{-1}. \tag{3.3.33}$$

By integrating this identity, we are therefore able to relate I_{2m+2}, I_{2m}, and I_{2m-2}, and thence reduce I_{2m} to a linear combination of I_0 and I_2. I_0 is an elliptic integral of the first kind already studied. It remains to calculate

$$I_2 = \int t^2\{(A_1 t^2 + B_1)(A_2 t^2 + B_2)\}^{-1/2}\, dt, \tag{3.3.34}$$

which is termed an *elliptic integral of the second kind*.

To reduce J_n, note that

$$\frac{d}{dt}\{t(t^2 + \gamma)^{-n+1}W\} = (t^2 + \gamma)^{-n}W^{-1}C(t^2), \qquad (3.3.35)$$

where $C(t^2)$ is a cubic in t^2. Rearranging C as a cubic in $(t^2 + \gamma)$, we show that

$$\frac{d}{dt}\{t(t^2 + \gamma)^{-n+1}W\} = [a(t^2 + \gamma)^{-n+3} + b(t^2 + \gamma)^{-n+2}$$

$$+ c(t^2 + \gamma)^{-n+1} + d(t^2 + \gamma)^{-n}]W^{-1}, \qquad (3.3.36)$$

where a, b, c, d are constants. Integration of this identity now provides a reduction formula relating J_n, J_{n-1}, J_{n-2}, and J_{n-3}. By repeated use of this, J_n can be expressed in terms of J_1, J_0, and J_{-1} (since $d = 0$ when $n = 1$, it is not possible to proceed further by expressing J_1 in terms of J_0, J_{-1}, and J_{-2}). J_0 is an integral of the first kind and J_{-1} is a sum of integrals of the first and second kinds. Thus, the only additional form we need consider is

$$J_1 = \int (t^2 + \gamma)^{-1}\{(A_1t^2 + B_1)(A_2t^2 + B_2)\}^{-1/2} dt. \qquad (3.3.37)$$

J_1 is called an *elliptic integral of the third kind*.

3.4. Elliptic Integrals of the Second Kind

One is I_2 given at (3.3.34).

The range of possibilities for the signs of A_1, B_1, A_2, B_2 has already been explored in connection with the integral of the first kind (*vide* the list of formulae (3.1.5), (3.2.3), (3.2.6)–(3.2.15)). In each case, the formula for the integral of the first kind can be appealed to to provide a change of variable which will reduce the corresponding integral of the second kind. For example, it follows from (3.2.7) that

$$\{(a^2 + t^2)(b^2 - t^2)\}^{-1/2} = \frac{1}{\sqrt{(a^2 + b^2)}} \frac{d}{dt} \text{sd}^{-1}\left[\frac{\sqrt{(a^2 + b^2)}t}{ab}, \frac{b}{\sqrt{(a^2 + b^2)}}\right] \qquad (3.4.1)$$

so that

$$\int t^2\{(a^2 + t^2)(b^2 - t^2)\}^{-1/2} dt = (a^2 + b^2)^{-1/2} \int t^2 \frac{d}{dt} \text{sd}^{-1}\left[\frac{\sqrt{(a^2 + b^2)}t}{ab}\right] dt,$$

$$= a^2b^2(a^2 + b^2)^{-3/2} \int \text{sd}^2u\, du, \qquad (3.4.2)$$

where $t = ab(\text{sd}u)/\sqrt{(a^2 + b^2)}$. An equivalent reduction of this integral of the second kind can be made by appeal to (3.2.3), viz.

$$\int t^2\{(a^2+t^2)(b^2-t^2)\}^{-1/2}\,dt = -b^2(a^2+b^2)^{-1/2}\int \mathrm{cn}^2 u\,du, \quad (3.4.3)$$

where, now, $t = b\,\mathrm{cn}\,u$ (the modulus is $b/\sqrt{(a^2+b^2)}$ in both cases).

We list below the remaining integrals of the second kind, reduced in this manner: With $a > b$ everywhere, except possibly (3.4.10) and (3.4.11), we find

$$\int t^2\{(a^2-t^2)(b^2-t^2)\}^{-1/2}\,dt = \frac{b^2}{a}\int \mathrm{sn}^2 u\,du, \qquad t = b\,\mathrm{sn}\,u, \quad k = b/a, \quad (3.4.4)$$

$$\int t^2\{(a^2-t^2)(b^2-t^2)\}^{-1/2}\,dt = -\frac{b^2}{a}\int \mathrm{cd}^2 u\,du, \qquad t = b\,\mathrm{cd}\,u, \quad k = b/a, \quad (3.4.5)$$

$$\int t^2\{(t^2-a^2)(t^2-b^2)\}^{-1/2}\,dt = a\int \mathrm{dc}^2 u\,du, \qquad t = a\,\mathrm{dc}\,u, \quad k = b/a, \quad (3.4.6)$$

$$\int t^2\{(t^2-a^2)(t^2-b^2)\}^{-1/2}\,dt = -a\int \mathrm{ns}^2 u\,du, \qquad t = a\,\mathrm{ns}\,u, \quad k = b/a, \quad (3.4.7)$$

$$\int t^2\{(a^2-t^2)(t^2-b^2)\}^{-1/2}\,dt = \frac{b^2}{a}\int \mathrm{nd}^2 u\,du, \qquad t = b\,\mathrm{nd}\,u, \quad k' = b/a, \quad (3.4.8)$$

$$\int t^2\{(a^2-t^2)(t^2-b^2)\}^{-1/2}\,dt = -a\int \mathrm{dn}^2 u\,du, \qquad t = a\,\mathrm{dn}\,u, \quad k' = b/a, \quad (3.4.9)$$

$$\int t^2\{(t^2-a^2)(t^2+b^2)\}^{-1/2}\,dt = \frac{a^2}{\sqrt{(a+b^2)}}\int \mathrm{nc}^2 u\,du, \qquad t = a\,\mathrm{nc}\,u,$$

$$k = \frac{b}{\sqrt{(a^2+b^2)}}, \quad (3.4.10)$$

$$\int t^2\{(t^2-a^2)(t^2+b^2)\}^{-1/2}\,dt = -\sqrt{(a^2+b^2)}\int \mathrm{ds}^2 u\,du,$$

$$t = \sqrt{(a^2+b^2)}\,\mathrm{ds}\,u, \quad k = \frac{b}{\sqrt{(a^2+b^2)}}, \quad (3.4.11)$$

$$\int t^2\{(t^2+a^2)(t^2+b^2)\}^{-1/2}\,dt = \frac{b^2}{a}\int \mathrm{sc}^2 u\,du, \qquad t = b\,\mathrm{sc}\,u, \quad k' = b/a, \quad (3.4.12)$$

$$\int t^2\{(t^2+a^2)(t^2+b^2)\}^{-1/2}\,dt = -a\int \operatorname{cs}^2 u\,du, \qquad t = a\operatorname{cs}u, \quad k'=b/a.$$

(3.4.13)

The integrals of the squares of the Jacobi elliptic functions can all be reduced to the integral of $\operatorname{dn}^2 u$ (with the same modulus). Thus, it can be verified by differentiation that

$$\int \operatorname{dn}^2 u\,du = u - k^2\int \operatorname{sn}^2 u\,du, \tag{3.4.14}$$

$$= k'^2 u + k^2\int \operatorname{cn}^2 u\,du, \tag{3.4.15}$$

$$= u - \operatorname{dn}u\operatorname{cs}u - \int \operatorname{ns}^2 u\,du, \tag{3.4.16}$$

$$= k'^2 u + \operatorname{dn}u\operatorname{sc}u - k'^2\int \operatorname{nc}^2 u\,du, \tag{3.4.17}$$

$$= k^2\operatorname{sn}u\operatorname{cd}u + k'^2\int \operatorname{nd}^2 u\,du, \tag{3.4.18}$$

$$= \operatorname{dn}u\operatorname{sc}u - k'^2\int \operatorname{sc}^2 u\,du, \tag{3.4.19}$$

$$= k'^2 u + k^2\operatorname{sn}u\operatorname{cd}u + k^2 k'^2\int \operatorname{sd}^2 u\,du, \tag{3.4.20}$$

$$= u + k^2\operatorname{sn}u\operatorname{cd}u - k^2\int \operatorname{cd}^2 u\,du, \tag{3.4.21}$$

$$= -\operatorname{dn}u\operatorname{cs}u - \int \operatorname{cs}^2 u\,du, \tag{3.4.22}$$

$$= k'^2 u - \operatorname{dn}u\operatorname{cs}u - \int \operatorname{ds}^2 u\,du, \tag{3.4.23}$$

$$= u + \operatorname{dn}u\operatorname{sc}u - \int \operatorname{dc}^2 u\,du. \tag{3.4.24}$$

We conclude that any integral of the second kind can always be reduced to an evaluation of Jacobi's epsilon function $E(u,k)$ defined by

$$E(u,k) = \int_0^u \operatorname{dn}^2 v\,dv = \int_0^\tau \sqrt{\left[\frac{1-k^2 t^2}{1-t^2}\right]}\,dt \qquad (t=\operatorname{sn}v,\ \tau=\operatorname{sn}u). \tag{3.4.25}$$

Legendre's form of this integral is derived by making the transformation $\operatorname{sn}v = \sin\theta$, $\operatorname{sn}u = \sin\phi$. Regarded as a function of ϕ and k, the integral will be denoted by $D(\phi, k)$.* Thus

$$D(\phi, k) = \int_0^\phi \sqrt{(1 - k^2 \sin^2 \theta)} \, d\theta. \tag{3.4.26}$$

Since, by equation (2.7.16), $\phi = \operatorname{am} u$, the relationship between the two notations is expressed by

$$E(u, k) = D(\operatorname{am} u, k). \tag{3.4.27}$$

Table D (p. 289) sets out values of $D(\phi, k)$ for $k = 0.1(0.1)0.9$.

Alternatively, we can take the canonical form of the integral of the second kind to be

$$\int \operatorname{sn}^2 u \, du = \int t^2 \{(1 - t^2)(1 - k^2 t^2)\}^{-1/2} \, dt. \tag{3.4.28}$$

3.5. Properties of the Function E(u)

$E(u)$ can be expressed in terms of theta functions thus: Differentiating the identity (1.4.19) twice with respect to y and then putting $y = 0$, we first show that

$$\frac{d}{dx}\left[\frac{\theta_4'(x)}{\theta_4(x)}\right] = \frac{\theta_4''(0)}{\theta_4(0)} - \theta_2^4(0) \operatorname{sn}^2 u, \tag{3.5.1}$$

where $u = x\theta_3^2(0)$ (see Exercise 16 in Chapter 2). Hence, recalling that $k = \theta_2^2(0)/\theta_3^2(0)$ (equations (2.1.7)), we calculate that

$$\operatorname{dn}^2 u = \frac{1}{\theta_3^2(0)} \frac{d}{du}\left[\frac{\theta_4'(x)}{\theta_4(x)}\right] + 1 - \frac{\theta_4''(0)}{\theta_3^4(0)\theta_4(0)}. \tag{3.5.2}$$

It now follows that

$$E(u) = \frac{1}{\theta_3^2(0)} \cdot \frac{\theta_4'(x)}{\theta_4(x)} + \left[1 - \frac{\theta_4''(0)}{\theta_3^4(0)\theta_4(0)}\right] u. \tag{3.5.3}$$

The *complete integral of the second kind* is denoted by E and is defined by the equation

$$E = E(k) = \int_0^K \operatorname{dn}^2 u \, du = E(K, k) = \int_0^{\pi/2} \sqrt{(1 - k^2 \sin^2 \theta)} \, d\theta = D(\tfrac{1}{2}\pi, k). \tag{3.5.4}$$

Putting $u = K$, $x = \tfrac{1}{2}\pi$ in equation (3.5.3), we deduce that

$$E = \left[1 - \frac{\theta_4''(0)}{\theta_3^4(0)\theta_4(0)}\right] K, \tag{3.5.5}$$

---

* The usual notation is $E(\phi, k)$, but this leads to confusion with the complete integral $E(k)$ defined in the next section.

since $\theta_4'(\tfrac{1}{2}\pi) = 0$. Equation (3.5.3) can now be written

$$E(u) = \frac{1}{\theta_3^2(0)} \cdot \frac{\theta_4'(x)}{\theta_4(x)} + Eu/K = \frac{d}{du}(\ln \theta_4) + Eu/K. \tag{3.5.6}$$

A table of values of $E(u, k)$ is set out in Table E (p. 293).
$E(u)$ is not periodic, but

$$E(u + 2K) - E(u) = \int_u^{u+2K} \mathrm{dn}^2 v \, dv = \int_0^{2K} \mathrm{dn}^2 v \, dv = E(2K), \tag{3.5.7}$$

since $\mathrm{dn}^2 v$ is periodic with period $2K$. Putting $u = -K$ in this result, we find
that $E(2K) = E(K) - E(-K) = 2E(K) = 2E$. Hence

$$E(u + 2K) = E(u) + 2E. \tag{3.5.8}$$

More generally, if n is an integer,

$$E(u + 2nK) = E(u) + 2nE. \tag{3.5.9}$$

A quasi-addition theorem for $E(u)$ can be established, as follows:
Logarithmic differentiation with respect to x of the identity (1.4.19) yields
the result

$$\frac{\theta_4'(x+y)}{\theta_4(x+y)} + \frac{\theta_4'(x-y)}{\theta_4(x-y)} = 2\frac{\theta_4(x)\theta_4'(x)\theta_4^2(y) - \theta_1(x)\theta_1'(x)\theta_1^2(y)}{\theta_4^2(x)\theta_4^2(y) - \theta_1^2(x)\theta_1^2(y)}. \tag{3.5.10}$$

It now follows that

$$2\frac{\theta_4'(x)}{\theta_4(x)} - \frac{\theta_4'(x+y)}{\theta_4(x+y)} - \frac{\theta_4'(x-y)}{\theta_4(x-y)} = 2\frac{\theta_1(x)\theta_1^2(y)\{\theta_1'(x)\theta_4(x) - \theta_1(x)\theta_4'(x)\}}{\theta_4(x)\{\theta_4^2(x)\theta_4^2(y) - \theta_1^2(x)\theta_1^2(y)\}}. \tag{3.5.11}$$

Making use of the identity (1.9.3), the last result reduces to

$$2\frac{\theta_4'(x)}{\theta_4(x)} - \frac{\theta_4'(x+y)}{\theta_4(x+y)} - \frac{\theta_4'(x-y)}{\theta_4(x-y)} = 2\frac{\theta_1(x)\theta_2(x)\theta_3(x)\theta_1^2(y)}{\theta_4(x)\theta_4(x+y)\theta_4(x-y)}. \tag{3.5.12}$$

Exchanging x and y in this identity and adding the result to the last
equation, we prove that

$$\frac{\theta_4'(x)}{\theta_4(x)} + \frac{\theta_4'(y)}{\theta_4(y)} - \frac{\theta_4'(x+y)}{\theta_4(x+y)}$$

$$= \frac{\theta_1(x)\theta_1(y)\{\theta_2(x)\theta_3(x)\theta_1(y)\theta_4(y) + \theta_2(y)\theta_3(y)\theta_1(x)\theta_4(x)\}}{\theta_4(x)\theta_4(y)\theta_4(x+y)\theta_4(x-y)}$$

$$= \theta_2(0)\theta_3(0)\frac{\theta_1(x)\theta_1(y)\theta_1(x+y)}{\theta_4(x)\theta_4(y)\theta_4(x+y)}, \tag{3.5.13}$$

the final step being permitted by appeal to identity (1.4.38).
The expression (3.5.6) for $E(u)$ in terms of theta functions now indicates that

the identity (3.5.13) is equivalent to

$$E(u + v) = E(u) + E(v) - k^2 \operatorname{sn} u \operatorname{sn} v \operatorname{sn}(u + v), \qquad (3.5.14)$$

where $u = x\theta_3^2(0)$, $v = y\theta_3^2(0)$ (refer to equations (2.1.1) and (2.1.7)).

Since $E(u + v)$ has not been expressed as an algebraic function of $E(u)$ and $E(v)$, this is not a true addition theorem for the E-function. It can be proved that such an expression for $E(u + v)$ does not exist.

In the extreme cases $k = 0, 1$, since $\operatorname{dn}(u, 0) = 1$ and $\operatorname{dn}(u, 1) = \operatorname{sech} u$, we conclude from equation (3.4.25) that

$$E(u, 0) = u, \qquad E(u, 1) = \tanh u. \qquad (3.5.15)$$

3.6. Jacobi's Zeta Function

It is convenient to introduce the zeta function $Z(u)$ defined by

$$Z(u) = \frac{1}{\theta_3^2(0)} \cdot \frac{\theta_4'(x)}{\theta_4(x)} = \frac{d}{du}(\ln \theta_4) = E(u) - Eu/K. \qquad (3.6.1)$$

Since θ_4 is an even function, $Z(u)$ must be odd. It follows immediately from (3.5.8) that $Z(u + 2K) = Z(u)$, i.e., that $Z(u)$ is periodic with period $2K$. Note that $Z(0) = Z(K) = 0$, so that $Z(nK) = 0$ for any integer n. Its quasi-addition theorem follows directly from (3.5.14), viz.

$$Z(u + v) = Z(u) + Z(v) - k^2 \operatorname{sn} u \operatorname{sn} v \operatorname{sn}(u + v). \qquad (3.6.2)$$

Clearly, both $E(u)$ and $Z(u)$ are defined for all complex values of the argument u, except those for which $\theta_4(x)$ vanishes, i.e., $x = m\pi + (n + \frac{1}{2})\pi\tau$ or $u = 2mK + (2n + 1)iK'$. Values of Z for real u and $k = 0.1(0.1)0.9$ will be found in Table E (p. 293).

Jacobi's transformation (section 1.7) can be used to relate $Z(iu, k)$ and $Z(u, k')$ thus:

Equations (1.7.15) and (2.1.2) show that

$$\operatorname{cn}(u, k') = \frac{\theta_4(0|\tau')}{\theta_2(0|\tau')}(-i\tau)^{1/2} e^{i\tau z'^2/\pi} \frac{\theta_4(\tau z'|\tau)}{\theta_4(z'|\tau')}, \qquad (3.6.3)$$

where $\tau' = -1/\tau$ and $u = z'\theta_3^2(0|\tau')$. Taking logarithmic derivatives with respect to z' of both members of this identity, we find that

$$-\operatorname{sc}(u, k') \operatorname{dn}(u, k')\theta_3^2(0|\tau') = \frac{2i}{\pi}\tau z' + \tau \frac{\theta_4'(\tau z'|\tau)}{\theta_4(\tau z'|\tau)} - \frac{\theta_4'(z'|\tau')}{\theta_4(z'|\tau')}. \qquad (3.6.4)$$

It follows from the transformation (1.7.12) and equations (2.2.3) that

$$\theta_3^2(0|\tau') = -i\tau\theta_3^2(0|\tau) = \frac{2}{\pi}K'(k) \qquad (3.6.5)$$

and, since $u = z\theta_3^2(0|\tau) = z'\theta_3^2(0|\tau')$, that

$$\tau z' = iz. \tag{3.6.6}$$

Hence, the identity (3.6.4) can be written in the form

$$\mathrm{sc}(u, k')\,\mathrm{dn}(u, k') = \frac{z}{K'} - \frac{i}{\theta_3^2(0|\tau)} \cdot \frac{\theta_4'(iz|\tau)}{\theta_4(iz|\tau)} + \frac{1}{\theta_3^2(0|\tau')} \cdot \frac{\theta_4'(z'|\tau')}{\theta_4(z'|\tau')},$$

$$= \frac{z}{K'} - iZ(iu, k) + Z(u, k'). \tag{3.6.7}$$

(Note: In future, wherever the modulus is not indicated explicitly, e.g., K and K', it will be assumed to be k.) Rearrangement of equation (3.6.7) after putting $z = \pi u/2K$ now yields the transformation equation

$$Z(iu, k) = i\,\mathrm{sc}(u, k')\,\mathrm{dn}(u, k') - iZ(u, k') - \frac{i\pi}{2KK'}u. \tag{3.6.8}$$

Indicating the modulus k explicitly in equation (3.6.1), this is written

$$Z(u, k) = E(u, k) - E(k)u/K(k). \tag{3.6.9}$$

Hence, using equation (2.6.5) and defining $E' = E(k')$, we have

$$Z(u, k') = E(u, k') - E'u/K'. \tag{3.6.10}$$

The identity (3.6.8) can now be put into the alternative form

$$E(iu, k) = i\,\mathrm{sc}(u, k')\,\mathrm{dn}(u, k') - iE(u, k') + \frac{iu}{KK'}(EK' + E'K - \tfrac{1}{2}\pi). \tag{3.6.11}$$

It will be proved in section 3.8 (equation (3.8.29)) that $EK' + E'K - KK' = \tfrac{1}{2}\pi$, thus permitting us to write the last identity in the simpler form

$$E(iu, k) = i\,\mathrm{sc}(u, k')\,\mathrm{dn}(u, k') - iE(u, k') + iu. \tag{3.6.12}$$

Since $K' = K(k')$, $\mathrm{sn}(2K', k') = 0$, and $E(2K', k') = 2E'$, putting $u = 2K'$ in equation (3.6.12), we find that

$$E(2iK') = 2i(K' - E'). \tag{3.6.13}$$

It now follows from the addition theorem (3.5.14) that

$$E(u + 2iK') = E(u) + 2i(K' - E'). \tag{3.6.14}$$

If we put $u = K'$ in the identity (3.6.12), since $\mathrm{sc}(K', k')$ is infinite, we conclude that $E(iK')$ is infinite (as remarked earlier). Thus, we cannot obtain an identity for $E(u + iK')$ directly from the addition theorem (3.5.14). Instead, from this theorem and the result (3.6.12), we first calculate that

$$E(u + iv, k) = E(u, k) - iE(v, k') + iv + i\,\mathrm{sc}(v, k')\,\mathrm{dn}(v, k')$$
$$- ik^2\,\mathrm{sn}(u, k)\,\mathrm{sc}(v, k')\,\mathrm{sn}(u + iv, k), \tag{3.6.15}$$

and then let $v \to K'$. After expansion of $\operatorname{sn}(u + iv, k)$ by (2.4.1) and some manipulation, it will be found that the last two terms are equal to

$$\frac{\operatorname{dn} u \operatorname{sn} v(k^2 \operatorname{sn} u \operatorname{cn} u \operatorname{sn} v + i \operatorname{dn} u \operatorname{dn} v \operatorname{cn} v)}{\operatorname{cn}^2 v + k^2 \operatorname{sn}^2 u \operatorname{sn}^2 v}, \tag{3.6.16}$$

where functions of u are to modulus k and functions of v to modulus k'. As $v \to K'$, the limiting value of this expression is $\operatorname{cs} u \operatorname{dn} u$ and the limiting form of equation (3.6.15) is accordingly

$$E(u + iK', k) = E(u, k) + i(K' - E') + \operatorname{cs} u \operatorname{dn} u$$

$$= E(u, k) + i(K' - E') + \frac{d}{du} \ln(\operatorname{sn} u). \tag{3.6.17}$$

Putting $v = K$ in equation (3.5.14), the associated identity

$$E(u + K) = E(u) + E - k^2 \operatorname{sn} u \operatorname{cd} u$$

$$= E(u) + E + \frac{d}{du} \ln(\operatorname{dn} u). \tag{3.6.18}$$

is obtained without difficulty.

Counterparts for the zeta function of the last two identities follow immediately; they are

$$Z(u + iK') = Z(u) - \frac{i\pi}{2K} + \operatorname{cs} u \operatorname{dn} u, \tag{3.6.19}$$

$$Z(u + K) = Z(u) - k^2 \operatorname{sn} u \operatorname{cd} u. \tag{3.6.20}$$

It then follows from (3.6.19) (or (3.6.14)) that

$$Z(u + 2iK') = Z(u) - i\pi/K. \tag{3.6.21}$$

Important individual values of $E(u)$ and $Z(u)$ have been collected together in the table below:

u	K	$2K$	$2iK'$	
$E(u)$	E	$2E$	$2i(K' - E')$	(3.6.22)
$Z(u)$	0	0	$-i\pi/K$	

3.7. Elliptic Integral of the Third Kind

This is the integral J_1 defined at (3.3.37).

The various possible combinations of signs of the constants A_1, B_1, A_2, B_2 lead to six distinct cases as for the integrals of the first two kinds. Each is treated by the same change of variable used to reduce the corresponding integral of the second kind.

For example,

$$\int (t^2 + \gamma)^{-1}\{(a^2 + t^2)(b^2 - t^2)\}^{-1/2}\,dt$$

$$= (a^2 + b^2)^{-1/2} \int \frac{a^2 + b^2 - b^2 \operatorname{sn}^2 u}{\gamma(a^2 + b^2) + b^2(a^2 - \gamma)\operatorname{sn}^2 u}\,du, \qquad (3.7.1)$$

under the transformation $t = ab(\operatorname{sd} u)/\sqrt{(a^2 + b^2)}$, the modulus of the function $\operatorname{sd} u$ being $b/\sqrt{(a^2 + b^2)}$ (cf. (3.4.2)).

In every case, the change of variable leads to an integral of the type

$$\int \frac{\alpha + \beta \operatorname{sn}^2 u}{\lambda + \mu \operatorname{sn}^2 u}\,du, \qquad (3.7.2)$$

where α, β, λ, μ are constants. If $\mu = 0$, $-\lambda$, $-k^2\lambda$, or $\lambda = 0$, this integral can be expressed in terms of integrals of the first two kinds. Otherwise, we have

$$\int \frac{\alpha + \beta \operatorname{sn}^2 u}{\lambda + \mu \operatorname{sn}^2 u}\,du = \frac{\alpha}{\lambda}u + \frac{1}{\lambda^2}(\beta\lambda - \alpha\mu)\int \frac{\operatorname{sn}^2 u}{1 + \nu \operatorname{sn}^2 u}\,du, \qquad (3.7.3)$$

where $\nu = \mu/\lambda$. The integral in the right-hand member of this equation can be evaluated in terms of theta functions, thus:

Define a parameter a by the equation

$$\operatorname{sn}^2 a = -\nu/k^2. \qquad (3.7.4)$$

(Note: a will be complex if $|\nu| > k^2$.) We then take as the canonical form for the integral to be evaluated, the expression

$$\Pi(u, a, k) = \int_0^u \frac{k^2 \operatorname{sn} a \operatorname{cn} a \operatorname{dn} a \operatorname{sn}^2 v}{1 - k^2 \operatorname{sn}^2 a \operatorname{sn}^2 v}\,dv. \qquad (3.7.5)$$

The identity (2.4.11) shows that the integrand can now be written in the form

$$\tfrac{1}{2}k^2 \operatorname{sn} a \operatorname{sn} v[\operatorname{sn}(v + a) + \operatorname{sn}(v - a)] = \tfrac{1}{2}[Z(v - a) - Z(v + a) + 2Z(a)], \quad (3.7.6)$$

having used the identity (3.6.2). Referring to equation (3.6.1), it now follows that

$$\Pi(u, a, k) = \frac{1}{2}\int_0^u [Z(v - a) - Z(v + a) + 2Z(a)]\,dv,$$

$$= \frac{1}{2}\int_{-a}^{u-a} Z(w)\,dw - \frac{1}{2}\int_a^{u+a} Z(w)\,dw + uZ(a),$$

$$= \tfrac{1}{2}[\ln\{\Theta(u - a)\} - \ln\{\Theta(-a)\} - \ln\{\Theta(u + a)\} + \ln\{\Theta(a)\}] + uZ(a), \tag{3.7.7}$$

where $\Theta(u) = \theta_4(x)(u = 2Kx/\pi)$. Since $\theta_4(x)$ is even, this result immediately reduces to

$$\Pi(u, a, k) = \tfrac{1}{2}\ln \frac{\Theta(u - a)}{\Theta(u + a)} + uZ(a). \qquad (3.7.8)$$

In practical calculations, a more useful canonical form is

$$\Lambda(u, \alpha, k) = \int_0^u \frac{dv}{1 - \alpha^2 \operatorname{sn}^2 v}. \tag{3.7.9}$$

Choosing a such that

$$\operatorname{sn} a = \alpha/k, \tag{3.7.10}$$

we find that

$$\Lambda(u, \alpha, k) = u + \operatorname{sc} a \operatorname{nd} a \Pi(u, a, k). \tag{3.7.11}$$

There are now four cases to consider, viz.:

$$\text{(i)} \quad 0 < \alpha < k < 1,$$
$$\text{(ii)} \quad 0 < k < \alpha < 1,$$
$$\text{(iii)} \quad \alpha > 1,$$
$$\text{(iv)} \quad \alpha^2 = -\beta^2, \qquad \beta > 0.$$

The cases $\alpha = 1$ and $\alpha = k$, of course, reduce to an integral of the second kind.

Case (i), $0 < \alpha < k < 1$: Equation (3.7.10) can be solved immediately for a in the interval $(0, K)$ to yield

$$a = \operatorname{sn}^{-1}(\alpha/k) = F(\theta, k), \tag{3.7.12}$$

where

$$\sin \theta = \alpha/k. \tag{3.7.13}$$

(Note: Refer to (3.1.8) for the definition of Legendre's incomplete elliptic integral of the first kind.) Now, introducing Legendre's integral of the second kind by equation (3.4.27), we have

$$E(a) = D(\operatorname{am} a) = D(\theta, k) \tag{3.7.14}$$

so that, by equation (3.6.1),

$$Z(a) = D(\theta, k) - \frac{E}{K} F(\theta, k). \tag{3.7.15}$$

Finally, we note that

$$\operatorname{sc} a = \alpha/\sqrt{(k^2 - \alpha^2)}, \qquad \operatorname{nd} a = 1/\sqrt{(1 - \alpha^2)}. \tag{3.7.16}$$

Equations (3.7.8), (3.7.11) now lead to the result

$$\Lambda(u, \alpha, k) = u + \frac{\alpha}{\sqrt{\{(k^2 - \alpha^2)(1 - \alpha^2)\}}} \left[\frac{1}{2} \ln \frac{\Theta\{u - F(\theta, k)\}}{\Theta\{u + F(\theta, k)\}} + u D(\theta, k) \right.$$
$$\left. - \frac{E}{K} u F(\theta, k) \right]. \tag{3.7.17}$$

Case (ii), $0 < k < \alpha < 1$: To solve equation (3.7.10), we shall put $a = K + ib$ and then, after substitution, we require that

$$\operatorname{sn}(K + ib) = \operatorname{cd}(ib) = \operatorname{nd}(b, k') = \alpha/k, \tag{3.7.18}$$

having used equations (2.2.17) and (2.6.12). Solving for b in the interval $(0, K')$, we obtain

$$b = \operatorname{dn}^{-1}(k/\alpha, k') = \operatorname{sn}^{-1}\{\sqrt{(\alpha^2 - k^2)/\alpha k'}, k'\} = F(\phi, k'), \tag{3.7.19}$$

where

$$\sin \phi = \sqrt{(\alpha^2 - k^2)/\alpha k'}. \tag{3.7.20}$$

Thus,

$$a = K + iF(\phi, k'). \tag{3.7.21}$$

The following derived results will be useful:

$$\operatorname{dn}(b, k') = k/\alpha, \qquad \operatorname{sn}(b, k') = \sqrt{(\alpha^2 - k^2)/\alpha k'},$$
$$\operatorname{cn}(b, k') = k\sqrt{(1 - \alpha^2)/\alpha k'}. \tag{3.7.22}$$

Referring to the identities (3.5.14) and (3.6.12), we next calculate that

$$\begin{aligned}
E(a) = E(K + ib) &= E + E(ib) - k^2 \operatorname{sn}(ib)\operatorname{sn}(K + ib) \\
&= E + i\operatorname{sc}(b, k')\operatorname{dn}(b, k') - iE(b, k') + ib - ik^2 \operatorname{sc}(b, k')\operatorname{nd}(b, k') \\
&= E + \frac{i}{\alpha}\sqrt{\{(\alpha^2 - k^2)(1 - \alpha^2)\}} - iD(\phi, k') + iF(\phi, k'). \tag{3.7.23}
\end{aligned}$$

Also,

$$\left.\begin{aligned}
\operatorname{sc} a = \operatorname{sc}(K + ib) &= -\frac{1}{k'}\operatorname{cs}(ib) = \frac{i}{k'}\operatorname{ns}(b, k') = \frac{i\alpha}{\sqrt{(\alpha^2 - k^2)}}, \\
\operatorname{nd} a = \operatorname{nd}(K + ib) &= \frac{1}{k'}\operatorname{dn}(ib) = \frac{1}{k'}\operatorname{dc}(b, k') = 1/\sqrt{(1 - \alpha^2)}.
\end{aligned}\right\} \tag{3.7.24}$$

Finally, putting $x = \pi u/2K$, $y = \pi F(\phi, k')/2K$, we note that

$$\left.\begin{aligned}
\Theta(u - a) &= \theta_4(x - iy - \tfrac{1}{2}\pi) = \theta_3(x - iy) = X + iY, \\
\Theta(u + a) &= \theta_4(x + iy + \tfrac{1}{2}\pi) = \theta_3(x + iy) = X - iY,
\end{aligned}\right\} \tag{3.7.25}$$

where, referring to equation (1.2.13), we calculate that

$$\left.\begin{aligned}
X &= 1 + 2 \sum_{n=0}^{\infty} q^{n^2} \cos 2nx \cosh 2ny, \\
Y &= 2 \sum_{n=0}^{\infty} q^{n^2} \sin 2nx \sinh 2ny,
\end{aligned}\right\} \tag{3.7.26}$$

taking $q = e^{i\pi\tau} = \exp(-\pi K'/K)$ (equation (2.2.3)). Clearly, $|\Theta(u - a)/\Theta(u + a)| = 1$, so that

$$\frac{1}{2}\ln\frac{\Theta(u - a)}{\Theta(u + a)} = i\tan^{-1}(Y/X). \tag{3.7.27}$$

We are now ready to substitute in equations (3.7.8) and (3.7.11). The result found is

$$\Lambda(u, \alpha, k) = \frac{\alpha}{\sqrt{\{(\alpha^2 - k^2)(1 - \alpha^2)\}}} \left[u \left\{ D(\phi, k') - F(\phi, k') + \frac{E}{K} F(\phi, k') \right\} \right.$$

$$\left. - \tan^{-1}(Y/X) \right]. \tag{3.7.28}$$

Case (iii), $\alpha > 1$: We now solve equation (3.7.10) by putting $a = c + iK'$. Upon substitution, we find that

$$\text{sn}(c + iK') = \frac{1}{k} \text{ns}\, c = \alpha/k. \tag{3.7.29}$$

Thus, we take c in the interval $(0, K)$ given by

$$c = \text{sn}^{-1}(1/\alpha) = F(\psi, k), \tag{3.7.30}$$

where

$$\sin \psi = 1/\alpha. \tag{3.7.31}$$

Then,

$$a = F(\psi, k) + iK'. \tag{3.7.32}$$

Making use of the identity (3.6.17), we now find that

$$E(a) = E(c + iK') = E(c) + i(K' - E') + \text{cs}\, c\, \text{dn}\, c,$$

$$= D(\psi, k) + i(K' - E') + \frac{1}{\alpha} \sqrt{\{(\alpha^2 - k^2)(\alpha^2 - 1)\}}. \tag{3.7.33}$$

Also,

$$\text{sc}\, a = \text{sc}(c + iK') = i\,\text{nd}\, c = i\alpha/\sqrt{(\alpha^2 - k^2)}, \\ \text{nd}\, a = \text{nd}(c + iK') = i\,\text{sc}\, c = i/\sqrt{(\alpha^2 - 1)}. \tag{3.7.34}$$

Finally, with $x = \pi u/2K, y = \pi F(\psi, k)/2K$,

$$\Theta(u - a) = \theta_4(x - y - \tfrac{1}{2}\pi\tau) = iq^{-1/4} e^{i(x-y)} \theta_1(y - x), \\ \Theta(u + a) = \theta_4(x + y + \tfrac{1}{2}\pi\tau) = iq^{-1/4} e^{-i(x+y)} \theta_1(x + y). \tag{3.7.35}$$

(Refer to equation (1.3.6).) Thus,

$$\frac{1}{2} \ln \frac{\Theta(u - a)}{\Theta(u + a)} = ix + \frac{1}{2} \ln \frac{\theta_1(y - x)}{\theta_1(y + x)}. \tag{3.7.36}$$

Substitution in equations (3.7.8) and (3.7.11) and use of the identity (3.8.29) now yields the result

$$\Lambda(u, \alpha, k) = \frac{\alpha}{\sqrt{\{(\alpha^2 - k^2)(\alpha^2 - 1)\}}} \left[\frac{1}{2} \ln \frac{\theta_1(y + x)}{\theta_1(y - x)} + u \left\{ \frac{E}{K} F(\psi, k) - D(\psi, k) \right\} \right]. \tag{3.7.37}$$

It should be noted that the integrand in (3.7.9) becomes infinite when $\operatorname{sn} v = 1/\alpha$, i.e., when $v = c = F(\psi, k)$, and the integral diverges when $u = F(\psi, k)$. We require, therefore, that $0 \leqslant u < F(\psi, k)$. When $u = F(\psi, k)$, then $x = y$ and $\theta_1(y - x) = 0$; thus, the formula (3.7.37) yields an infinite value for Λ, as expected.

Case (iv), $\alpha^2 = -\beta^2, \beta > 0$. In this case, equation (3.7.10) is solved by putting $a = id$, thus:

$$\operatorname{sn}(id) = i\operatorname{sc}(d, k') = i\beta/k. \tag{3.7.38}$$

We, therefore, require that $\operatorname{sc}(d, k') = \beta/k$, or

$$d = \operatorname{sn}^{-1}(\beta/\sqrt{(\beta^2 + k^2)}, k') = F(\omega, k'), \tag{3.7.39}$$

where

$$\sin \omega = \beta/\sqrt{(\beta^2 + k^2)}. \tag{3.7.40}$$

Thus

$$a = iF(\omega, k'). \tag{3.7.41}$$

The following results will be needed shortly:

$$\operatorname{sn}(d, k') = \beta/\sqrt{(\beta^2 + k^2)}, \qquad \operatorname{cn}(d, k') = k/\sqrt{(\beta^2 + k^2)},$$
$$\operatorname{dn}(d, k') = k\sqrt{(\beta^2 + 1)}/\sqrt{(\beta^2 + k^2)}. \tag{3.7.42}$$

Next, appealing to the identity (3.6.12), we obtain

$$\begin{aligned}E(a) = E(id) &= i\beta\sqrt{(\beta^2 + 1)}/\sqrt{(\beta^2 + k^2)} - iE(d, k') + ia\\ &= i\beta\sqrt{(\beta^2 + 1)}/\sqrt{(\beta^2 + k^2)} - iD(\omega, k') + iF(\omega, k'). \end{aligned}\tag{3.7.43}$$

Putting $x = \pi u/2K, y = \pi F(\omega, k')/2K$, we have

$$\left.\begin{aligned}\Theta(u - a) = \theta_4(x - iy) = X + iY,\\ \Theta(u + a) = \theta_4(x + iy) = X - iY,\end{aligned}\right\} \tag{3.7.44}$$

where

$$\left.\begin{aligned}X = 1 + 2 \sum_{n=1}^{\infty} (-1)^n q^{n^2} \cos 2nx \cosh 2ny,\\[2mm] Y = 2 \sum_{n=1}^{\infty} (-1)^n q^{n^2} \sin 2nx \sinh 2ny,\end{aligned}\right\} \tag{3.7.45}$$

taking $q = \exp(-\pi K'/K)$. Thus

$$\frac{1}{2}\ln\frac{\Theta(u - a)}{\Theta(u + a)} = i\tan^{-1}(Y/X). \tag{3.7.46}$$

Substitution of these results in equations (3.7.8) and (3.7.11) leads to the formula

$$\begin{aligned}\Lambda(u, i\beta, k) = \frac{k^2}{\beta^2 + k^2}u + \frac{\beta}{\sqrt{\{(\beta^2 + k^2)(\beta^2 + 1)\}}}\Bigg[u\Big\{D(\omega, k') - F(\omega, k')\\ + \frac{E}{K}F(\omega, k')\Big\} - \tan^{-1}(Y/X)\Bigg].\end{aligned}\tag{3.7.47}$$

3.8. Complete Elliptic Integrals

The complete elliptic integral of the first kind has been defined at (3.1.3). It can be expressed in the form

$$K(k) = \int_0^{\pi/2} (1 - k^2 \sin^2\theta)^{-1/2} \, d\theta. \tag{3.8.1}$$

Since $K'(k) = K(k')$ (equation (2.6.5)), we also have

$$K'(k) = \int_0^{\pi/2} (1 - k'^2 \sin^2\theta)^{-1/2} \, d\theta. \tag{3.8.2}$$

Referring to equations (2.2.3), we note that $\tau = iK'/K$. Thus, we are now able to exhibit τ (and therefore q) as a function of k and so supply an "inversion theorem" for the relationship (2.1.12). Later (section 8.12) we shall extend the definitions of $K(k)$ and $K'(k)$ to all complex values of k (except for singularities at $k = 0, \pm 1$; they prove to be multivalued functions) and the inversion of (2.1.12) will then be complete.

The complete elliptic integral of the second kind was defined at (3.5.4). This is equivalent to the result

$$E(k) = \int_0^{\pi/2} (1 - k^2 \sin^2\theta)^{1/2} \, d\theta \tag{3.8.3}$$

(see equation (3.4.26)).

Since we have defined $E'(k) = E(k')$, then

$$E'(k) = \int_0^{\pi/2} (1 - k'^2 \sin^2\theta)^{1/2} \, d\theta. \tag{3.8.4}$$

Having assumed $0 < k < 1$, the integrands in (3.8.1) and (3.8.3) are expansible in ascending powers of k^2 and can then be integrated term by term to yield

$$K = \tfrac{1}{2}\pi \left[1 + \left(\frac{1}{2}\right)^2 k^2 + \left(\frac{1.3}{2.4}\right)^2 k^4 + \left(\frac{1.3.5}{2.4.6}\right)^2 k^6 + \cdots \right], \tag{3.8.5}$$

$$E = \tfrac{1}{2}\pi \left[1 - \left(\frac{1}{2}\right)^2 k^2 - \frac{1}{3}\left(\frac{1.3}{2.4}\right)^2 k^4 - \frac{1}{5}\left(\frac{1.3.5}{2.4.6}\right)^2 k^6 - \cdots \right]. \tag{3.8.6}$$

Differentiating under the integral sign in (3.8.3) with respect to k, we find that

$$\frac{dE}{dk} = -k \int_0^{\pi/2} \frac{\sin^2\theta}{\sqrt{(1 - k^2 \sin^2\theta)}} \, d\theta = \frac{1}{k}(E - K). \tag{3.8.7}$$

A similar differentiation of (3.8.1) gives

$$\frac{dK}{dk} = k \int_0^{\pi/2} \sin^2\theta (1 - k^2 \sin^2\theta)^{-3/2} \, d\theta. \tag{3.8.8}$$

This integral, also, can be expressed in terms of E and K. The reader should first verify that

$$\frac{k'^2 \sin^2\theta}{(1 - k^2 \sin^2\theta)^{3/2}} = \frac{\cos^2\theta}{\sqrt{(1 - k^2 \sin^2\theta)}} - \frac{d}{d\theta} \frac{\sin\theta\cos\theta}{\sqrt{(1 - k^2 \sin^2\theta)}}. \tag{3.8.9}$$

Then, integrating this identity over the range $0 \leqslant \theta \leqslant \frac{1}{2}\pi$, we can show that

$$k'^2 \int_0^{\pi/2} \frac{\sin^2\theta}{(1 - k^2 \sin^2\theta)^{3/2}} \, d\theta = \int_0^{\pi/2} \frac{\cos^2\theta}{\sqrt{(1 - k^2 \sin^2\theta)}} \, d\theta. \tag{3.8.10}$$

But

$$E - k'^2 K = k^2 \int_0^{\pi/2} \frac{\cos^2\theta}{\sqrt{(1 - k^2 \sin^2\theta)}} \, d\theta. \tag{3.8.11}$$

It now follows from equations (3.8.8), (3.8.10), and (3.8.11) that

$$\frac{dK}{dk} = \frac{1}{kk'^2}(E - k'^2 K). \tag{3.8.12}$$

Exchanging primed for unprimed symbols and vice versa in the identities (3.8.7) and (3.8.12), we obtain further identities, viz.

$$\frac{dE'}{dk'} = \frac{1}{k'}(E' - K'), \qquad \frac{dK'}{dk'} = \frac{1}{k^2 k'}(E' - k^2 K'). \tag{3.8.13}$$

These are easily seen to be equivalent to

$$\frac{dE'}{dk} = \frac{k}{k'^2}(K' - E'), \qquad \frac{dK'}{dk} = \frac{1}{kk'^2}(k^2 K' - E'). \tag{3.8.14}$$

If we put

$$J = K - E = k^2 \int_0^{\pi/2} \frac{\sin^2\theta}{\sqrt{(1 - k^2 \sin^2\theta)}} \, d\theta = k^2 \int_0^K \mathrm{sn}^2 u \, du, \tag{3.8.15}$$

$(\sin\theta = \mathrm{sn}\, u)$ the identities (3.8.7) and (3.8.12) can be written in the alternative forms

$$\frac{dJ}{dk} = \frac{k}{k'^2}(K - J), \qquad \frac{dK}{dk} = \frac{1}{kk'^2}(k^2 K - J). \tag{3.8.16}$$

Comparison of these equations with the equations (3.8.14) indicates that J and K satisfy the same differential equations as E' and K' respectively. If we eliminate K' between the equations (3.8.14), we find that E' (and hence J) satisfies the second-order linear differential equation

$$k(1 - k^2)\frac{d^2 w}{dk^2} - (1 + k^2)\frac{dw}{dk} + kw = 0. \tag{3.8.17}$$

The other linearly independent solution must be J and its general solution is accordingly

$$w = AE' + BJ = AE' + B(K - E). \qquad (3.8.18)$$

Similarly, by eliminating E' between the equations (3.8.14), we can show that the equation

$$k(1 - k^2)\frac{d^2w}{dk^2} + (1 - 3k^2)\frac{dw}{dk} - kw = 0 \qquad (3.8.19)$$

has the general solution

$$w = AK' + BK. \qquad (3.8.20)$$

Similarly, defining

$$J' = K' - E' = k'^2 \int_0^{\pi/2} \frac{\sin^2\theta}{\sqrt{(1 - k'^2\sin^2\theta)}}\,d\theta, \qquad (3.8.21)$$

it follows from equations (3.8.14) that

$$\frac{dJ'}{dk} = \frac{1}{k}(J' - K'), \qquad \frac{dK'}{dk} = \frac{1}{kk'^2}(J' - k'^2K'), \qquad (3.8.22)$$

showing that J' and K' satisfy the same equations as E and K respectively. Hence, E and J' are independent solutions of the equation

$$k(1 - k^2)\frac{d^2w}{dk^2} + (1 - k^2)\frac{dw}{dk} + kw = 0 \qquad (3.8.23)$$

and K', K are independent solutions of the equation (3.8.19) (as already proved).

It also follows from equations (3.8.7), (3.8.12), and (3.8.14) that

$$\frac{d}{dk}(EK' + E'K - KK') = 0, \qquad (3.8.24)$$

showing that $(EK' + E'K - KK')$ is a constant.

To find the value of this constant, we let $k \to 0$. Clearly

$$\lim_{k \to 0} E = \lim_{k \to 0} K = \tfrac{1}{2}\pi, \qquad \lim_{k \to 0} E' = 1. \qquad (3.8.25)$$

As $k \to 0$, the nome q tends to zero also. Hence, using equations (2.1.12) and (2.2.8),

$$\lim_{k \to 0} \{K' + \ln(\tfrac{1}{4}k)\} = \lim_{q \to 0} \{(-2q + O(q^2))\ln q - 4q + O(q^2)\} = 0. \qquad (3.8.26)$$

It now follows that

$$\lim_{k \to 0} (E - K)K' = \lim_{k \to 0} \{(E - K)(K' + \ln(\tfrac{1}{4}k)) - (E - K)\ln(\tfrac{1}{4}k)\}$$

$$= 0, \qquad (3.8.27)$$

since the expansions (3.8.5), (3.8.6) show that $(E - K)$ is $O(k^2)$ for small k. We conclude that

$$\lim_{k \to 0} \{(E - K)K' + E'K\} = \tfrac{1}{2}\pi \tag{3.8.28}$$

and, hence, that

$$EK' + E'K - KK' = \tfrac{1}{2}\pi. \tag{3.8.29}$$

The complete elliptic integral of the third kind is

$$\Pi(K, a, k) = \int_0^K \frac{k^2 \operatorname{sn} a \operatorname{cn} a \operatorname{dn} a \operatorname{sn}^2 u}{1 - k^2 \operatorname{sn}^2 a \operatorname{sn}^2 u} \, du$$

$$= \frac{1}{2} \ln \frac{\Theta(K - a)}{\Theta(K + a)} + KZ(a), \tag{3.8.30}$$

after putting $u = K$ in equation (3.7.8). But, by the definition of the Θ-function,

$$\left. \begin{aligned} \Theta(K - a) &= \theta_4(\tfrac{1}{2}\pi - \pi a/2K) = \theta_3(\pi a/2K), \\ \Theta(K + a) &= \theta_4(\tfrac{1}{2}\pi + \pi a/2K) = \theta_3(\pi a/2K). \end{aligned} \right\} \tag{3.8.31}$$

Hence, equation (3.8.30) reduces to

$$\Pi(K, a, k) = KZ(a) = \frac{1}{2}\pi \cdot \frac{\theta_4'(\pi a/2K)}{\theta_4(\pi a/2K)}, \tag{3.8.32}$$

after using (3.6.1).

If we prefer to make use of the canonical form (3.7.9), by substituting $u = K$ into the formulae (3.7.17), (3.7.28), (3.7.37), and (3.7.47), we find that

(i) $\quad \Lambda(K, \alpha, k) = K + \dfrac{\alpha}{\sqrt{\{(k^2 - \alpha^2)(1 - \alpha^2)\}}} [KD(\theta, k) - EF(\theta, k)],$

$$0 < \alpha < k < 1, \tag{3.8.33}$$

(ii) $\quad \Lambda(K, \alpha, k) = \dfrac{\alpha}{\sqrt{\{(\alpha^2 - k^2)(1 - \alpha^2)\}}} [KD(\phi, k') - KF(\phi, k') + EF(\phi, k')],$

$$0 < k < \alpha < 1, \tag{3.8.34}$$

(iii) $\quad \Lambda(K, \alpha, k) = \dfrac{\alpha}{\sqrt{\{(\alpha^2 - k^2)(\alpha^2 - 1)\}}} [EF(\psi, k) - KD(\psi, k)],$

$$\alpha > 1, \tag{3.8.35}$$

(iv) $\quad \Lambda(K, i\beta, k) = \dfrac{k^2}{\beta^2 + k^2} K + \dfrac{\beta}{\sqrt{\{(\beta^2 + k^2)(\beta^2 + 1)\}}} [KD(\omega, k') - KF(\omega, k')$

$$+ EF(\omega, k')], \qquad \beta > 0. \tag{3.8.36}$$

The angles θ, ϕ, ψ, and ω are defined at equations (3.7.13), (3.7.20), (3.7.31), and (3.7.40), respectively.

3.9. Further Transformations of the Modulus

The effect on the Jacobian elliptic functions of transforming from a modulus k to the modulus $k' = \sqrt{(1 - k^2)}$ has already been calculated in section 2.6. In this section, two further transformations of this type will be studied, viz.

$$\kappa = 1/k \quad \text{and} \quad k_1 = \frac{1 - k'}{1 + k'}. \tag{3.9.1}$$

Equation (3.1.1) is valid for all values of x and k, real or complex, for which $\mathrm{sn}^{-1}(x, k)$ is defined. Equation (3.1.2) is, accordingly, also valid in this extended domain, provided it is recognized that, in general, the variable of integration t is complex and that the integral must be taken over a curve in the t-plane connecting the origin to the point with affix x. Since the integrand possesses singularities at $t = 1$ and $t = 1/k$, the value of the integral will depend upon the position of the contour in relation to these singularities and $\mathrm{sn}^{-1}(x, k)$ is therefore multivalued, as already noted. In equation (3.1.2), it is assumed that the contour of integration is the real axis and there is no ambiguity; it always yields the value of $\mathrm{sn}^{-1}(x, k)$ lying in the range $(0, K)$.

The general form taken by equation (3.1.2), when x and u are complex, will be studied in Chapter 8. However, it will here be noted that the equation yields a definite value for $\mathrm{sn}^{-1}(x, k)$ when $k > 1$ and x lies in the range $(0, 1/k)$. It therefore provides values (real) for $x = \mathrm{sn}(u, k)$ when the modulus is greater than unity, for real values of u corresponding to values of x lying in the range $(0, 1/k)$. We shall show that these values are simply related to the values of the function sn taken with modulus $1/k < 1$.

For, changing the variable by $s = kt$ $(k > 1)$ in equation (3.1.2), we find

$$u = \mathrm{sn}^{-1}(x, k) = \frac{1}{k} \int_0^{kx} \{(1 - s^2)(1 - s^2/k^2)\}^{-1/2} \, ds = \frac{1}{k} \mathrm{sn}^{-1}(kx, 1/k).$$

$$\tag{3.9.2}$$

Thus

$$x = \mathrm{sn}(u, k) = \frac{1}{k} \mathrm{sn}(ku, 1/k). \tag{3.9.3}$$

Putting $\kappa = 1/k$ and replacing u by u/k, this transformation can be expressed in the form

$$\mathrm{sn}(u, \kappa) = k \, \mathrm{sn}(u/k, k). \tag{3.9.4}$$

It now follows by use of the identities (2.1.9)–(2.1.11) that

$$\mathrm{cn}(u, \kappa) = \mathrm{dn}(u/k, k), \tag{3.9.5}$$

$$\mathrm{dn}(u, \kappa) = \mathrm{cn}(u/k, k). \tag{3.9.6}$$

In these transformation equations, we have assumed $0 < \kappa < 1$ and $k > 1$. However, replacing u by ku and then exchanging the symbols k and κ, they can be proved valid for $\kappa > 1$, $0 < k < 1$, also. Indeed, since sn, cn, and dn are analytic functions of both their arguments, the principle of analytic continuation implies that these identities are valid for all complex values of u and k for which the functions are defined.

Next, suppose k_1 is related to k by the second of the transformations (3.9.1). We shall assume $0 < k < 1$, so that k' and k_1 both lie in the interval (0, 1). Then, by equation (3.1.8),

$$\text{sn}^{-1}(\sin \phi_1, k_1) = \int_0^{\phi_1} (1 - k_1^2 \sin^2 \theta_1)^{-1/2} \, d\theta_1, \qquad (3.9.7)$$

where $0 \leqslant \phi_1 \leqslant \frac{1}{2}\pi$. Changing the variable of integration to θ by the transformation

$$2\theta = \theta_1 + \sin^{-1}(k_1 \sin \theta_1), \qquad (3.9.8)$$

where the inverse sine is to be taken in the range $(0, \frac{1}{2}\pi)$, we find that

$$\text{sn}^{-1}(\sin \phi_1, k_1) = (1 + k') \int_0^{\phi} (1 - k^2 \sin^2 \theta)^{-1/2} \, d\theta = (1 + k') \, \text{sn}^{-1}(\sin \phi, k),$$

$$(3.9.9)$$

where ϕ is given by

$$2\phi = \phi_1 + \sin^{-1}(k_1 \sin \phi_1). \qquad (3.9.10)$$

(Note: Make use of the identity

$$(1 + k_1)^2 (1 - k^2 \sin^2 \theta) = [k_1 \cos \theta_1 + \sqrt{(1 - k_1^2 \sin^2 \theta_1)}]^2.)$$

Rearranging equation (3.9.10) into the form

$$\cot \phi_1 = \cot 2\phi + k_1 \csc 2\phi \qquad (3.9.11)$$

and then squaring and adding 1 to both sides, we calculate that

$$\csc^2 \phi_1 = \frac{4}{(1 + k')^2} \csc^2 2\phi \, (1 - k^2 \sin^2 \phi). \qquad (3.9.12)$$

Thus,

$$\sin \phi_1 = (1 + k') \sin \phi \cos \phi \, (1 - k^2 \sin^2 \phi)^{-1/2}. \qquad (3.9.13)$$

Putting $u = \text{sn}^{-1}(\sin \phi, k)$, $u_1 = \text{sn}^{-1}(\sin \phi_1, k_1)$, we have

$$\sin \phi = \text{sn} \, u, \qquad \cos \phi = \text{cn} \, u, \qquad \sin \phi_1 = \text{sn} \, u_1, \qquad \cos \phi_1 = \text{cn} \, u_1.$$

$$(3.9.14)$$

Also, equation (3.9.9) shows that $u_1 = (1 + k')u$. It then follows from equation (3.9.13) that

$$\text{sn}(u_1, k_1) = (1 + k') \text{sn}(u, k) \text{cd}(u, k), \qquad (3.9.15)$$

where

$$u_1 = (1 + k')u. \qquad (3.9.16)$$

This is *Landen's transformation*. It has been proved for $0 \leqslant \phi_1 \leqslant \frac{1}{2}\pi$, or $0 \leqslant u_1 \leqslant K(k_1)$, or $0 \leqslant u \leqslant K(k_1)/(1 + k')$, but may be extended to all complex values of u and k by analytic continuation.

The transformation can also be established with general complex values of u and k directly, by application of the Landen transformation for the theta functions (section 1.8).

Thus, if k, k_1 are the moduli corresponding to parameters $\tau, 2\tau$, then division of equations (1.8.5), (1.8.6) and reference to the equations (2.1.7) shows that

$$k_1 = \frac{1 - k'}{1 + k'}. \tag{3.9.17}$$

Further, equation (1.8.6) yields the result

$$\theta_3^2(0|2\tau) = \tfrac{1}{2}(1 + k')\theta_3^2(0|\tau). \tag{3.9.18}$$

Next, dividing the identities (1.8.8) and (1.8.11) and replacing ratios of theta functions by Jacobian elliptic functions, we arrive immediately at the transformation

$$k_1^{1/2}\,\mathrm{sn}(u_1, k_1) = k\,\mathrm{sn}(u, k)\mathrm{cd}(u, k), \tag{3.9.19}$$

where $u = x\theta_3^2(0|\tau), u_1 = 2x\theta_3^2(0|2\tau)$. Appeal to equations (3.9.17), (3.9.18) now shows that

$$u_1 = (1 + k')u, \qquad k/k_1^{1/2} = 1 + k'. \tag{3.9.20}$$

Equation (3.9.19) is accordingly Landen's transformation.

Again, dividing (1.8.9) by (1.8.11) and introducing elliptic functions, we find that

$$\mathrm{cn}(u_1, k_1) = \{1 - (1 + k')\mathrm{sn}^2(u, k)\}/\mathrm{dn}(u, k). \tag{3.9.21}$$

Similarly, division of (1.8.10) by (1.8.11) yields the third transformation equation, viz.

$$\mathrm{dn}(u_1, k_1) = \{1 - (1 - k')\mathrm{sn}^2(u, k)\}/\mathrm{dn}(u, k). \tag{3.9.22}$$

Alternatively, this last pair of transformation equations can be derived from (3.9.15) by use of the fundamental identities.

Multiplying equations (3.9.21), (3.9.22) by $(1 - k')$ and $(1 + k')$ respectively and adding, we get

$$\mathrm{dn}\,u = (\mathrm{dn}\,u_1 + k_1\,\mathrm{cn}\,u_1)/(1 + k_1), \tag{3.9.23}$$

it being understood that functions of u are to modulus k and functions of u_1 are to modulus k_1. Multiplication by the same factors followed by subtraction leads to the result

$$\mathrm{nd}\,u = (\mathrm{dn}\,u_1 - k_1\,\mathrm{cn}\,u_1)/(1 - k_1). \tag{3.9.24}$$

Then, using the identity (2.4.7), we calculate that

$$\frac{1 - k_1 \operatorname{sn}^2\frac{1}{2}u_1}{1 + k_1 \operatorname{sn}^2\frac{1}{2}u_1} = \frac{1 - k_1 + \operatorname{dn} u_1 + k_1 \operatorname{cn} u_1}{1 + k_1 + \operatorname{dn} u_1 - k_1 \operatorname{cn} u_1},$$

$$= \frac{1 - k_1 + (1 + k_1)\operatorname{dn} u}{1 + k_1 + (1 - k_1)\operatorname{nd} u},$$

$$= \operatorname{dn} u. \tag{3.9.25}$$

Thus,

$$\operatorname{dn}(u, k) = \frac{1 - k_1 \operatorname{sn}^2(u_1', k_1)}{1 + k_1 \operatorname{sn}^2(u_1', k_1)}, \tag{3.9.26}$$

where

$$u_1' = \tfrac{1}{2}u_1 = u/(1 + k_1). \tag{3.9.27}$$

The following expressions for $\operatorname{sn} u$, $\operatorname{cn} u$ can now be obtained:

$$\operatorname{sn}(u, k) = \frac{(1 + k_1)\operatorname{sn}(u_1', k_1)}{1 + k_1 \operatorname{sn}^2(u_1', k_1)}, \tag{3.9.28}$$

$$\operatorname{cn}(u, k) = \frac{\operatorname{cn}(u_1', k_1)\operatorname{dn}(u_1', k_1)}{1 + k_1 \operatorname{sn}^2(u_1', k_1)}. \tag{3.9.29}$$

These relationships constitute the inverse of Landen's transformation and are referred to as *Gauss's transformation*.

It is easily verified that, if $0 < k < 1$, then $k_1 < k$. Thus, Gauss's transformation expresses Jacobi's functions to modulus k in terms of functions to a smaller modulus. Repeated application of the transformation accordingly yields functions to moduli $k_1, k_2, \ldots$, where $k_{n+1} < k_n$. Since k_n is decreasing and $k_n > 0$, it must tend to a limit $l \geqslant 0$ as $n \to \infty$. Letting $n \to \infty$ in the recurrence relationship

$$k_{n+1} = \frac{1 - k_n'}{1 + k_n'}, \tag{3.9.30}$$

we show that

$$l = \frac{1 - \sqrt{(1 - l^2)}}{1 + \sqrt{(1 - l^2)}}. \tag{3.9.31}$$

This last equation has real roots 0 and 1; since $l < 1$, we conclude $l = 0$. Hence, by repeated application of the Gauss transformation, we can reduce the modulus upon which the elliptic functions depend to as small a value as we please. Such elliptic functions are then readily computed by the use of expansions such as (3.1.11).

Let us consider more carefully the transformation (3.9.8). As θ_1 increases from 0 to π, θ increases monotonically from 0 to $\frac{1}{2}\pi$. Thus putting $\phi = \frac{1}{2}\pi$, $\phi_1 = \pi$ in equation (3.9.9), we obtain the result

$$\int_0^\pi (1 - k_1^2 \sin^2 \theta_1)^{-1/2} \, d\theta_1 = (1 + k') \int_0^{\pi/2} (1 - k^2 \sin^2 \theta)^{-1/2} \, d\theta. \tag{3.9.32}$$

Putting $k' = b/a(0 < b < a)$, the last equation will be found to be equivalent to

$$\int_0^{\pi/2} \frac{d\theta_1}{\sqrt{(a_1^2 \cos^2 \theta_1 + b_1^2 \sin^2 \theta_1)}} = \int_0^{\pi/2} \frac{d\theta}{\sqrt{(a^2 \cos^2 \theta + b^2 \sin^2 \theta)}}, \quad (3.9.33)$$

where

$$a_1 = \tfrac{1}{2}(a + b), \qquad b_1 = \sqrt{(ab)}. \quad (3.9.34)$$

Thus, equation (3.9.33) indicates that it is permissible to replace the parameters a and b, in the right-hand integral, by their arithmetic and geometric means respectively.

Suppose this transformation is applied to this integral repeatedly. After n applications, the parameters will take the values a_n, b_n, where

$$a_{n+1} = \tfrac{1}{2}(a_n + b_n), \qquad b_{n+1} = \sqrt{(a_n b_n)}. \quad (3.9.35)$$

As $n \to \infty$, a_n steadily decreases and b_n steadily increases toward the same limit m. This follows from the inequalities

$$\left. \begin{array}{l} a_{n+1} - b_{n+1} = \tfrac{1}{2}(\sqrt{a_n} - \sqrt{b_n})^2 > 0, \\ a_n - a_{n+1} = \tfrac{1}{2}(a_n - b_n) > 0, \\ b_{n+1} - b_n = \sqrt{b_n}(\sqrt{a_n} - \sqrt{b_n}) > 0. \end{array} \right\} \quad (3.9.36)$$

Thus,

$$\lim_{n \to \infty} \int_0^{\pi/2} \frac{d\theta}{\sqrt{(a_n^2 \cos^2 \theta + b_n^2 \sin^2 \theta)}} = \int_0^{\pi/2} \frac{d\theta}{\sqrt{(m^2 \cos^2 \theta + m^2 \sin^2 \theta)}} = \pi/(2m). \quad (3.9.37)$$

Referring to equation (3.8.1), we note that

$$K(k) = \int_0^{\pi/2} \frac{d\theta}{\sqrt{(\cos^2 \theta + k'^2 \sin^2 \theta)}}. \quad (3.9.38)$$

Hence, the result (3.9.37) implies that

$$K(k) = \pi/\{2m(1, k')\}. \quad (3.9.39)$$

$m(1, k')$ is readily calculable using a computer and this equation accordingly provides us with a powerful means of constructing a table of values of $K(k)$.

3.10. k-Derivatives of Elliptic Functions

Since it is possible to define Jacobi's functions as the solutions of the differential equations (2.5.2)–(2.5.4) which satisfy the initial conditions $\operatorname{sn} 0 = 0$, $\operatorname{cn} 0 = \operatorname{dn} 0 = 1$, the derivatives of these functions with respect to the modulus k must be calculable from these equations.

We first note that the fundamental identities (2.1.9)–(2.1.11) are first integrals of the equations (2.5.2)–(2.5.4) (the reader may verify this by differentiation of these identities) and will also be used in the calculation of the k-derivatives.

Writing

$$\frac{\partial}{\partial k}\operatorname{sn} u = \sigma, \qquad \frac{\partial}{\partial k}\operatorname{cn} u = \gamma, \qquad \frac{\partial}{\partial k}\operatorname{dn} u = \delta, \tag{3.10.1}$$

by differentiating equations (2.5.2)–(2.5.4) partially with respect to k, we find that

$$\frac{\partial\sigma}{\partial u} = \gamma\operatorname{dn} u + \delta\operatorname{cn} u, \tag{3.10.2}$$

$$\frac{\partial\gamma}{\partial u} = -\sigma\operatorname{dn} u - \delta\operatorname{sn} u, \tag{3.10.3}$$

$$\frac{\partial\delta}{\partial u} = -2k\operatorname{sn} u\operatorname{cn} u - k^2\sigma\operatorname{cn} u - k^2\gamma\operatorname{sn} u. \tag{3.10.4}$$

Further, differentiation of the identities (2.1.9), (2.1.10) with respect to k gives

$$\gamma\operatorname{cn} u + \sigma\operatorname{sn} u = 0, \tag{3.10.5}$$

$$\delta\operatorname{dn} u + k\operatorname{sn}^2 u + k^2\sigma\operatorname{sn} u = 0. \tag{3.10.6}$$

We can now eliminate γ and δ between the equations (3.10.2), (3.10.5), and (3.10.6) to arrive at the equation

$$\frac{\partial\sigma}{\partial u} + \frac{\operatorname{sn} u}{\operatorname{cn} u\operatorname{dn} u}(\operatorname{dn}^2 u + k^2\operatorname{cn}^2 u) = -\frac{k\operatorname{sn}^2 u\operatorname{cn} u}{\operatorname{dn} u}. \tag{3.10.7}$$

After multiplication through by an integrating factor $1/(\operatorname{cn} u\operatorname{dn} u)$, this equation can be expressed in the form

$$\frac{\partial}{\partial u}\left[\frac{\sigma}{\operatorname{cn} u\operatorname{dn} u}\right] = -k\operatorname{sd}^2 u, \tag{3.10.8}$$

which integrates immediately to yield

$$\frac{\partial}{\partial k}\operatorname{sn} u = \sigma = \frac{1}{k}u\operatorname{cn} u\operatorname{dn} u + \frac{k}{k'^2}\operatorname{sn} u\operatorname{cn}^2 u - \frac{1}{kk'^2}E(u)\operatorname{cn} u\operatorname{dn} u. \tag{3.10.9}$$

(Note: An arbitrary function of k appears in the integral, but can be shown to be identically zero since, for $u = 0$,

$$\sigma = \frac{\partial}{\partial k}\operatorname{sn}(0, k) = 0.) \tag{3.10.10}$$

γ and δ now follow from equations (3.10.5), (3.10.6), thus:

$$\frac{\partial}{\partial k}\operatorname{cn} u = \gamma = -\frac{1}{k}u\operatorname{sn} u\operatorname{dn} u - \frac{k}{k'^2}\operatorname{sn}^2 u\operatorname{cn} u + \frac{1}{kk'^2}E(u)\operatorname{sn} u\operatorname{dn} u, \tag{3.10.11}$$

$$\frac{\partial}{\partial k}\operatorname{dn} u = \delta = -\frac{k}{k'^2}\operatorname{sn}^2 u\operatorname{dn} u - ku\operatorname{sn} u\operatorname{cn} u + \frac{k}{k'^2}E(u)\operatorname{sn} u\operatorname{cn} u. \qquad (3.10.12)$$

The *k*-derivative of $E(u)$ can now be found thus:

$$\frac{\partial}{\partial k}E(u) = \frac{\partial}{\partial k}\int_0^u \operatorname{dn}^2 v\, dv$$

$$= 2\int_0^u \operatorname{dn} v\left(-\frac{k}{k'^2}\operatorname{sn}^2 v\operatorname{dn} v - kv\operatorname{sn} v\operatorname{cn} v + \frac{k}{k'^2}E(v)\operatorname{sn} v\operatorname{cn} v\right)dv$$

$$= \int_0^u\left[\frac{\partial}{\partial v}\left(\frac{k}{k'^2}E(v) - kv\right)\operatorname{sn}^2 v - \frac{3k}{k'^2}\operatorname{sn}^2 v\operatorname{dn}^2 v + k\operatorname{sn}^2 v\right]dv$$

$$= \left(\frac{k}{k'^2}E(u) - ku\right)\operatorname{sn}^2 u + \frac{1}{k}\int_0^u(1 - \operatorname{dn}^2 v)\left(1 - \frac{3}{k'^2}\operatorname{dn}^2 v\right)dv. \qquad (3.10.13)$$

Since

$$\left.\begin{aligned}\int_0^u \operatorname{dn}^2 v\, dv &= E(u),\\[2mm]\int_0^u \operatorname{dn}^4 v\, dv &= \tfrac{1}{3}[2(2 - k^2)E(u) - k'^2 u + k^2\operatorname{sn} u\operatorname{cn} u\operatorname{dn} u]\end{aligned}\right\} \qquad (3.10.14)$$

(the second integral may be verified by differentiation or calculated as suggested in Exercise 15), we have finally

$$\frac{\partial}{\partial k}E(u) = \frac{k}{k'^2}\operatorname{sn} u\operatorname{cn} u\operatorname{dn} u - ku\operatorname{sn}^2 u - \frac{k}{k'^2}E(u)\operatorname{cn}^2 u. \qquad (3.10.15)$$

Next, we note that

$$\operatorname{cn}[K(k), k] = 0 \qquad (3.10.16)$$

is an identity in *k*. Differentiation then leads to

$$\frac{\partial}{\partial K}\operatorname{cn}(K, k)\frac{dK}{dk} + \frac{\partial}{\partial k}\operatorname{cn}(K, k) = 0. \qquad (3.10.17)$$

But

$$\frac{\partial}{\partial K}\operatorname{cn}(K, k) = -\operatorname{sn}(K, k)\operatorname{dn}(K, k) = -k' \qquad (3.10.18)$$

and, putting $u = K$ in the result (3.10.11),

$$\frac{\partial}{\partial k}\operatorname{cn}(K, k) = -\frac{k'}{k}K + \frac{1}{kk'}E. \qquad (3.10.19)$$

It follows from equation (3.10.17) that

$$\frac{dK}{dk} = \frac{1}{kk'^2}(E - k'^2 K). \qquad (3.10.20)$$

Thus, equation (3.8.12) has been derived by an alternative argument.

We now differentiate the identity

$$E(k) = E(K(k), k) \tag{3.10.21}$$

to give

$$\frac{dE}{dk} = \frac{\partial E}{\partial K}\frac{dK}{dk} + \frac{\partial E}{\partial k}. \tag{3.10.22}$$

But

$$\partial E/\partial K = \mathrm{dn}^2 K = k'^2 \tag{3.10.23}$$

and, putting $u = K$ in equation (3.10.15),

$$\partial E/\partial k = -kK. \tag{3.10.24}$$

Thus,

$$\frac{dE}{dk} = k'^2\frac{dK}{dk} - kK = \frac{1}{k}(E - K), \tag{3.10.25}$$

by use of equations (3.10.22) and (3.10.20). This is equation (3.8.7) again.

EXERCISES

1. Show that

$$\int_0^2 \{(2x - x^2)(4x^2 + 9)\}^{-1/2}\,dx = \frac{2}{\sqrt{15}}K(1/\sqrt{5}).$$

2. If

$$u = \int_x^\infty \{(t + 1)(t^2 + t + 1)\}^{-1/2}\,dt,$$

   show that

$$x = \frac{2\,\mathrm{cn}\,u}{1 - \mathrm{cn}\,u},$$

   where $k = \sqrt{3}/2$.

3. Show that, if $0 \leqslant x \leqslant 1$, then

$$\int_0^x dt/\sqrt{(1 + t^2 - 2t^4)} = \frac{1}{\sqrt{3}}(K - \mathrm{cn}^{-1}x),$$

   with $k^2 = 2/3$.

4. Show that

$$\int_1^x \{(t^2 + 1)(2t^2 - 3t + 2)\}^{-1/2}\,dt = \sqrt{\tfrac{2}{7}}\,\mathrm{sc}^{-1}\left[\sqrt{7}\frac{x-1}{x+1}, \sqrt{\frac{6}{7}}\right].$$

5. Using the change of variable

$$\operatorname{cn}(u, 1/\sqrt{2}) = \frac{t^2 - 1}{t^2 + 1},$$

show that

$$\int_x^\infty \frac{dt}{\sqrt{(t^4 + 1)}} = \tfrac{1}{2}\operatorname{cn}^{-1}\left(\frac{x^2 - 1}{x^2 + 1}\right).$$

6. Show that

$$\int_x^1 \frac{dt}{\sqrt{(1 - t^4)}} = \frac{1}{\sqrt{2}}\operatorname{cn}^{-1}(x, 1/\sqrt{2}),$$

and deduce that

$$\int_1^x \frac{dt}{\sqrt{(t^4 - 1)}} = \frac{1}{\sqrt{2}}\operatorname{cn}^{-1}(1/x, 1/\sqrt{2}).$$

7. Show that the change of variable

$$t = 1 + \sqrt{3}\frac{1 - s}{1 + s}$$

brings the integral

$$\int_1^x \frac{dt}{\sqrt{(t^3 - 1)}} \qquad (x \geqslant 1)$$

to canonical form and deduce that its value is

$$3^{-1/4}\operatorname{cn}^{-1}\left(\frac{\sqrt{3} + 1 - x}{\sqrt{3} - 1 + x}\right),$$

with modulus $k = (\sqrt{3} - 1)/2\sqrt{2}$.

8. Changing the variable by

$$t = 1 - \sqrt{3}\frac{1 - s}{1 + s},$$

show that

$$\int_x^1 \frac{dt}{\sqrt{(1 - t^3)}} = 3^{-1/4}\operatorname{cn}^{-1}\left(\frac{\sqrt{3} - 1 + x}{\sqrt{3} + 1 - x}\right),$$

with $k = (\sqrt{3} + 1)/2\sqrt{2}$. Deduce that

$$\int_{-1}^x \frac{dt}{\sqrt{(t^3 + 1)}} = 3^{-1/4}\operatorname{cn}^{-1}\left(\frac{\sqrt{3} - 1 - x}{\sqrt{3} + 1 + x}\right).$$

9. Show that

$$\int_1^\infty \frac{dt}{\sqrt{(t^3 - 1)}} = \frac{1}{2\sqrt{(3\pi)}} \Gamma(\tfrac{1}{6})\Gamma(\tfrac{1}{3}).$$

Deduce, using the result of Exercise 7, that

$$K = \frac{1}{4.3^{1/4}\sqrt{\pi}} \Gamma(\tfrac{1}{6})\Gamma(\tfrac{1}{3})$$

when $k = (\sqrt{3} - 1)/2\sqrt{2}$.

10. Show that

$$\int_{-\infty}^1 \frac{dt}{\sqrt{(1 - t^3)}} = \frac{1}{2\sqrt{\pi}} \Gamma(\tfrac{1}{6})\Gamma(\tfrac{1}{3}).$$

(Hint: Split into integrals over the ranges $(-\infty, 0)$ and $(0, 1)$.) Deduce, using the result of Exercise 8, that

$$K = \frac{3^{1/4}}{4\sqrt{\pi}} \Gamma(\tfrac{1}{6})\Gamma(\tfrac{1}{3}),$$

when $k = (\sqrt{3} + 1)/2\sqrt{2}$. Hence verify that $K' = \sqrt{3}K$ when $k = (\sqrt{3} - 1)/2\sqrt{2}$, and $K' = K/\sqrt{3}$ when $k = (\sqrt{3} + 1)/2\sqrt{2}$. (Hint: Use Exercise 9.)

11. Changing the variable by

$$t = 1 + \sqrt{3}\frac{1 - \mathrm{cn}(u, k)}{1 + \mathrm{cn}(u, k)},$$

with $k = (\sqrt{3} - 1)/2\sqrt{2}$, prove that

$$\int_1^\infty \frac{(t - 1)^2}{(t^3 - 1)^{3/2}} dt = \frac{2}{3^{3/4}}[4E - (2 + \sqrt{3})K].$$

12. Integrating by parts, show that

$$\int_1^\infty \frac{(t - 1)^2}{(t^3 - 1)^{3/2}} dt = \frac{4}{3}\int_1^\infty \frac{t - 1}{t^3\sqrt{(t^3 - 1)}} dt.$$

Hence, evaluate the integral in terms of gamma functions and verify that

$$K[2\sqrt{3}E - (\sqrt{3} + 1)K] = \tfrac{1}{2}\pi,$$

for $k = (\sqrt{3} - 1)/2\sqrt{2}$. (Hint: Use the results of Exercises 9 and 11.)

13. Changing the variable by

$$t = 1 - \sqrt{3}\frac{1 - \mathrm{cn}\, u}{1 + \mathrm{cn}\, u}$$

with $k = (\sqrt{3} + 1)/2\sqrt{2}$, prove that

$$\int_{-\infty}^1 \frac{(1 - t)^2}{(1 - t^3)^{3/2}} dt = \frac{2}{3^{3/4}}[4E' - (2 - \sqrt{3})K'],$$

where K' and E' are evaluated with $k = (\sqrt{3} - 1)/2\sqrt{2}$. Splitting the range of integration into two parts, viz. $(-\infty, 0)$ and $(0, 1)$, and expressing each part integral in terms of gamma functions, verify that, for $k = (\sqrt{3} - 1)/2\sqrt{2}$,

$$K'[2\sqrt{3}E' - (\sqrt{3} - 1)K'] = \tfrac{3}{2}\pi.$$

(Hint: Use a result from Exercise 10.)

14. Show that, if m is an integer,

$$\frac{d^2}{du^2} sn^m u = m(m - 1) sn^{m-2} u - m^2(1 + k^2) sn^m u + m(m + 1)k^2 sn^{m+2} u.$$

If

$$I_m = \int sn^m u \, du = \int \frac{t^m \, dt}{\sqrt{\{(1 - t^2)(1 - k^2 t^2)\}}},$$

obtain the recurrence relationship

$$(m + 1)k^2 I_{m+2} = m(1 + k^2)I_m - (m - 1)I_{m-2} + sn^{m-1} u \operatorname{cn} u \operatorname{dn} u$$

and deduce that

$$\int sn^3 u \, du = \frac{1}{2k^3}[(1 + k^2)\ln(\operatorname{dn} u - k \operatorname{cn} u) + k \operatorname{cn} u \operatorname{dn} u],$$

$$\int sn^4 u \, du = \frac{1}{3k^4}[(2 + k^2)u - 2(1 + k^2)E(u) + k^2 sn u \operatorname{cn} u \operatorname{dn} u],$$

$$\int ns^3 u \, du = -\tfrac{1}{2}(1 + k^2)\ln(\operatorname{ds} u + \operatorname{cs} u) - \tfrac{1}{2}\operatorname{cs} u \operatorname{ds} u,$$

$$\int ns^4 u \, du = \tfrac{1}{3}[(2 + k^2)u - 2(1 + k^2)E(u) - \operatorname{dn} u \operatorname{cs} u(2 + 2k^2 + ns^2 u)].$$

15. If

$$J_m = \int cn^m u \, du, \quad K_m = \int dn^m u \, du,$$

obtain the recurrence relationships

$$(m + 1)k^2 J_{m+2} = m(2k^2 - 1)J_m + (m - 1)k'^2 J_{m-2} + cn^{m-1} u \operatorname{sn} u \operatorname{dn} u,$$
$$(m + 1)K_{m+2} = m(2 - k^2)K_m - (m - 1)k'^2 K_{m-2} + k^2 dn^{m-1} u \operatorname{sn} u \operatorname{cn} u.$$

Deduce the results

$$\int cn^3 u \, du = \frac{1}{2k^3}[(2k^2 - 1)\sin^{-1}(k \operatorname{sn} u) + k \operatorname{sn} u \operatorname{dn} u],$$

$$\int cn^4 u \, du = \frac{1}{3k^4}[(2 - 3k^2)k'^2 u + 2(2k^2 - 1)E(u) + k^2 \operatorname{sn} u \operatorname{cn} u \operatorname{dn} u],$$

$$\int dn^3 u\, du = \tfrac{1}{2}[(2-k^2)\sin^{-1}(sn\,u) + k^2\,sn\,u\,cn\,u],$$

$$\int dn^4 u\, du = \tfrac{1}{3}[2(2-k^2)E(u) - k'^2 u + k^2\,sn\,u\,cn\,u\,dn\,u]$$

$$\int nc^3 u\, du = \frac{1}{2k'^3}[(1-2k^2)\ln(dc\,u + k'\,sc\,u) + k'\,sc\,u\,dc\,u],$$

$$\int nc^4 u\, du = \frac{1}{3k'^4}[k'^2(2-3k^2)u + 2(1-2k^2)(dn\,u\,sc\,u - E(u))$$
$$+ k'^2\,sc\,u\,dc\,u\,nc\,u],$$

$$\int nd^3 u\, du = \frac{1}{2k'^3}[(2-k^2)\sin^{-1}(cd\,u) - k^2 k'\,sd\,u\,cd\,u],$$

$$\int nd^4 u\, du = \frac{1}{3k'^4}[2(2-k^2)(E(u) - k^2\,sn\,u\,cd\,u) - k'^2 u - k^2 k'^2\,sd\,u\,cd\,u\,nd\,u].$$

16. Using the identity $sn(u+K) = cd\,u$ and one of the results of Exercise 14, show that

$$\int cd^4 u\, du = \frac{1}{3k^4}[(2+k^2)u - 2(1+k^2)E(u)$$
$$+ k^2\,sn\,u\,cd\,u(2 + 2k^2 - k'^2\,nd^2 u)].$$

Integrate the third and fourth powers of $dc\,u$, $sd\,u$, and $ds\,u$ by a similar method.

17. Obtain the following integrals:

(i) $\displaystyle\int_0^{2K} u\,nd^2 u\, du = 2KE/k'^2,$

(ii) $\displaystyle\int_0^{2K} u\,nd^3 u\, du = \frac{\pi}{2k'^3}(1+k'^2)K,$

(iii) $\displaystyle\int_0^{2K} u\,nd^4 u\, du = \frac{2K}{3k'^4}\{2(1+k'^2)E - k'^2 K\}.$

(Hint: $dn(u-K) = k'\,nd\,u$.)

18. Show that

$$\int \frac{cn\,u}{1+cn\,u}\, du = E(u) - \frac{sn\,u\,dn\,u}{1+cn\,u}.$$

(Hint: Multiply integrand by $(1-c)/(1-c)$.) Prove, similarly, that

$$\int \frac{sn\,u}{1+sn\,u}\, du = \frac{1}{k'^2}\left[E(u) + \frac{cn\,u\,dn\,u}{1+sn\,u}\right],$$

$$\int \frac{dn\,u}{1+dn\,u}\, du = \frac{1}{k^2}\left[E(u) - k'^2 u - \frac{k^2\,sn\,u\,cn\,u}{1+dn\,u}\right].$$

Deduce expressions for the integrals of $(1+sn\,u)^{-1}$, $(1+cn\,u)^{-1}$, and $(1+dn\,u)^{-1}$.

19. Prove that, for small u,

   (i) $E(u) = u - \frac{1}{3}k^2 u^3 + \frac{1}{15}k^2(1 + k^2)u^5 + O(u^7)$,

   (ii) $E(u + iK') = \frac{1}{u} + i(K' - E') + \frac{1}{3}(2 - k^2)u - \frac{1}{45}(1 - k^2 + k^4)u^3 + O(u^5)$.

20. Show that, if $x > 9$, then

$$\int_x^\infty \{(t-1)(t-4)(t-6)(t-9)\}^{-1/2}\,dt = \frac{1}{4}\,\mathrm{ns}^{-1}\{\frac{1}{4}(x-5), \frac{1}{4}\}.$$

   (Hint: Pair $(t-1)$ with $(t-9)$ and $(t-4)$ with $(t-6)$.)

21. Show that

$$E(u + K + iK') = E(u) + E + i(K' - E') - \mathrm{sc}\,u\,\mathrm{dn}\,u$$

   and deduce that $E(K + iK') = E + i(K' - E')$.

22. Using the identity $\mathrm{dn}(u + iK') = -i\,\mathrm{cs}\,u$ and results from Exercise 15, show that

$$\int \mathrm{cs}^3 u\,du = \frac{1}{2}[(2 - k^2)\ln(\mathrm{ns}\,u + \mathrm{ds}\,u) - \mathrm{ns}\,u\,\mathrm{ds}\,u],$$

$$\int \mathrm{cs}^4 u\,du = \frac{1}{3}[2(2 - k^2)E(u) - k'^2 u + \mathrm{dn}\,u\,\mathrm{cs}\,u(4 - 2k^2 - \mathrm{ns}^2 u)].$$

   Obtain the integrals of $\mathrm{sc}^3 u$ and $\mathrm{sc}^4 u$, similarly.

23. Prove $\lim_{k \to 0}(K - E)/k^2 = \frac{1}{4}\pi$.

24. With modulus $k = 1/\sqrt{2}$, show that

$$K = \{\Gamma(\tfrac{1}{4})\}^2/4\sqrt{\pi}, \qquad E = (2K^2 + \pi)/4K.$$

   (Hint: Put $t = \cos\theta$ in equations (3.8.1) and (3.8.3).)

25. With modulus $k = 1/\sqrt{2}$, show that $k' = k$, $K' = K$ and deduce that $2^{-1/4}\theta_3(0) = \theta_2(0) = \theta_4(0)$, $q = e^{-\pi}$. Hence, prove that

$$\sum_{n=0}^\infty \exp\{-n(n+1)\pi\} = \frac{e^{\pi/4}\Gamma(\tfrac{1}{4})}{2^{7/4}\pi^{3/4}},$$

$$\sum_{n=0}^\infty \exp\{-(2n+1)^2\pi\} = \frac{(2^{1/4} - 1)\Gamma(\tfrac{1}{4})}{2^{11/4}\pi^{3/4}}.$$

26. Transforming the variable of integration by $t = \sqrt{2}\,\mathrm{sn}\,u\,\mathrm{dn}\,u$ with modulus $1/\sqrt{2}$, show that

$$\int_0^x \frac{dt}{(1 - t^2)^{1/4}} = \sqrt{2}\{2E(y) - y\},$$

   where $\mathrm{cn}\,y = (1 - x^2)^{1/4}$.

27. Prove

   (i) $\Pi(u, v) - \Pi(v, u) = uZ(v) - vZ(u)$,

(ii) $\Pi(2K, a) = 2KZ(a)$,

(iii) $\Pi(2iK', a) = 2iK'Z(a) + i\pi a/K$.

(Note: The dependence of Π on k is not explicitly indicated.)

28. Establish the identity

$$\theta_1(x')\theta_1(y')\theta_1(z')\theta_1(w') + \theta_4(x')\theta_4(y')\theta_4(z')\theta_4(w')$$
$$= \theta_1(x)\theta_1(y)\theta_1(z)\theta_1(w) + \theta_4(x)\theta_4(y)\theta_4(z)\theta_4(w),$$

where $x' = s - x$, etc. $(2s = x + y + z + w)$. (Hint: Use Exercise 5 in Chapter 1.)
Putting $x = u, y = v, z = \pm a, w = u + v \pm a$, deduce that

$$\Pi(u, a) + \Pi(v, a) - \Pi(u + v, a) = \frac{1}{2}\ln\frac{1 - k^2\,\text{sn}\,a\,\text{sn}\,u\,\text{sn}\,v\,\text{sn}(u + v - a)}{1 + k^2\,\text{sn}\,a\,\text{sn}\,u\,\text{sn}\,v\,\text{sn}(u + v + a)}.$$

Deduce, further, that

$$\Pi(u, a) + \Pi(u, b) - \Pi(u, a + b) = \frac{1}{2}\ln\frac{1 - k^2\,\text{sn}\,a\,\text{sn}\,b\,\text{sn}\,u\,\text{sn}(a + b - u)}{1 + k^2\,\text{sn}\,a\,\text{sn}\,b\,\text{sn}\,u\,\text{sn}(a + b + u)}$$
$$+ k^2 u\,\text{sn}\,a\,\text{sn}\,b\,\text{sn}(a + b).$$

(Note: These are quasi-addition theorems for Jacobi's integral of the third kind.)

29. Show that $\Pi(iu, ia + K, k) = \Pi(u, a + K', k')$.

30. Legendre's equation of order n is

$$(1 - z^2)\frac{d^2w}{dz^2} - 2z\frac{dw}{dz} + n(n + 1)w = 0.$$

Show that $K(k)$ and $K'(k)$ satisfy the equation with $n = -\frac{1}{2}, z = 1 - 2k^2$.

31. If $c = k^2, c' = k'^2$, show that $E - k'^2 K$ satisfies the equation

$$4cc'\frac{d^2u}{dc^2} = u.$$

Deduce that $E' - k^2 K'$ is a second solution of this equation. (Hint: Exchange c and c'.)

32. Replacing k by ik in the integrals for $\text{sn}^{-1}(x, k)$, $\text{cn}^{-1}(x, k)$, and $\text{dn}^{-1}(x, k)$, obtain the following transformation equations:

(i) $\text{sn}(u, ik) = \dfrac{1}{h}\text{sd}(hu, k/h)$,

(ii) $\text{cn}(u, ik) = \text{cd}(hu, k/h)$,

(iii) $\text{dn}(u, ik) = \text{nd}(hu, k/h)$,

where $h = \sqrt{(1 + k^2)}$.

33. Obtain the transformations

(i) $\text{sn}(k'u, ik/k') = k'\,\text{sd}(u, k)$,

(ii) $\text{cn}(k'u, ik/k') = \text{cd}(u, k)$,

(iii) $\text{dn}(k'u, ik/k') = \text{nd}(u, k)$.

34. Obtain the transformations

   (i) $\mathrm{sn}(ik'u, 1/k') = ik'\,\mathrm{sc}(u, k)$,
   (ii) $\mathrm{cn}(ik'u, 1/k') = \mathrm{dc}(u, k)$,
   (iii) $\mathrm{dn}(ik'u, 1/k') = \mathrm{nc}(u, k)$.

35. Obtain the transformations

   (i) $\mathrm{sn}\{i(1 + k)u, (1 - k)/(1 + k)\} = i(1 + k)\mathrm{sc}(u, k)\mathrm{nd}(u, k)$,
   (ii) $\mathrm{sn}\{2\sqrt{k}u, (1 + k)/2\sqrt{k}\} = 2\sqrt{k}\,\mathrm{sn}(u, k)/\{1 + k\,\mathrm{sn}^2(u, k)\}$,
   (iii) $\mathrm{sn}\{i(1 + k')u, 2\sqrt{k'}/(1 + k')\}$
         $= i(1 + k')\mathrm{sn}(u, k)\mathrm{cn}(u, k)/\{1 - (1 + k')\mathrm{sn}^2(u, k)\}$
   (iv) $\mathrm{sn}\{(k' + ik)u, 2\sqrt{(ikk')}/(k' + ik)\}$
         $= (k' + ik)\mathrm{sn}(u, k)\mathrm{dn}(u, k)/\{1 + k(ik' - k)\mathrm{sn}^2(u, k)\}$.

36. If K, K' are the quarter-periods for $\mathrm{sn}(u, k), \mathrm{cn}(u, k)$, and $\mathrm{dn}(u, k)$, show that the quarter-periods of $\mathrm{sn}(u_1, k_1), \mathrm{cn}(u_1, k_1)$, and $\mathrm{dn}(u_1, k_1)$ obtained by Landen's transformation are $\Lambda = \frac{1}{2}(1 + k')K$ and $\Lambda' = (1 + k')K'$. Taking $k = 1/\sqrt{2}$, deduce that quarter-periods for elliptic functions with modulus $3 - 2\sqrt{2}$ are given by

$$\Lambda = \tfrac{1}{2}\Lambda' = \frac{\sqrt{2} + 1}{8\sqrt{(2\pi)}}\{\Gamma(\tfrac{1}{4})\}^2.$$

(Hint: Refer to Exercise 24.)

37. Prove that $kE(ku, 1/k) = E(u, k) - k'^2 u$.

38. Prove that

$$\frac{\partial}{\partial k}Z(u, k) = \frac{k}{k'^2}\,\mathrm{sn}\,u\,\mathrm{cn}\,u\,\mathrm{dn}\,u + \frac{1}{k}u\,\mathrm{dn}^2 u - \frac{k}{k'^2}E(u)\mathrm{cn}^2 u$$

$$+ \frac{1}{kk'^2}\cdot\frac{E}{K^2}(E - 2k'^2 K)u.$$

39. Show that

$$\Pi(u, a, k) = uE(a) - \tfrac{1}{2}\int_{u-a}^{u+a} E(v)\,dv.$$

40. Show that

$$\frac{\partial}{\partial k}\int_0^u E(v)\,dv = \frac{k}{2k'^2}\,\mathrm{sn}^2 u - \frac{1}{2k}u^2 + \frac{1}{k}uE(u) - \frac{1}{2kk'^2}E^2(u).$$

41. Deduce from the results of the last two exercises and equation (3.10.15) that

$$\frac{\partial}{\partial k}\Pi(u, a, k) = \left[\frac{k}{k'^2}\,\mathrm{sn}\,a\,\mathrm{cn}\,a\,\mathrm{dn}\,a - ka\,\mathrm{sn}^2 a - \frac{k}{k'^2}E(a)\mathrm{cn}^2 a\right]u$$

$$- \frac{k}{4k'^2}\{\mathrm{sn}^2(u + a) - \mathrm{sn}^2(u - a)\} + \frac{1}{k}au$$

$$-\frac{1}{2k}\{(u+a)E(u+a)-(u-a)E(u-a)\}$$

$$+\frac{1}{4kk'^2}\{E^2(u+a)-E^2(u-a)\}.$$

42. Prove that

$$\int_0^u \frac{1-\beta^2\,\mathrm{sn}^2 v}{1-\alpha^2\,\mathrm{sn}^2 v}\,dv=\frac{1}{\alpha^2}[\beta^2 u+(\alpha^2-\beta^2)\Lambda(u,\alpha,k)].$$

Hence evaluate

$$\int_0^u \frac{\mathrm{cn}^2 v}{1-\alpha^2\,\mathrm{sn}^2 v}\,dv,\qquad \int_0^u \frac{\mathrm{dn}^2 v}{1-\alpha^2\,\mathrm{sn}^2 v}\,dv.$$

43. Prove that

$$\int_0^u \frac{\mathrm{sd}^2 v}{1-\alpha^2\,\mathrm{sn}^2 v}\,dv=\frac{1}{k'^2(k^2-\alpha^2)}[E(u)-k^2\,\mathrm{sn}\,u\,\mathrm{cd}\,u-k'^2\Lambda(u,\alpha,k)].$$

44. Prove that $E(\tfrac{1}{2}K,k)=\tfrac{1}{2}E+k^2/\{2(1+k')\}$.

45. Prove that

$$\int \frac{\mathrm{dn}\,u}{1+\alpha^2\,\mathrm{sn}^2 u}\,du=\frac{1}{\sqrt{(1+\alpha^2)}}\tan^{-1}\{\sqrt{(1+\alpha^2)}\,\mathrm{sc}\,u\}.$$

(Hint: Put $s=\mathrm{sn}\,u$.)

46. Prove that

(i) $\displaystyle\int \frac{du}{1\pm\mathrm{sn}\,u}=u+\frac{1}{k'^2}\{\mathrm{dc}\,u(\mathrm{sn}\,u\mp 1)-E(u)\}$,

(ii) $\displaystyle\int \frac{du}{1\pm k\,\mathrm{sn}\,u}=\frac{1}{k'^2}\{E(u)-k\,\mathrm{cd}\,u(k\,\mathrm{sn}\,u\mp 1)\}$.

(Hint: Change denominator to $\mathrm{cn}^2 u$ or $\mathrm{dn}^2 u$.)

47. Prove that, if α^2 is not 1 or k^2 (see previous exercise), then

$$\int \frac{du}{1+\alpha\,\mathrm{sn}\,u}=\Lambda(u,\alpha,k)-\frac{\alpha}{\sqrt{\{(\alpha^2-1)(\alpha^2-k^2)\}}}\tanh^{-1}\!\left(\sqrt{\frac{\alpha^2-k^2}{\alpha^2-1}}\,\mathrm{cd}\,u\right),$$

$$\alpha^2>1,$$

$$=\Lambda(u,\alpha,k)+\frac{\alpha}{\sqrt{\{(\alpha^2-k^2)(1-\alpha^2)\}}}\tan^{-1}\!\left(\sqrt{\frac{\alpha^2-k^2}{1-\alpha^2}}\,\mathrm{cd}\,u\right),$$

$$k^2<\alpha^2<1,$$

$$=\Lambda(u,\alpha,k)+\frac{\alpha}{\sqrt{\{(k^2-\alpha^2)(1-\alpha^2)\}}}\tanh^{-1}\!\left(\sqrt{\frac{k^2-\alpha^2}{1-\alpha^2}}\,\mathrm{cd}\,u\right),$$

$$\alpha^2<k^2.$$

48. Prove that

(i) $\displaystyle\int \frac{du}{1 \pm \mathrm{dn}\, u} = \frac{1}{k^2}\{u - \mathrm{cs}\, u(\mathrm{dn}\, u \mp 1) - E(u)\},$

(ii) $\displaystyle\int \frac{du}{\mathrm{dn}\, u \pm k'} = \frac{1}{k^2}\,\mathrm{sc}\, u \pm \frac{1}{k^2 k'}\{E(u) - k'^2 u - \mathrm{sc}\, u\, \mathrm{dn}\, u\}.$

49. Provided $\alpha^2 \neq 1$ or $1/k'^2$, prove that

$$\int \frac{du}{1 + \alpha\, \mathrm{dn}\, u} = \frac{1}{1-\alpha^2}\Lambda\!\left(u, \frac{i k \alpha}{\sqrt{(1-\alpha^2)}}, k\right) - \frac{\alpha}{\sqrt{\{(1-\alpha^2)(1-k'^2\alpha^2)\}}}$$

$$\times \tan^{-1}\!\left(\sqrt{\frac{1-k'^2\alpha^2}{1-\alpha^2}}\,\mathrm{sc}\, u\right), \qquad \alpha^2 < 1,$$

$$= -\frac{1}{\alpha^2-1}\Lambda\!\left(u, \frac{k\alpha}{\sqrt{(\alpha^2-1)}}, k\right) + \frac{\alpha}{\sqrt{\{(\alpha^2-1)(1-k'^2\alpha^2)\}}}$$

$$\times \tanh^{-1}\!\left(\sqrt{\frac{1-k'^2\alpha^2}{\alpha^2-1}}\,\mathrm{sc}\, u\right), \qquad 1 < \alpha^2 < 1/k'^2,$$

$$= -\frac{1}{\alpha^2-1}\Lambda\!\left(u, \frac{k\alpha}{\sqrt{(\alpha^2-1)}}, k\right) + \frac{\alpha}{\sqrt{\{(\alpha^2-1)(k'^2\alpha^2-1)\}}}$$

$$\times \tan^{-1}\!\left(\sqrt{\frac{k'^2\alpha^2-1}{\alpha^2-1}}\,\mathrm{sc}\, u\right), \qquad \alpha^2 > 1/k'^2.$$

50. Prove that

$$\int \frac{du}{1 \pm \mathrm{cn}\, u} = u - \mathrm{ds}\, u(\mathrm{cn}\, u \mp 1) - E(u).$$

51. If $\alpha^2 \neq 1$, show that

$$\int \frac{du}{1 + \alpha\, \mathrm{cn}\, u} = \frac{1}{1-\alpha^2}\Lambda\!\left(u, \frac{i\alpha}{\sqrt{(1-\alpha^2)}}, k\right) - \frac{\alpha}{\sqrt{\{(1-\alpha^2)(k^2 + k'^2\alpha^2)\}}}$$

$$\times \tan^{-1}\!\left(\sqrt{\frac{k^2 + k'^2\alpha^2}{1-\alpha^2}}\,\mathrm{sd}\, u\right), \qquad \alpha^2 < 1,$$

$$= -\frac{1}{\alpha^2-1}\Lambda\!\left(u, \frac{\alpha}{\sqrt{(\alpha^2-1)}}, k\right) + \frac{\alpha}{\sqrt{\{(\alpha^2-1)(k^2 + k'^2\alpha^2)\}}}$$

$$\times \tanh^{-1}\!\left(\sqrt{\frac{k^2 + k'^2\alpha^2}{\alpha^2-1}}\,\mathrm{sd}\, u\right), \qquad \alpha^2 > 1.$$

52. Obtain a transformation of the type (3.3.15) which reduces the cubic $11x^2 - 8x^3 - 3$ to the form (3.3.16) with real factors and deduce that

$$\int (11x^2 - 8x^3 - 3)^{-1/2}\, dx = \frac{2}{\sqrt{21}}\,\mathrm{dn}^{-1}\!\left(\sqrt{\frac{5}{21}}\cdot\frac{3-2x}{2x-1}\right),$$

with $k^2 = 16/21$. (Hint: $\lambda = \frac{1}{2}$ satisfies the sextic (3.3.20).)

53. Prove that

$$\int_0^{\pi/2} \frac{\sin^2\theta}{\sqrt{(1-k^2\sin^2\theta)}}\,d\theta = \frac{1}{k^2}(K-E).$$

By differentiating $\sin\theta\cos\theta\sqrt{(1-k^2\sin^2\theta)}$ and integrating the result between the limits 0 and $\frac{1}{2}\pi$, deduce that

$$\int_0^{\pi/2} \frac{\sin^4\theta}{\sqrt{(1-k^2\sin^2\theta)}}\,d\theta = \frac{1}{3k^4}\{(k^2+2)K - 2(k^2+1)E\}.$$

54. Prove that

$$\int_0^{\pi/2} \frac{k^4\sin^2\theta\cos^2\theta}{(1-k^2\sin^2\theta)^{3/2}}\,d\theta = (2-k^2)K - 2E.$$

55. If $c = k^2, c' = k'^2$, prove that

(i) $\displaystyle\int_0^c K\,dc = 2(E - c'K),$

(ii) $\displaystyle\int_0^c E\,dc = \frac{2}{3}\{(1+c)E - (1-c)K\},$

(iii) $\displaystyle\int_0^k (E/k'^2)\,dk = kK.$

(Hint: Verify by differentiation.)

56. Show that $D(\phi, k)$ and $F(\phi, k)$ satisfy

(i) $\displaystyle\frac{\partial D}{\partial k} = \frac{1}{k}(D - F),$

(ii) $\displaystyle\frac{\partial F}{\partial k} = \frac{1}{kk'^2}(D - k'^2 F) - \frac{k}{k'^2}\sin\phi\cos\phi.$

57. Putting $\operatorname{sn}(u, k) = x$ and $\operatorname{sn}(iu, k') = y$ in the transformation $\operatorname{sn}(iu, k') = i\operatorname{sc}(u, k)$, show that the change of variable $y = ix/\sqrt{(1-x^2)}$ generates the transformation

$$[(1-y^2)(1-k'^2 y^2)]^{-1/2}\,dy = i[(1-x^2)(1-k^2 x^2)]^{-1/2}\,dx.$$

Similarly, starting from the Gauss transformation (3.9.28), show that the change of variable $y = (1+k)x/(1+kx^2)$ generates the transformation

$$[(1-y^2)(1-\lambda^2 y^2)]^{-1/2}\,dy = (1+k)[(1-x^2)(1-k^2 x^2)]^{-1/2}\,dx$$

where $\lambda = 2\sqrt{k}/(1+k)$.

CHAPTER 4

Geometrical Applications

4.1. Geometry of the Ellipse

Taking an ellipse to have parametric equations

$$x = a \sin \theta, \qquad y = b \cos \theta, \tag{4.1.1}$$

where $a > b$ and the eccentric angle θ is measured from the minor axis, if s is the arc length parameter measured clockwise around the curve from the end B of the minor axis, then

$$ds^2 = \sqrt{(dx^2 + dy^2)} = \sqrt{(a^2 \cos^2 \theta + b^2 \sin^2 \theta)} \, d\theta = a\sqrt{(1 - e^2 \sin^2 \theta)} \, d\theta, \tag{4.1.2}$$

where $e = \sqrt{(1 - b^2/a^2)}$ is the eccentricity. Thus, the length of arc from B to any point P where $\theta = \phi$ is given by

$$s = a \int_0^\phi \sqrt{(1 - e^2 \sin^2 \theta)} \, d\theta = aE(u, e), \tag{4.1.3}$$

where $\phi = \text{am}(u, e)$ or $\text{sn}(u, e) = \sin \phi$ (*vide* equation (3.4.27)). Note that the modulus equals the eccentricity.

Putting $\phi = \frac{1}{2}\pi, u = K$ gives aE for the length of a quadrant of the ellipse.

If we introduce an alternative parameter u by putting $\theta = \text{am}(u, e)$, then by equations (2.7.16) and (2.7.17), parametric equations for the ellipse can be reexpressed in the form

$$x = a \, \text{sn} \, u, \qquad y = b \, \text{cn} \, u. \tag{4.1.4}$$

Now suppose that P and Q are two points on the ellipse with eccentric angles $\theta = \text{am}(u + \alpha)$ and $\phi = \text{am}(u - \alpha)$ respectively. We shall permit u to

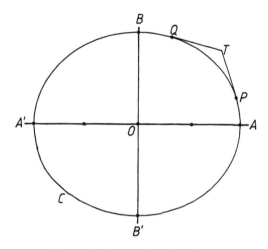

Figure 4.1. Properties of an ellipse.

vary, but regard α as a constant such that $0 < \alpha < K$. Then, the equations of the tangents PT, QT (see Fig. 4.1) are

$$\left.\begin{array}{l} \dfrac{x}{a}\operatorname{sn}(u+\alpha) + \dfrac{y}{b}\operatorname{cn}(u+\alpha) = 1, \\[3mm] \dfrac{x}{a}\operatorname{sn}(u-\alpha) + \dfrac{y}{b}\operatorname{cn}(u-\alpha) = 1. \end{array}\right\} \tag{4.1.5}$$

Adding and subtracting these equations and using the identities (2.4.11), (2.4.12), and (2.4.14), we find that the coordinates of their point of intersection T satisfy

$$\left.\begin{array}{l} \dfrac{x}{a}\operatorname{sn} u\,\operatorname{cn}\alpha\,\operatorname{dn}\alpha + \dfrac{y}{b}\operatorname{cn} u\,\operatorname{cn}\alpha = 1 - k^2\operatorname{sn}^2 u\,\operatorname{sn}^2\alpha, \\[3mm] \dfrac{x}{a}\operatorname{cn} u\,\operatorname{dn} u\,\operatorname{sn}\alpha - \dfrac{y}{b}\operatorname{sn} u\,\operatorname{dn} u\,\operatorname{sn}\alpha\,\operatorname{dn}\alpha = 0. \end{array}\right\} \tag{4.1.6}$$

Solving these equations for the coordinates of T, we find that

$$x = a\,\operatorname{dc}\alpha\,\operatorname{sn} u, \qquad y = b\,\operatorname{nc}\alpha\,\operatorname{cn} u. \tag{4.1.7}$$

We conclude that the locus of T is an ellipse with semiaxes $a\,\operatorname{dc}\alpha$ and $b\,\operatorname{nc}\alpha$. The foci of this ellipse lie at the points $(\pm f, 0)$, where

$$f^2 = a^2\operatorname{dc}^2\alpha - b^2\operatorname{nc}^2\alpha = a^2 e^2, \tag{4.1.8}$$

since $b^2 = a^2(1 - e^2) = a^2 e'^2$, where e' is the complementary modulus. Thus, for all values of α, the locus of T is a confocal ellipse.

By differentiating the parametric equations (4.1.4), we obtain direction ratios for the tangent at the point u, viz.

$$dx = a\,\mathrm{cn}\,u\,\mathrm{dn}\,u\,du, \qquad dy = -b\,\mathrm{sn}\,u\,\mathrm{dn}\,u\,du. \tag{4.1.9}$$

Removing the factor $\mathrm{dn}\,u\,du$, we derive direction ratios $(a\,\mathrm{cn}\,u, -b\,\mathrm{sn}\,u)$ and, since

$$a^2\,\mathrm{cn}^2u + b^2\,\mathrm{sn}^2u = a^2(\mathrm{cn}^2u + e'^2\,\mathrm{sn}^2u) = a^2\,\mathrm{dn}^2u, \tag{4.1.10}$$

the direction cosines are found to be $(\mathrm{cd}\,u, -e'\,\mathrm{sd}\,u)$. Thus, if t is distance measured along the tangent at u from the point of contact,

$$x = a\,\mathrm{sn}\,u + t\,\mathrm{cd}\,u, \qquad y = b\,\mathrm{cn}\,u - te'\,\mathrm{sd}\,u \tag{4.1.11}$$

are parametric equations for the tangent.

For the tangent PT, we have

$$x = a\,\mathrm{sn}(u + \alpha) + t\,\mathrm{cd}(u + \alpha). \tag{4.1.12}$$

At $T, x = a\,\mathrm{dc}\,\alpha\,\mathrm{sn}\,u$ and, hence,

$$t = a(\mathrm{dc}\,\alpha\,\mathrm{sn}\,u - \mathrm{sn}(u + \alpha))\mathrm{dc}(u + \alpha),$$
$$= -a\,\mathrm{sc}\,\alpha\,\mathrm{dn}\,u\,\mathrm{dn}(u + \alpha), \tag{4.1.13}$$

by application of the identities (2.4.1), (2.4.2) and some manipulation. Thus

$$PT = a\,\mathrm{sc}\,\alpha\,\mathrm{dn}\,u\,\mathrm{dn}(u + \alpha). \tag{4.1.14}$$

Replacing α by $-\alpha$, we find similarly that

$$QT = a\,\mathrm{sc}\,\alpha\,\mathrm{dn}\,u\,\mathrm{dn}(u - \alpha). \tag{4.1.15}$$

Next, it follows from equation (4.1.3) that the arc PQ of the ellipse is given by

$$\begin{aligned}
\mathrm{arc}\,PQ &= a\{E(u + \alpha) - E(u - \alpha)\} \\
&= 2aE(\alpha) - ae^2\,\mathrm{sn}\,u\,\mathrm{sn}\,\alpha\{\mathrm{sn}(u + \alpha) + \mathrm{sn}(u - \alpha)\} \\
&= 2aE(\alpha) - \frac{2ae^2\,\mathrm{sn}^2u\,\mathrm{sn}\,\alpha\,\mathrm{cn}\,\alpha\,\mathrm{dn}\,\alpha}{1 - e^2\,\mathrm{sn}^2u\,\mathrm{sn}^2\alpha},
\end{aligned} \tag{4.1.16}$$

after making use of the identities (3.5.14) and (2.4.11). Also

$$\begin{aligned}
PT + QT &= a\,\mathrm{sc}\,\alpha\,\mathrm{dn}\,u\{\mathrm{dn}(u + \alpha) + \mathrm{dn}(u - \alpha)\} \\
&= \frac{2a\,\mathrm{sc}\,\alpha\,\mathrm{dn}^2u\,\mathrm{dn}\,\alpha}{1 - e^2\,\mathrm{sn}^2u\,\mathrm{sn}^2\alpha}.
\end{aligned} \tag{4.1.17}$$

Thus,

$$PT + QT - \mathrm{arc}\,PQ = 2a\{\mathrm{sc}\,\alpha\,\mathrm{dn}\,\alpha - E(\alpha)\}. \tag{4.1.18}$$

We have now proved that, as T describes a confocal ellipse, $PT + QT - \mathrm{arc}\,PQ$ remains constant.

Now suppose that a string is wrapped around the ellipse $x^2/a^2 + y^2/b^2 = 1$

and, that, after being pulled taut so that it is in contact with the whole ellipse except for the arc PQ, its two ends are knotted at T. The length of the loop is then $PT + QT + p -$ arc PQ, where p is the perimeter of the ellipse. Thus, if the string is permitted to slip around the ellipse, T will describe a confocal ellipse.

In the special case which arises when $b \rightarrow 0, e \rightarrow 1$, the ellipse degenerates into the segment $A'A$ and the foci become the points A and A'. The string, when taut, now forms a triangle $TA'A$ and the locus of T is well known to be an ellipse with its foci at A and A'. Thus, the theorem we have proved is revealed to be a generalization of a familiar property of the ellipse.

Next, suppose P and Q are any two points on the ellipse with parameters u, v respectively, where $u + v = K$. Referring to the results (2.2.17), (2.2.18), we have

$$\operatorname{sn} v = \operatorname{cd} u, \qquad \operatorname{cn} v = e' \operatorname{sd} u. \tag{4.1.19}$$

Hence

$$\operatorname{cn} u \operatorname{cn} v = e' \operatorname{sn} u \operatorname{sn} v. \tag{4.1.20}$$

Thus, if θ, ϕ are the eccentric angles of P and Q,

$$\cot \theta \cot \phi = e', \tag{4.1.21}$$

from which it follows that

$$\sin \phi = 1 / \sqrt{(1 + e'^2 \tan^2 \theta)}. \tag{4.1.22}$$

The equation of the normal to the ellipse at P is

$$ax \cos \theta - by \sin \theta = (a^2 - b^2) \sin \theta \cos \theta, \tag{4.1.23}$$

and the distance of the normal from the center O is accordingly given by

$$d = \frac{(a^2 - b^2) \sin \theta \cos \theta}{\sqrt{(a^2 \cos^2 \theta + b^2 \sin^2 \theta)}} = \frac{ae^2 \sin \theta}{\sqrt{(1 + e'^2 \tan^2 \theta)}},$$

$$= ae^2 \sin \theta \sin \phi. \tag{4.1.24}$$

The distance of the normal at Q from the center is calculated similarly also to be d.

Using the addition theorem (3.5.14), we find that

$$E = E(K) = E(u + v) = E(u) + E(v) - e^2 \operatorname{sn} u \operatorname{sn} v$$

$$= E(u) + E(v) - e^2 \sin \theta \sin \phi. \tag{4.1.25}$$

Hence,

$$d = aE(u) + aE(v) - aE = \operatorname{arc} BQ - \operatorname{arc} AP. \tag{4.1.26}$$

We have proved *Fagnano's theorem* that, if AP and BQ are two arcs of the quadrant AB of an ellipse and the normals at the extremities P, Q lie at the same distance d from the center, then the length difference of the arcs AP and BQ is also d.

If P and Q coincide, then $u = v = \frac{1}{2}K$ and the coordinates of this point are

$$x = a \operatorname{sn}\tfrac{1}{2}K = a/\sqrt{(1+e')} = \left(\frac{a^3}{a+b}\right)^{1/2},$$

$$y = b \operatorname{cn}\tfrac{1}{2}K = b\sqrt{e'}/\sqrt{(1+e')} = \left(\frac{b^3}{a+b}\right)^{1/2}. \tag{4.1.27}$$

(See equations (2.4.10).) This is called *Fagnano's point*. Thus, if F is this point, then

$$d = \operatorname{arc} BF - \operatorname{arc} AF = ae^2 \operatorname{sn}^2\tfrac{1}{2}K = a - b. \tag{4.1.28}$$

Another problem involving elliptic functions is to calculate the curve on which an ellipse can roll so that its center describes a straight line.

In Fig. 4.2, AC is a semimajor axis of an ellipse and the arc AP of the ellipse has rolled on the arc YP of the curve, while the center has moved from O to C along a straight line which we take to be horizontal and the x-axis. The y-axis is taken to be OY, vertically downward in the figure. Since C must move horizontally and P is the instantaneous center of rotation, CP must be vertical.

Let (r, θ) be polar coordinates of P as indicated, so that the polar equation of the ellipse is

$$\frac{1}{r^2} = \frac{1}{a^2}\cos^2\theta + \frac{1}{b^2}\sin^2\theta. \tag{4.1.29}$$

Then, if ϕ is the angle between the tangent to the ellipse at P and the radius r,

$$\tan \phi = r\, d\theta/dr = -\frac{ab}{\sqrt{\{(a^2 - r^2)(r^2 - b^2)\}}}. \tag{4.1.30}$$

But, the tangent to the ellipse is also the tangent to the curve on which it rolls and, hence, this tangent makes an angle $(\frac{3}{2}\pi - \phi)$ with the positive x-axis.

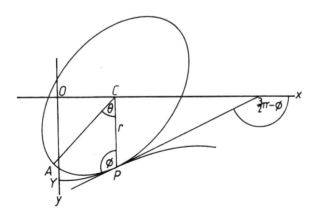

Figure 4.2. Rolling an ellipse on a curve.

It follows that, if P is a point (x, y) on this curve, then

$$\frac{dx}{dy} = \cot(\tfrac{3}{2}\pi - \phi) = \tan\phi = -\frac{ab}{\sqrt{\{(a^2 - y^2)(y^2 - b^2)\}}}, \qquad (4.1.31)$$

since $y = PC = r$. Integrating this equation using the result (3.2.11), we obtain

$$x = b\,dn^{-1}(y/a) + \text{constant}, \qquad (4.1.32)$$

the modulus being $\sqrt{(a^2 - b^2)}/a = e$. Initially, $x = 0$ and $y = a$, so that the constant vanishes.

We conclude that the equation of the curve on which the ellipse rolls is

$$y = a\,dn(x/b, e). \qquad (4.1.33)$$

4.2. Area of an Ellipsoid

We shall calculate the area of the surface of the ellipsoid with equation

$$\frac{x^2}{a^2} + \frac{y^2}{b^2} + \frac{z^2}{c^2} = 1, \qquad (4.2.1)$$

where $a > b > c$, which lies above the plane $z = 0$.

If (α, β, γ) are the direction angles of the normal to this surface at the point (x, y, z), then

$$\cos\gamma = 1 \Big/ \sqrt{\left[1 + \frac{c^4 x^2}{a^4 z^2} + \frac{c^4 y^2}{b^4 z^2}\right]}. \qquad (4.2.2)$$

Hence, the curve along which $\cos\gamma = h$ (constant) has equations

$$1 + \frac{c^4 x^2}{a^4 z^2} + \frac{c^4 y^2}{b^4 z^2} = h^2, \qquad \frac{x^2}{a^2} + \frac{y^2}{b^2} + \frac{z^2}{c^2} = 1. \qquad (4.2.3)$$

Eliminating z between these equations, we calculate that the projection of the curve on the xy-plane has equations

$$\frac{x^2\{a^2 - h^2(a^2 - c^2)\}}{a^4(1 - h^2)} + \frac{y^2\{b^2 - h^2(b^2 - c^2)\}}{b^4(1 - h^2)} = 1, \qquad z = 0, \qquad (4.2.4)$$

which determine an ellipse.

We now define parameters k, v by the equations

$$1 - c^2/a^2 = \sin^2 v, \qquad 1 - c^2/b^2 = k^2 \sin^2 v. \qquad (4.2.5)$$

Clearly, we can assume $0 < k < 1, 0 < v < \tfrac{1}{2}\pi$. The semiaxes of the ellipse (4.2.4) then have lengths

$$a\sqrt{\left[\frac{1 - h^2}{1 - h^2\sin^2 v}\right]}, \qquad b\sqrt{\left[\frac{1 - h^2}{1 - h^2 k^2 \sin^2 v}\right]} \qquad (4.2.6)$$

and its area is given by

$$A(h) = \frac{\pi ab(1 - h^2)}{\sqrt{\{(1 - h^2 \sin^2 v)(1 - h^2 k^2 \sin^2 v)\}}}. \tag{4.2.7}$$

Consider the area dA of the xy-plane lying between the ellipses with constants h and $h + dh$. This is the projection of a strip of the ellipsoid, all of whose normals make an angle $\cos^{-1} h$ with the z-axis. The area of this strip is accordingly $dA/\cos \gamma = dA/h$. It follows that the area of the ellipsoid lying between the curves $h = 1$ (the end of the semiaxis c) and $h = \eta$ is given by

$$S(\eta) = \int_1^\eta \frac{1}{h} \frac{dA}{dh} \, dh = |A/h|_1^\eta + \int_1^\eta h^{-2} A \, dh$$

$$= \eta^{-1} A(\eta) - \int_\eta^1 h^{-2} A \, dh, \tag{4.2.8}$$

having integrated by parts. By letting $\eta \to 0$, we shall obtain a formula for the area of the hemiellipsoid.

To calculate the integral appearing in equation (4.2.8), we substitute for A from (4.2.7) and then change the variable of integration by the transformation $t = h \sin v$ to give

$$\int_\eta^1 h^{-2} A \, dh = \pi ab \sin v I_1 - \pi ab \cosec v I_2, \tag{4.2.9}$$

where

$$I_1 = \int_{\eta \sin v}^{\sin v} t^{-2} \{(1 - t^2)(1 - k^2 t^2)\}^{-1/2} \, dt, \tag{4.2.10}$$

$$I_2 = \int_{\eta \sin v}^{\sin v} \{(1 - t^2)(1 - k^2 t^2)\}^{-1/2} \, dt. \tag{4.2.11}$$

It remains to calculate this pair of elliptic integrals.

Putting $t = \text{sn}(u, k)$, we calculate that

$$I_1 = \int_{\text{sn}^{-1}(\eta \sin v)}^{\text{sn}^{-1}(\sin v)} \text{ns}^2 u \, du = \left[u - \text{dn } u \text{ cs } u - \int \text{dn}^2 u \, du \right]_{\text{sn}^{-1}(\eta \sin v)}^{\text{sn}^{-1}(\sin v)}, \tag{4.2.12}$$

having used equation (3.4.16). Substituting the limits and introducing the E-function (equation (3.4.25)), we arrive at the result

$$I_1 = \text{sn}^{-1}(\sin v) - \text{sn}^{-1}(\eta \sin v) - \cot v \sqrt{(1 - k^2 \sin^2 v)}$$
$$+ \eta^{-1} \cosec v \sqrt{\{(1 - k^2 \eta^2 \sin^2 v)(1 - \eta^2 \sin^2 v)\}}$$
$$- E(\text{sn}^{-1}(\sin v), k) + E(\text{sn}^{-1}(\eta \sin v), k). \tag{4.2.13}$$

Similarly, we calculate that

$$I_2 = \text{sn}^{-1}(\sin v) - \text{sn}^{-1}(\eta \sin v). \tag{4.2.14}$$

We can now identify the terms in the expression (4.2.8) for $S(\eta)$ which become infinite as $\eta \to 0$. They are

$$\eta^{-1}A(\eta) - \pi ab\eta^{-1}\sqrt{\{(1 - k^2\eta^2 \sin^2 v)(1 - \eta^2 \sin^2 v)\}}. \qquad (4.2.15)$$

Substituting for $A(\eta)$ from equation (4.2.7), it is easy to verify that this expression tends to zero as $\eta \to 0$.

Putting $\eta = 0$ in the remaining terms, we now find that the area of the upper half of the ellipsoid is given by

$$S(0) = \pi ab[\cos v\sqrt{(1 - k^2 \sin^2 v)} + \cos v \cot v \, \text{sn}^{-1}(\sin v)$$
$$+ \sin v E(\text{sn}^{-1}(\sin v), k)]. \qquad (4.2.16)$$

Changing to Legendre's notation by equations (3.1.8) and (3.4.27), the area of the whole ellipsoid can now be expressed in the form

$$2S(0) = 2\pi ab[\cos v\sqrt{(1 - k^2 \sin^2 v)} + \cos v \cot v F(v, k) + \sin v D(v, k)]$$
$$= 2\pi c^2 + \frac{2\pi ab}{\sin v}\left[\cos^2 v F(v, k) + \sin^2 v D(v, k)\right], \qquad (4.2.17)$$

where

$$\cos v = c/a \quad \text{and} \quad k^2 = \frac{a^2(b^2 - c^2)}{b^2(a^2 - c^2)}. \qquad (4.2.18)$$

The special cases of an ellipsoid of revolution are (i) $a = b$ (oblate spheroid) and (ii) $b = c$ (prolate spheroid). In the first case $k = 1$ and $F(v, k) = \frac{1}{2}\ln\{(1 + \sin v)/(1 - \sin v)\}$, $D(v, k) = \sin v$; thus, the area is

$$2\pi a^2 + \frac{\pi c^2}{\sin v}\ln\left[\frac{1 + \sin v}{1 - \sin v}\right]. \qquad (4.2.19)$$

In the second case $k = 0$ and $F(v, k) = D(v, k) = v$; this reduces the area to

$$\frac{2\pi ab}{\sin v}(v + \sin v \cos v). \qquad (4.2.20)$$

4.3. The Lemniscate

This is the curve whose polar equation is

$$r^2 = a^2 \cos 2\theta. \qquad (4.3.1)$$

Its shape is that of a figure-8, with the two branches of the curve intersecting at an angle of $\frac{1}{2}\pi$ at the pole.

The length of arc connecting the points $\theta = 0$, $r = a$, and (r, θ) is

$$s = \int_r^a \sqrt{\{r^2(d\theta/dr)^2 + 1\}}\, dr = a^2 \int_r^a (a^4 - r^4)^{-1/2}\, dr. \qquad (4.3.2)$$

This is an elliptic integral of the first kind of the type which has been evaluated at (3.2.3); putting $a = b$ in this result, we find that

$$s = \frac{a}{\sqrt{2}} \operatorname{cn}^{-1}(r/a, 1/\sqrt{2}),$$
(4.3.3)

or

$$r = a \operatorname{cn}(\sqrt{2}s/a, 1/\sqrt{2}).$$
(4.3.4)

As s increases from zero, r decreases from its initial value a, first becoming zero when $\sqrt{2}s/a = K(1/\sqrt{2})$. Hence, the total length of the lemniscate is $2\sqrt{2}aK(1/\sqrt{2})$. But, from equation (4.3.2), this total length must be

$$4a \int_0^1 (1 - t^4)^{-1/2} \, dt = a \int_0^1 u^{-3/4}(1 - u)^{-1/2} \, du$$

$$= aB(\tfrac{1}{4}, \tfrac{1}{2})$$

$$= a \frac{\Gamma(\tfrac{1}{4})\Gamma(\tfrac{1}{2})}{\Gamma(\tfrac{3}{4})}$$

$$= \frac{a}{\sqrt{(2\pi)}}[\Gamma(\tfrac{1}{4})]^2,$$
(4.3.5)

making use of well-known properties of the beta and gamma functions (viz. $B(x, y) = \Gamma(x)\Gamma(y)/\Gamma(x + y)$, $\Gamma(x)\Gamma(1 - x) = \pi/\sin(\pi x)$, $\Gamma(\tfrac{1}{2}) = \sqrt{\pi}$). It now follows that

$$K(1/\sqrt{2}) = \tfrac{1}{4}\pi^{-1/2}[\Gamma(\tfrac{1}{4})]^2.$$
(4.3.6)

Note also that $K'(k) = K(k')$ implies that $K' = K$ for modulus $1/\sqrt{2}$.

From equations (4.3.1) and (4.3.4), we deduce that

$$\sin \theta = \frac{1}{\sqrt{2}} \operatorname{sn}(\sqrt{2}s/a)$$
(4.3.7)

and, hence,

$$\cos \theta = \sqrt{(1 - \sin^2\theta)} = \operatorname{dn}(\sqrt{2}s/a).$$
(4.3.8)

Parametric equations for the lemniscate can now be written down, thus:

$$\left.\begin{array}{l} x = r \cos \theta = a \operatorname{cn}(\sqrt{2}s/a) \operatorname{dn}(\sqrt{2}s/a), \\[2mm] y = r \sin \theta = \dfrac{a}{\sqrt{2}} \operatorname{cn}(\sqrt{2}s/a) \operatorname{sn}(\sqrt{2}s/a). \end{array}\right\}$$
(4.3.9)

4.4. Formulae of Spherical Trigonometry

We shall only consider the case of a spherical triangle, all of whose sides and angles are acute.

Adopting the usual notation, let a, b, c represent the sides and A, B, C the respective opposite angles of the triangle. Then the following relationships are established in most texts devoted to this topic:

$$\frac{\sin a}{\sin A} = \frac{\sin b}{\sin B} = \frac{\sin c}{\sin C} = k \quad \text{(sine rule)}, \tag{4.4.1}$$

$$\cos c = \cos a \cos b + \sin a \sin b \cos C \quad \text{(cosine rule)}, \tag{4.4.2}$$

$$\cos C = -\cos A \cos B + \sin A \sin B \cos c \quad \text{(polar cosine rule)}, \tag{4.4.3}$$

$$\sin A \cos b = \cos B \sin C + \sin B \cos C \cos a \quad \text{(five parts formula)}, \tag{4.4.4}$$

$$\cos C \cos b = \sin b \cot a - \sin C \cot A \quad \text{(four parts formula)}. \tag{4.4.5}$$

It follows from formula (4.4.3) that

$$\cos C < \sin A \sin B \cos c < \cos c \tag{4.4.6}$$

and hence that $C > c$. Thus, in equation (4.4.1), $k < 1$.

We now define *elliptic measures* u, v, w of the angles A, B, C by means of the equations

$$\operatorname{sn} u = \sin A, \qquad \operatorname{sn} v = \sin B, \qquad \operatorname{sn} w = \sin C, \tag{4.4.7}$$

the modulus of the elliptic functions being k. Clearly, we can choose these measures all to lie in the interval $(0, K)$. If we use Jacobi's notation (see equation (2.7.16)), then $A = \operatorname{am} u$, $B = \operatorname{am} v$, $C = \operatorname{am} w$. It now follows that

$$\operatorname{cn} u = \cos A, \qquad \operatorname{cn} v = \cos B, \qquad \operatorname{cn} w = \cos C. \tag{4.4.8}$$

Using the equations (4.4.1), we also deduce that

$$\sin a = k \operatorname{sn} u, \qquad \sin b = k \operatorname{sn} v, \qquad \sin c = k \operatorname{sn} w, \tag{4.4.9}$$

from which it follows that

$$\cos a = \operatorname{dn} u, \qquad \cos b = \operatorname{dn} v, \qquad \cos c = \operatorname{dn} w. \tag{4.4.10}$$

Substitution in the formulae (4.4.2) and (4.4.3) now leads to the equations

$$\operatorname{dn} w = \operatorname{dn} u \operatorname{dn} v + k^2 \operatorname{sn} u \operatorname{sn} v \operatorname{cn} w, \tag{4.4.11}$$

$$\operatorname{cn} w = -\operatorname{cn} u \operatorname{cn} v + \operatorname{sn} u \operatorname{sn} v \operatorname{dn} w. \tag{4.4.12}$$

Solving these equations for $\operatorname{cn} w$ and $\operatorname{dn} w$, we find (using the addition theorems (2.4.2) and (2.4.3)) that

$$\operatorname{cn} w = -\operatorname{cn}(u + v), \qquad \operatorname{dn} w = \operatorname{dn}(u + v). \tag{4.4.13}$$

Since $0 < u < K$, etc., these equations have the unique solution $w = 2K - (u + v)$, i.e.,

$$u + v + w = 2K. \tag{4.4.14}$$

Thus, we have proved that the elliptic measures of the angles of a spherical triangle always sum to $2K$—this result is the counterpart for spherical triangles of the angle sum theorem for plane triangles (to which it reduces in the limit as the triangle sides a, b, c become small and $k \to 0$).

Substitution in the five parts formula leads to the equation

$$\operatorname{sn} u \operatorname{dn} v = \operatorname{cn} v \operatorname{sn} w + \operatorname{sn} v \operatorname{cn} w \operatorname{dn} u, \tag{4.4.15}$$

or

$$\operatorname{cn} v \operatorname{sn}(u + v) = \operatorname{sn} u \operatorname{dn} v + \operatorname{dn} u \operatorname{sn} v \operatorname{cn}(u + v). \tag{4.4.16}$$

This identity also can be verified by application of the addition theorems.

Alternatively, we can establish the theorem (4.4.14) by an independent method and then use the formulae of spherical trigonometry to derive the addition theorems for the elliptic functions.

Thus, suppose c and C are kept constant and a, b, A, B are varied. Then k is constant by equations (4.4.1). Differentiating equation (4.4.3), we find that

$$(\sin A \cos B + \cos A \sin B \cos c)\, dA$$
$$+ (\cos A \sin B + \sin A \cos B \cos c)\, dB = 0. \tag{4.4.17}$$

Using the five parts formula, this reduces to

$$dA \cos b + dB \cos a = 0. \tag{4.4.18}$$

Differentiating equations (4.4.7), we get

$$\operatorname{cn} u \operatorname{dn} u \, du = \cos A \, dA = \operatorname{cn} u \, dA, \tag{4.4.19}$$

i.e., $dA = \operatorname{dn} u \, du = du \cos a$. Similarly, $dB = dv \cos b$. Hence, equation (4.4.18) is equivalent to

$$du + dv = 0. \tag{4.4.20}$$

Integrating, we conclude that

$$u + v = \text{constant} \tag{4.4.21}$$

for this variation.

In the extreme situation where $a = 0$, $b = c$, then $A = 0$, $B = \pi - C$, and thus $u = 0$, $v = 2K - w$. Thus, the constant in equation (4.4.21) must be $2K - w$ and this equation is identical with equation (4.4.14).

The equations (4.4.11), (4.4.12) now follow as before from the formulae (4.4.2), (4.4.3). Putting $w = 2K - (u + v)$ in these equations and solving for $\operatorname{cn}(u + v)$, $\operatorname{dn}(u + v)$, we can derive the addition theorems (2.4.2) and (2.4.3).

4.5. Seiffert's Spiral

This is a curve traced on the surface of a globe, having the property that increments in longitude are everywhere proportional to distances measured along the track.

If (θ, ϕ) are colatitude and longitude respectively, then along the curve

$$\phi = ks + \text{constant}, \tag{4.5.1}$$

where k is constant. Taking the sphere to have unit radius, the metric for its surface is

$$ds^2 = d\theta^2 + \sin^2\theta \, d\phi^2. \tag{4.5.2}$$

Substituting $d\phi = k \, ds$, we deduce that

$$ds/d\theta = (1 - k^2 \sin^2\theta)^{-1/2}. \tag{4.5.3}$$

If $0 < k < 1$, integration now leads to the equation

$$s = \text{sn}^{-1}(\sin\theta, k) + \text{constant}, \tag{4.5.4}$$

or

$$\sin\theta = \text{sn}(s, k), \tag{4.5.5}$$

choosing to measure s from the point on the curve where $\theta = 0$, i.e., the North pole. Employing Jacobi's notation, we can write

$$\theta = \text{am}(s, k). \tag{4.5.6}$$

Equations (4.5.1) and (4.5.6) now determine a family of spirals winding out from the North pole and in to the South pole. They cross the equator where $\theta = \frac{1}{2}\pi$ and, hence, $s = K$. These curves differ from one another only by a rotation about the polar axis; thus, taking the constant in equation (4.5.1) to be zero, we can eliminate s to yield

$$\theta = \text{am}(\phi/k, k) \tag{4.5.7}$$

as the equation of *Seiffert's spiral*.

If ψ is the angle between the spiral and the meridian at any point, then

$$\cos\psi = d\theta/ds = \text{dn}(s, k). \tag{4.5.8}$$

This shows that ψ takes its smallest value where $s = K$ and the curve then meets the equator; at this point $\text{dn } s = k'$ and, therefore, $\psi = \sin^{-1}k$.

If $k = 1$, then

$$\sin\theta = \tanh s, \qquad \phi = s, \tag{4.5.9}$$

and the spiral approaches the equator as $s \to \infty$, but never crosses it.

In case $k > 1$, the integral of equation (4.5.3) can be written

$$s = \frac{1}{k} \int \{(1 - t^2)(k^{-2} - t^2)\}^{-1/2} \, dt, \tag{4.5.10}$$

where $t = \sin\theta$. Referring to the standard integral (3.1.7), we now deduce that

$$\sin\theta = \frac{1}{k} \text{sn}(ks, 1/k). \tag{4.5.11}$$

When $ks = K(1/k)$, i.e.,

$$s = s_0 = \int_0^{\pi/2} \frac{d\theta}{\sqrt{(k^2 - \sin^2\theta)}}, \tag{4.5.12}$$

$sn(ks, 1/k)$ takes its maximum value of unity and θ its maximum value of $\sin^{-1}(1/k)$. Thereafter, θ decreases, becoming zero again when $s = 2s_0$. In this case, therefore, the curve is closed and passes through the pole.

Differentiating equation (4.5.11), we obtain

$$\cos\theta \frac{d\theta}{ds} = cn(ks, \kappa) dn(ks, \kappa), \tag{4.5.13}$$

where $\kappa = 1/k$. But

$$\cos\theta = \sqrt{(1 - \sin^2\theta)} = \sqrt{\{1 - \kappa^2 sn^2(ks, \kappa)\}} = dn(ks, \kappa). \tag{4.5.14}$$

Hence

$$\cos\psi = \frac{d\theta}{ds} = cn(ks, \kappa) \tag{4.5.15}$$

gives the angle made by the curve with the meridian. When $ks = K(\kappa)$ (i.e., $s = s_0$), $\psi = \frac{1}{2}\pi$ and the curve has reached its most southerly point.

4.6. Orthogonal Systems of Cartesian Ovals

We consider the conformal transformation

$$z = sn^2(w, k) \tag{4.6.1}$$

mapping a period parallelogram in the w-plane onto the z-plane. If $w = u + iv$, equations (2.4.1), (2.6.12) show that

$$sn(w, k) = \frac{sn(u, k) dn(v, k') + i cn(u, k) dn(u, k) sn(v, k') cn(v, k')}{1 - dn^2(u, k) sn^2(v, k')}. \tag{4.6.2}$$

It follows that, if $0 < k < 1$, then

$$sn^*(w, k) = sn(w^*, k). \tag{4.6.3}$$

Now let r, r', r'' be distances measured from the points $0, 1$, and $1/k^2$ respectively on the real axis in the z-plane to the point z. Then

$$r = |z| = \sqrt{(zz^*)} = \sqrt{(sn^2w \, sn^2w^*)} = sn\,w\,sn\,w^*, \tag{4.6.4}$$

$$r' = |1 - z| = \sqrt{\{(1 - z)(1 - z^*)\}} = \sqrt{(cn^2w\,cn^2w^*)} = cn\,w\,cnw^*, \tag{4.6.5}$$

$$r'' = |1/k^2 - z| = k^{-2}\sqrt{\{(1 - kz)(1 - kz^*)\}} = k^{-2}\sqrt{(dn^2w\,dn^2w^*)}$$
$$= k^{-2}\,dn\,w\,dn\,w^*. \tag{4.6.6}$$

Application of the identities (2.4.22), (2.4.25)–(2.4.28) now shows that

$$r = \frac{\text{cn } 2iv - \text{cn } 2u}{\text{dn } 2iv + \text{dn } 2u} = \frac{1}{k^2} \frac{\text{dn } 2iv - \text{dn } 2u}{\text{cn } 2iv + \text{cn } 2u}, \tag{4.6.7}$$

$$r' = \frac{\text{cn } 2u \text{ dn } 2iv + \text{dn } 2u \text{ cn } 2iv}{\text{dn } 2iv + \text{dn } 2u} = \frac{k'^2}{k^2} \frac{\text{dn } 2iv - \text{dn } 2u}{\text{cn } 2iv \text{ dn } 2u - \text{cn } 2u \text{ dn } 2iv}, \tag{4.6.8}$$

$$k^2 r'' = \frac{\text{cn } 2u \text{ dn } 2iv + \text{dn } 2u \text{ cn } 2iv}{\text{cn } 2iv + \text{cn } 2u} = k'^2 \frac{\text{cn } 2iv - \text{cn } 2u}{\text{cn } 2iv \text{ dn } 2u - \text{cn } 2u \text{ dn } 2iv}. \tag{4.6.9}$$

Eliminating first v and then u between the equations (4.6.7), (4.6.8), we arrive at the equations

$$\left. \begin{array}{l} r' - r \text{ dn } 2u = \text{cn } 2u, \\ r' + r \text{ dn } 2iv = \text{cn } 2iv. \end{array} \right\} \tag{4.6.10}$$

These equations determine the curves in the z-plane which map the straight lines $u = $ constant, $v = $ constant in the w-plane. Since the two families of straight lines intersect at right angles and the transformation is conformal, the two families of curves determined by equations (4.6.10) must be orthogonal.

Each point in the z-plane has associated with it a unique pair of values of r and r'. Then (r, r') are referred to as the *bipolar coordinates* of the point with respect to the poles at 0 and 1. Not all positive pairs of values of these coordinates can be identified with real points since, by the triangle inequality, it is necessary that $r + r' > 1$, $r + 1 > r'$, and $r' + 1 > r$. However, any pair of positive values of r and r' satisfying these inequalities specifies two points, symmetrically positioned with respect to the real axis. Thus, a relationship $f(r, r') = 0$ between r and r', which can be satisfied by admissible pairs of values of these bipolar coordinates, determines a curve in the z-plane which is symmetrical about the real axis.

In particular, a linear equation of the type

$$\alpha r + \beta r' = \gamma \tag{4.6.11}$$

specifies a closed curve called a *Cartesian oval*. Without loss of generality, we can assume $\gamma \geqslant 0$. Then, one at least of the coefficients α, β must be positive and this is normally taken to be α (it may be necessary to switch poles to abide by this convention). However, it is not convenient to subject the coefficients to these conditions in this section. If $\alpha = \beta$, the sum $(r + r')$ is constant and the curve is an ellipse with its foci at the poles. If $\alpha = -\beta$, the difference $(r - r')$ is constant and the curve is one branch of a hyperbola with its foci at the poles— this is the only case where the curve is not closed (if the difference is zero, the hyperbola degenerates to a straight line). If any one of the constants α, β, γ vanishes, the curve is a circle. The two ovals $\alpha r \pm \beta r' = \gamma$ are said to be *conjugate* and possess a number of interesting properties (see E.H. Lockwood, *A Book of Curves*, Cambridge University Press, 1963).

We can now identify the two families of curves determined by the equations

(4.6.10) as Cartesian ovals, each member of one set cutting each member of the other set orthogonally.

Eliminating first v and then u between equations (4.6.8), (4.6.9), the following pair of equations is obtained:

$$\left.\begin{array}{l} k^2 r'' \,\mathrm{dn}\,2u - k^2 r' \mathrm{cn}\,2u = k'^2, \\ k^2 r'' \,\mathrm{dn}\,2iv - k^2 r' \,\mathrm{cn}\,2iv = k'^2. \end{array}\right\} \qquad (4.6.12)$$

These equations determine two families of Cartesian ovals based upon the poles 1 and $1/k^2$. These families must, of course, be identical with those already introduced earlier and we conclude that a Cartesian oval possesses three poles and is specified by an equation of the type (4.6.11) with respect to any two.

Thus, eliminating u and v between equations (4.6.7), (4.6.9), we are led to the following equations:

$$\left.\begin{array}{l} k^2 r'' - k^2 r\,\mathrm{cn}\,2u = \mathrm{dn}\,2u, \\ k^2 r'' + k^2 r\,\mathrm{cn}\,2iv = \mathrm{dn}\,2iv, \end{array}\right\} \qquad (4.6.13)$$

representing the two families when using the poles 0 and $1/k^2$.

As an exercise, the reader should show by simple geometry that an oval whose equation is (4.6.11) when poles 0 and 1 are used possesses a third pole at $z = (\gamma^2 - \beta^2)/(\alpha^2 - \beta^2)$ $(|\alpha| \neq |\beta|)$. In the case of the first of the families (4.6.10), $\alpha = \mathrm{dn}\,2u$, $\beta = 1$, $\gamma = \mathrm{cn}\,2u$, so the third pole is at $z = (\mathrm{cn}^2 2u - 1)/(\mathrm{dn}^2 2u - 1) = 1/k^2$.

EXERCISES

1. An ellipse has semimajor axis a and eccentricity $(\sqrt{3} - 1)/2\sqrt{2}$. Show that its perimeter is

$$a\left(\frac{\pi}{\sqrt{3}}\right)^{1/2}\left[(1 + 1/\sqrt{3})\frac{\Gamma(\tfrac{1}{3})}{\Gamma(\tfrac{5}{6})} + \frac{2\Gamma(\tfrac{5}{6})}{\Gamma(\tfrac{1}{3})}\right].$$

2. If (x, y) are Cartesian coordinates and

$$x = a\,\mathrm{dc}\,u\,\mathrm{sn}\,v, \qquad y = b\,\mathrm{nc}\,u\,\mathrm{cn}\,v,$$

show that the curves $u = $ constant are ellipses and the curves $v = $ constant are hyperbolae.

A point moves on an ellipse $u = $ const. Show that its polar with respect to the ellipse $x^2/a^2 + y^2/b^2 = 1$ envelopes the ellipse

$$\frac{x^2}{a^2}\mathrm{dn}^2 u + \frac{y^2}{b^2} = \mathrm{cn}^2 u.$$

Another point moves on a hyperbola $v = $ const. Show that its polar with respect to the same ellipse envelopes the hyperbola

$$\frac{k^2 x^2}{a^2}\mathrm{sn}^2 v - \frac{k^2 y^2}{k'^2 b^2}\mathrm{cn}^2 v = 1.$$

In the special case when the modulus k is taken to be the eccentricity e of the ellipse $x^2/a^2 + y^2/b^2 = 1$, show that the conics $u = $ const., $v = $ const. are confocal with this ellipse.

3. P and Q are variable points on the ellipse $x = a\cos\theta$, $y = b\sin\theta$, with $\theta = 2\,\mathrm{am}(u+\alpha)$, $2\,\mathrm{am}(u-\alpha)$ respectively, u being variable and α a constant. Show that the chord PQ has equation

$$\frac{x}{a}\{k^2\,\mathrm{sn}^2\alpha + (1+\mathrm{dn}^2\alpha)\cos\psi\} + \frac{2y}{b}\,\mathrm{dn}\,\alpha\sin\psi$$

$$= 2(\mathrm{cn}^2\alpha - \mathrm{sn}^2\alpha) + k^2\,\mathrm{sn}^2\alpha(1-\cos\psi),$$

where $\psi = 2\,\mathrm{am}(u,k)$. Deduce that, as u varies, the chord envelopes the ellipse

$$\left(\frac{x}{a} + k^2\,\mathrm{cd}^2\alpha\,\mathrm{sn}^2\alpha\right)^2 + \frac{y^2}{b^2} = (\mathrm{cd}^2\alpha - \mathrm{sn}^2\alpha)^2.$$

If $\alpha = \frac{1}{2}K$, show that the chords are all concurrent at the point $x = a(k'-1)/(k'+1)$, $y = 0$. Show that, if the original ellipse is a circle, then so is the envelope.

4. With the notation of Exercise 2 above, let T be a point which moves on the hyperbola $v = \beta$ and let P, Q be the points of contact of tangents from T to the ellipse $x^2/a^2 + y^2/b^2 = 1$. If $k = e$, prove that

$$PT - \mathrm{arc}\,PX = QT - \mathrm{arc}\,QX,$$

where X is the point on the ellipse between P and Q where it is intersected by the hyperbola.

5. Show that

$$x = a\,\mathrm{dc}(u,k), \qquad y = bk'\,\mathrm{sc}(u,k),$$

are parametric equations for the hyperbola $x^2/a^2 - y^2/b^2 = 1$. If $k = a/\sqrt{(a^2+b^2)}$, show that the arc length along the curve from $u = 0$ is given by

$$s = \frac{a}{k}\{k'^2 u + \mathrm{dn}\,u\,\mathrm{sc}\,u - E(u)\}.$$

6. Using the notation of the previous exercise, show that the tangents to the hyperbola $x^2/a^2 - y^2/b^2 = 1$ at the points P, Q with parameters $u \pm v$ intersect at the point T ($a\,\mathrm{dc}\,u\,\mathrm{cd}\,v$, $bk'\,\mathrm{sc}\,u\,\mathrm{nd}\,v$). Prove that

$$PT = bk'\,\mathrm{nc}\,u\,\mathrm{sd}\,v\,\mathrm{nc}(u+v), \qquad QT = bk'\,\mathrm{nc}\,u\,\mathrm{sd}\,v\,\mathrm{nc}(u-v).$$

7. Show that the locus of the point T (defined in the previous exercise) when v is constant is a hyperbola confocal with the hyperbola $x^2/a^2 - y^2/b^2 = 1$. Show, also, that as T describes its locus, the polar of T (viz. PQ) envelopes the hyperbola

$$\frac{x^2}{a^2}\,\mathrm{cn}^2 v - \frac{y^2}{b^2} = \mathrm{dn}^2 v.$$

8. Show that the locus of the point T (defined in Exercise 6) when u is constant is an ellipse confocal with the hyperbola $x^2/a^2 - y^2/b^2 = 1$. Show, also, that as T

describes its locus, the polar of T envelopes the ellipse

$$\frac{x^2}{k^2a^2}dc^2u + \frac{y^2}{b^2}sc^2u = 1.$$

9. Show that, as T describes the hyperbola $v = $ const. (see Exercise 7),

$$PT + QT - \text{arc } PQ = \frac{2a}{k}\{E(v) - k^2 \text{ sd } v \text{ cn } v - k'^2v\}.$$

10. Show that, as T describes the ellipse $u = $ const. (see Exercise 8),

$$\text{arc } AP + \text{arc } AQ - PT + QT = \frac{2a}{k}\{k'^2u + \text{dn } u \text{ sc } u - E(u)\},$$

where A is the point $u = 0$ on the hyperbola $x^2/a^2 - y^2/b^2 = 1$. Deduce that

$$PT - \text{arc } PX = QT - \text{arc } QX,$$

X being the point where the ellipse $u = $ const. intersects the hyperbola $x^2/a^2 - y^2/b^2 = 1$.

11. The concave side of a branch of a hyperbola having transverse axis $2a$ and eccentricity e rolls on a curve whose Cartesian equation is $y = f(x)$. Initially, the point of contact lies on the transverse axis of the hyperbola and on the positive y-axis. Thereafter, the center of the hyperbola describes the x-axis. Show that $f(x) = a \text{ nc}(ex/b)$, where $b^2 = a^2(e^2 - 1)$ and the modulus k is given by $k^2 = 1 - 1/e^2$.

12. Show that the length of the arc of the curve, whose equation is (4.1.33), which lies between the points where $x = 0$ and $x = \xi$ is $a[E - E\{\text{sn}^{-1}(\text{cn}(\xi/b))\}]$ and verify that this is also the length of the arc of the ellipse (4.1.29) from $r = a$ to $r = y = a \text{ dn}(\xi/b)$.

13. Show that the surface area of the elliptical paraboloid

$$\frac{x^2}{a^2} + \frac{y^2}{b^2} = 2z, \qquad a > b,$$

lying between the planes $z = 0$ and $z = h$ is

$$\frac{4}{3}\left[\frac{2ab^2h}{\sqrt{(b^2 + 2h)}}K + 2ah\sqrt{(b^2 + 2h)}E + a^2b^2\{(E - K)F(\phi, k')\right.$$
$$\left. + KD(\phi, k') - \tfrac{1}{2}\pi\}\right],$$

where

$$k^2 = \frac{2h(a^2 - b^2)}{a^2(b^2 + 2h)}, \qquad \sin\phi = \frac{a}{\sqrt{(a^2 + 2h)}},$$

and K, E are calculated to modulus k.

14. Show that the lengths of the arcs of the curves

(i) $y = \frac{2k}{k'^2}\text{sn } x,$

(ii) $y = 2k' \operatorname{nc} x$,

measured from $x = 0$ are

(i) $\dfrac{2}{k'^2}E(x) - x$,   (ii) $2 \operatorname{dn} x \operatorname{sc} x + x - 2E(x)$,

respectively.

15. The elliptical cylinder

$$\frac{x^2}{\sin^2\alpha} + \frac{y^2}{\sin^2\beta} = 1, \qquad \alpha > \beta,$$

is said to intersect the unit sphere $x^2 + y^2 + z^2 = 1$ in a *sphero-ellipse* whose semiaxes are α and β. If $(1, \theta, \phi)$ are spherical polar coordinates of a point on the sphere, show that parametric equations for the spheroellipse can be taken in the form

$$\cos\theta = \cos\beta \operatorname{dn} u, \qquad \tan\phi = \frac{\sin\alpha}{\sin\beta}\operatorname{sc} u,$$

where $k' = \cos\alpha/\cos\beta$ and ϕ is measured from the yz-plane. Show, also, that the area of the sector of the sphere enclosed by the sphero ellipse and the great circle arcs (semidiameters) from the pole (center) to the points u_1, u_2 on the curve is

$$\int_{u_1}^{u_2} (1 - \cos\theta)\frac{d\phi}{du}\,du = \left|\phi + u\sin\alpha\tan\beta - \frac{\sin\alpha}{\sin\beta\cos\beta}\Lambda(u, i\gamma, k)\right|_{u_1}^{u_2}$$

where $\gamma^2 = (\sin^2\alpha/\sin^2\beta) - 1$.

16. (r, r') are bipolar coordinates of a point. If (r, r') are plotted as Cartesian coordinates of a point P in a plane, show that P must lie in an infinite rectangular region bounded by the lines $r + r' = 1$, $r = r' + 1$, $r' = r + 1$. Deduce that the equation $r/a + r'/a' = 1$ defines a real Cartesian oval if, and only if, a and a' satisfy one of the six sets of conditions (i) $a > 1$, $0 < a' \leqslant a$, (ii) $a' > 1$, $0 < a \leqslant a'$, (iii) $0 < a < 1$, $-1 < a' < 0$, (iv) $0 < a' < 1$, $-1 < a < 0$, (v) $a \leqslant -1$, $0 < a' < -a$, (vi) $a' \leqslant -1$, $0 < a < -a'$.

17. Prove that, for all values of k and k' such that $k^2 + k'^2 = 1$, the family of ovals

$$r - r'\sqrt{(k'^2 + k^2\lambda^2)} = \lambda \qquad (-1 \leqslant \lambda \leqslant 1)$$

is orthogonal to the family

$$\mu r + r'\sqrt{(k^2 + k'^2\mu^2)} = 1 \qquad (-1 \leqslant \mu \leqslant 1),$$

the distance between the poles being the unit.

18. Prove that, provided both ovals exist, the curves $\alpha r \pm \beta r' = \gamma$ are inverses of one another with respect to the pole $r = 0$. (Hint: Obtain the polar equation for the two ovals.) If only one of the ovals exists, show that it is its own inverse.

19. Show that the length of an arc of the sine curve $y = a\sin(x/b)$ measured from O is $\sqrt{(a^2 + b^2)}D(x/b, k)$, where $k^2 = a^2/(a^2 + b^2)$.

20. Show that the volume bounded by the two elliptical cylinders

$$\frac{x^2}{a^2} + \frac{z^2}{c^2} = 1, \qquad \frac{y^2}{b^2} + \frac{z^2}{c'^2} = 1 \qquad (c < c')$$

is

$$\frac{8ab}{3c}\{(c'^2 + c^2)E - (c'^2 - c^2)K\},$$

with $k = c/c'$. (Hint: The integrals given in Exercise 53 in Chapter 3 should prove helpful.)

CHAPTER 5

Physical Applications

5.1. The Simple Pendulum

Let l be the length of the suspension, g the gravitational acceleration, and m the mass of the bob. Then, if θ is the angle made by the string with the downward vertical and v is the velocity of the bob at any time t, its energy is conserved provided

$$\tfrac{1}{2}mv^2 - mgl\cos\theta = \text{constant.} \qquad (5.1.1)$$

Since $v = l\dot{\theta}$, putting $\omega^2 = g/l$, this equation may be written

$$\dot{\theta}^2 - 2\omega^2\cos\theta = \text{constant.} \qquad (5.1.2)$$

If, now, α is the amplitude of the swing, then $\dot{\theta} = 0$ when $\theta = \alpha$ and, hence

$$\dot{\theta}^2 = 2\omega^2(\cos\theta - \cos\alpha) = 4\omega^2(\sin^2\tfrac{1}{2}\alpha - \sin^2\tfrac{1}{2}\theta). \qquad (5.1.3)$$

A further integration now leads to the result

$$\omega t = \tfrac{1}{2}\int_0^\theta \frac{d\theta}{\sqrt{(\sin^2\tfrac{1}{2}\alpha - \sin^2\tfrac{1}{2}\theta)}}, \qquad (5.1.4)$$

where t is the time taken for the bob to move from its lowest position to a position in which the string is inclined at θ to the downward vertical.

Changing the variable of integration in (5.1.4) to ϕ, where

$$\sin\tfrac{1}{2}\theta = \sin\tfrac{1}{2}\alpha\sin\phi, \qquad (5.1.5)$$

we find

$$\omega t = \int_0^\phi \frac{d\phi}{\sqrt{(1 - \sin^2\tfrac{1}{2}\alpha\sin^2\phi)}} = \text{sn}^{-1}(\sin\phi, \sin\tfrac{1}{2}\alpha), \qquad (5.1.6)$$

having used the standard form (3.1.8) of the elliptic integral of the first kind. Inverting, we get

$$\sin \phi = \text{sn}(\omega t, \sin \tfrac{1}{2}\alpha), \tag{5.1.7}$$

or, in Jacobi's notation,

$$\phi = \text{am}(\omega t, \sin \tfrac{1}{2}\alpha). \tag{5.1.8}$$

Thus

$$\sin \tfrac{1}{2}\theta = \sin \tfrac{1}{2}\alpha \, \text{sn}(\omega t, k), \tag{5.1.9}$$

where $k = \sin \tfrac{1}{2}\alpha$, showing that $\sin \tfrac{1}{2}\theta$ oscillates with amplitude $\sin \tfrac{1}{2}\alpha$ and period

$$T = 4K/\omega, \tag{5.1.10}$$

where

$$K = \int_0^{\pi/2} \frac{d\phi}{\sqrt{(1 - \sin^2\tfrac{1}{2}\alpha \sin^2\phi)}}. \tag{5.1.11}$$

Note that

$$\cos \tfrac{1}{2}\theta = \sqrt{(1 - \sin^2\tfrac{1}{2}\alpha \, \text{sn}^2\omega t)} = \sqrt{(1 - k^2 \, \text{sn}^2\omega t)} = \text{dn } \omega t. \tag{5.1.12}$$

Differentiating equation (5.1.9), we get

$$\tfrac{1}{2}\dot\theta \cos \tfrac{1}{2}\theta = \omega \sin \tfrac{1}{2}\alpha \, \text{cn } \omega t \, \text{dn } \omega t \tag{5.1.13}$$

and then, using the previous equation, this reduces to the result

$$\dot\theta = 2\omega \sin \tfrac{1}{2}\alpha \, \text{cn } \omega t. \tag{5.1.14}$$

Many of these quantities can be illustrated geometrically, thus: If O is the point of suspension and the bob P moves on a circle center O radius l (Fig. 5.1), let B be one extreme of the swing and let A, E be the lowest and highest points respectively on the circle. Construct perpendiculars BD, PN from B, P on to the vertical diameter AE. Then $\angle AOB = \alpha$, $\angle AEB = \angle ABD = \tfrac{1}{2}\alpha$, $\angle AOP = \theta$, $\angle AEP = \angle APN = \tfrac{1}{2}\theta$. C is the midpoint of AD. Construct a circle, center C, through A and D, and let Q be the point where it meets PN.

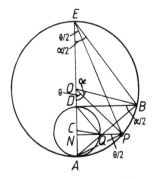

Figure 5.1. Simple pendulum.

Then $AD = l(1 - \cos\alpha) = 2l\sin^2\frac{1}{2}\alpha$ and, hence, $CQ = l\sin^2\frac{1}{2}\alpha$. Also, $AN = l(1 - \cos\theta) = 2l\sin^2\frac{1}{2}\theta$; thus, $CN = l\sin^2\frac{1}{2}\alpha - 2l\sin^2\frac{1}{2}\theta$. It now follows that

$$\cos\angle ACQ = CN/CQ = 1 - \frac{2\sin^2\frac{1}{2}\theta}{\sin^2\frac{1}{2}\alpha} = 1 - 2\sin^2\phi = \cos 2\phi, \quad (5.1.15)$$

using equation (5.1.5). We conclude that $\angle ACQ = 2\phi$ and, hence, $\angle ADQ = \phi$. Clearly, as the bob P oscillates between its extreme points, Q oscillates between A and D, and ϕ oscillates between $-\pi$ and π in accordance with equation (5.1.7).

The modulus k is given by any one of the ratios AD/AB, AB/AE, BD/BE. We also note that

$$AP = AE\sin\tfrac{1}{2}\theta = AE\sin\tfrac{1}{2}\alpha \, \mathrm{sn}\,\omega t = AB\,\mathrm{sn}\,\omega t, \quad (5.1.16)$$

$$PE = AE\cos\tfrac{1}{2}\theta = AE\,\mathrm{dn}\,\omega t. \quad (5.1.17)$$

Thus, the distances of the bob from the ends of the vertical diameter oscillate like $\mathrm{sn}\,\omega t$ and $\mathrm{dn}\,\omega t$ respectively.

If the amplitude of the swing is changed from α to $\pi - \alpha$ (replacing the string by a light wire to maintain the circular motion of the bob), then $\sin\frac{1}{2}\alpha$ must be replaced by $\cos\frac{1}{2}\alpha$ and the modulus k by k'. The period then becomes $4K(k')/\omega = 4K'/\omega$.

In the extreme case $\alpha = \pi$, we have $k = 1$ and sn becomes tanh and dn becomes sech. Thus

$$AP = AE\tanh\omega t, \qquad PE = AE\,\mathrm{sech}\,\omega t. \quad (5.1.18)$$

Evidently, the bob now approaches the point E, but takes an infinite time to reach this point.

Next, consider the case where the bob is projected horizontally from A with sufficient velocity V to permit it to execute a complete circle about O. The energy equation (5.1.2) now takes the form

$$\tfrac{1}{2}mv^2 - mgl\cos\theta = \tfrac{1}{2}mV^2 - mgl, \quad (5.1.19)$$

or

$$\dot\theta^2 = 2\omega^2\left(\frac{V^2}{2gl} - 1 + \cos\theta\right). \quad (5.1.20)$$

We require $\dot\theta^2 > 0$ when $\theta = \pi$ for circular motion to occur and, hence, $V^2 > 4gl$. Putting $\phi = \frac{1}{2}\theta$ and $k^2 = 4gl/V^2$ ($k < 1$), this equation transforms to.

$$\dot\phi^2 = \frac{\omega^2}{k^2}(1 - k^2\sin^2\phi). \quad (5.1.21)$$

We can now integrate as in the previous case to give

$$\omega t = k\,\mathrm{sn}^{-1}(\sin\phi, k) \quad (5.1.22)$$

or

$$\sin\tfrac{1}{2}\theta = \sin\phi = \mathrm{sn}(\omega t/k, k). \quad (5.1.23)$$

Thus

$$\theta = 2\,\mathrm{am}(\omega t/k, k),$$

(5.1.24)

$$\cos\tfrac{1}{2}\theta = \mathrm{cn}(\omega t/k, k).$$

(5.1.25)

Differentiation of equation (5.1.23) yields

$$\tfrac{1}{2}\dot{\theta}\cos\tfrac{1}{2}\theta = \frac{\omega}{k}\,\mathrm{cn}(\omega t/k)\mathrm{dn}(\omega t/k)$$

(5.1.26)

which, in view of equation (5.1.25), implies that

$$\dot{\theta} = \frac{2\omega}{k}\,\mathrm{dn}(\omega t/k).$$

(5.1.27)

Thus, $\dot{\theta}$ is always positive and oscillates between the values $2\omega/k$ (at A) and $2\omega k'/k = \sqrt{(V^2 - 4gl)}/l$ (at E).

The time T for a complete revolution ($\theta = 0$ to $\theta = 2\pi$) can be found from equation (5.1.23); it is given by

$$T = 2kK/\omega = 4lK/V.$$

(5.1.28)

5.2. Duffing's Equation

This is the equation governing the oscillations of a mass attached to the end of a spring whose tension (or compression) T is related to its extension x by an equation of the form

$$T = \alpha x + \beta x^3.$$

(5.2.1)

α is always positive. If $\beta = 0$, the spring obeys Hooke's law and the oscillations are simple harmonic. If $\beta > 0$, the tension increases with the extension more rapidly than required by Hooke's law and the spring is said to be *hard*. If $\beta < 0$, the tension increases less rapidly than required by the law and the spring is said to be *soft*.

By a suitable choice of the unit of time, the equation of motion of the mass can be put into the form

$$\ddot{x} + x + \varepsilon x^3 = 0,$$

(5.2.2)

which is the canonical form of *Duffing's equation*.

Consider, first, the case of a hard spring, for which $\varepsilon > 0$. Suppose that initially, $t = 0$, $x = a$, $\dot{x} = 0$. Since $\ddot{x} = d(\tfrac{1}{2}\dot{x}^2)/dx$, we can integrate with respect to x to give

$$\tfrac{1}{2}\dot{x}^2 + \tfrac{1}{2}x^2 + \tfrac{1}{4}\varepsilon x^4 = \tfrac{1}{2}a^2 + \tfrac{1}{4}\varepsilon a^4$$

(5.2.3)

or

$$\dot{x}^2 = (a^2 - x^2)(1 + \tfrac{1}{2}\varepsilon a^2 + \tfrac{1}{2}\varepsilon x^2).$$

(5.2.4)

Integrating to obtain t, we find

$$t = \sqrt{\frac{2}{\varepsilon}} \int_x^a \frac{dx}{\sqrt{\{(a^2 - x^2)(2/\varepsilon + a^2 + x^2)\}}}$$

$$= \frac{1}{\sqrt{(1 + \varepsilon a^2)}} \mathrm{cn}^{-1}\left[\frac{x}{a}, \sqrt{\left(\frac{\varepsilon a^2}{2 + 2\varepsilon a^2}\right)}\right], \qquad (5.2.5)$$

having referred to the standard form (3.2.3). Inversion now shows that

$$x = a\,\mathrm{cn}\{\sqrt{(1 + \varepsilon a^2)}\,t\}, \qquad (5.2.6)$$

with modulus given by

$$k^2 = \frac{\varepsilon a^2}{2 + 2\varepsilon a^2}. \qquad (5.2.7)$$

The period of oscillation determined by (5.2.6) is given by

$$T = \frac{4K}{\sqrt{(1 + \varepsilon a^2)}}. \qquad (5.2.8)$$

If the departure from Hooke's law is small, we can work to the first order in ε. Thus, $k^2 = \frac{1}{2}\varepsilon a^2$ and equation (3.8.5) shows that $K = \frac{1}{2}\pi(1 + \frac{1}{8}\varepsilon a^2)$. Hence, to $O(\varepsilon)$

$$T = 2\pi(1 - \tfrac{3}{8}\varepsilon a^2), \qquad (5.2.9)$$

indicating that, as the amplitude of the oscillation increases, the period decreases and the frequency therefore increases.

For a soft spring, $\varepsilon < 0$ and we shall write $\varepsilon = -\eta$. Since $x - \eta x^3$ has a maximum at $x = 1/\sqrt{(3\eta)}$, theoretically the tension should decrease with increasing extension for sufficiently large values of x; this is unrealistic and we shall accordingly assume $a < 1/\sqrt{(3\eta)}$.

It will be convenient to measure t from an instant when the mass is at the center of oscillation ($x = 0$) and x is increasing. Thus, $x = 0$ at $t = 0$, and the equation for the time is

$$t = \sqrt{\left(\frac{2}{\eta}\right)} \int_0^x \frac{dx}{\sqrt{\{(a^2 - x^2)(2/\eta - a^2 - x^2)\}}}$$

$$= \sqrt{\left(\frac{2}{2 - \eta a^2}\right)} \mathrm{sn}^{-1}\left[\frac{x}{a}, \sqrt{\left(\frac{\eta a^2}{2 - \eta a^2}\right)}\right], \qquad (5.2.10)$$

after reference to the standard form (3.1.7) (to apply this result, we must assume $2/\eta - a^2 > a^2$, i.e., $a < 1/\sqrt{\eta}$, which is guaranteed by $a < 1/\sqrt{(3\eta)}$). Inverting the last equation, we obtain

$$x = a\,\mathrm{sn}\{\sqrt{(1 - \tfrac{1}{2}\eta a^2)}\,t\}, \qquad (5.2.11)$$

where the modulus is determined by

$$k^2 = \frac{\eta a^2}{2 - \eta a^2}. \tag{5.2.12}$$

Thus, the period of oscillation is given by

$$T = \frac{4K}{\sqrt{(1 - \tfrac{1}{2}\eta a^2)}} \tag{5.2.13}$$

and, for small η, this reduces to

$$T = 2\pi(1 + \tfrac{3}{8}\eta a^2) \tag{5.2.14}$$

to $O(\eta)$, in agreement with (5.2.9). For a soft spring, therefore, the frequency of oscillation decreases as the amplitude increases.

5.3. Orbits under a μ/r^4 Law of Attraction

Suppose a particle of unit mass is attracted towards a center O by a force μ/r^4, r and θ being its polar coordinates at time t in the plane of motion. Then, since the particle's energy E and angular momentum h about O will be conserved, we can write down the equations of motion

$$\tfrac{1}{2}(\dot{r}^2 + r^2\dot{\theta}^2) - \frac{\mu}{3r^3} = E, \qquad r^2\dot{\theta} = h. \tag{5.3.1}$$

Putting $r = 1/u$ and eliminating t between these equations, we arrive at the equation

$$\alpha\left(\frac{du}{d\theta}\right)^2 = u^3 - \alpha u^2 + \beta = f(u), \tag{5.3.2}$$

where

$$\alpha = 3h^2/2\mu, \qquad \beta = 3E/\mu. \tag{5.3.3}$$

Equation (5.3.2) determines the polar equation of the orbit. Clearly, $\alpha > 0$ (we ignore the case of rectilinear motion), but β may take any real value. We shall always assume the sense of the motion to be such that θ increases (i.e., $h > 0$).

By sketching the graph of $u^3 - \alpha u^2$, the reader should verify that there are five cases to consider: (i) If $\beta < 0$, then $f(u)$ has one real zero greater than α and two complex zeros whose real parts are negative (since the sum of the zeros is α). (ii) If $\beta = 0$, $f(u)$ has a double zero at $u = 0$ and a simple zero at $u = \alpha$. (iii) If $0 < \beta < 4\alpha^3/27$, $f(u)$ has three real zeros u_1, u_2, u_3, satisfying $u_1 < 0 < u_2 < 2\alpha/3 < u_3 < \alpha$. (iv) If $\beta = 4\alpha^3/27$, $f(u)$ has a pair of coincident zeros at $u = 2\alpha/3$ and a simple zero at $u = -\alpha/3$. (v) If $\beta > 4\alpha^3/27$, $f(u)$ has a real zero with negative u and two complex zeros with positive real parts. Since $\alpha(du/d\theta)^2 > 0$, by consideration of the sign of $f(u)$ over the range of all positive values of u, it is possible to establish the character of each of the possible orbits in these cases without further integration. However, we shall first calculate the orbital equations and determine their shapes from these.

The intermediate cases (ii) and (iv) can be analyzed without the introduction of elliptic functions. Thus, if $\beta = 0$, we calculate that

$$\alpha^{-1/2}\theta = \int \frac{du}{u\sqrt{(u-\alpha)}} = \alpha^{-1/2}\cos^{-1}\left(\frac{2\alpha}{u}-1\right). \qquad (5.3.4)$$

(Use the substitution $u = 1/v$.) We have ignored the constant of integration, since this can always be eliminated by suitable choice of the line $\theta = 0$. The polar equation of the orbit is now found to be

$$r = \frac{1}{2\alpha}(1 + \cos\theta), \qquad (5.3.5)$$

which is a cardioid. Thus, the particle recedes to a maximum distance $1/2\alpha$ from the pole and then falls into the center of attraction.

If $\beta = 4\alpha^3/27$, then

$$\alpha^{-1/2}\theta = \int \frac{du}{(u - 2\alpha/3)\sqrt{(u + \alpha/3)}} = -\alpha^{-1/2}\cosh^{-1}\left|\frac{u + 4\alpha/3}{u - 2\alpha/3}\right|. \qquad (5.3.6)$$

(Put $u - 2\alpha/3 = 1/v$ if $u > 2\alpha/3$ and $2\alpha/3 - u = 1/v$ if $u < 2\alpha/3$.) The equation of the orbit is accordingly

$$\begin{aligned}
r &= \frac{3}{2\alpha} \cdot \frac{\cosh\theta - 1}{\cosh\theta + 2}, \quad \text{if } u > 2\alpha/3, \\
&= \frac{3}{2\alpha} \cdot \frac{\cosh\theta + 1}{\cosh\theta - 2}, \quad \text{if } u < 2\alpha/3.
\end{aligned} \right\} \qquad (5.3.7)$$

In the first case, for positive values of θ, the orbit spirals outward from the center of attraction, approaching the circle $r = 3/2\alpha$ asymptotically from inside the circle. In the second case, for $\theta > \cosh^{-1}2$, the orbit spirals inward from infinity, approaching the circle $r = 3/2\alpha$ asymptotically from outside the circle. Reflecting these orbits in the line $\theta = 0$, we obtain the orbits for negative values of θ, which represent similar trajectories, traversed in the opposite sense, i.e., diverging inward and outward from the circular motion. The circular orbit represents a possible motion, but our analysis demonstrates that it is unstable.

Now consider case (i), where

$$f(u) = (u - a)\{(u + b)^2 + c^2\}, \qquad (5.3.8)$$

a, b, c being all positive and $a > \alpha$. Clearly, we need $u \geqslant a$ to make $f(u)$ positive. Applying the method described in section 3.3, this cubic can be written as a pair of quadratic factors

$$\begin{aligned}
S_1 &= \tfrac{1}{2}(p+q)^{-1}[(u+p)^2 - (u-q)^2], \\
S_2 &= (p+q)^{-1}[(q+b)(u+p)^2 + (p-b)(u-q)^2],
\end{aligned} \right\} \qquad (5.3.9)$$

where p and q are positive numbers given by

$$p = \sqrt{\{(a+b)^2 + c^2\}} - a, \qquad q = \sqrt{\{(a+b)^2 + c^2\}} + a, \qquad (5.3.10)$$

as the reader can verify directly. Then, integrating equation (5.3.2) we find

$$\alpha^{-1/2}\theta$$

$$= \sqrt{2(p+q)} \int \frac{du}{\sqrt{[(u+p)^2 - (u-q)^2][(q+b)(u+p)^2 + (p-b)(u-q)^2]}}.$$

$$(5.3.11)$$

We now make the substitution

$$x = \frac{u-q}{u+p}, \qquad (5.3.12)$$

where x increases monotonically for increasing u. This gives

$$\alpha^{-1/2}\theta = \sqrt{\frac{2}{p-b}} \int \frac{dx}{\sqrt{\{(1-x^2)(d^2 + x^2)\}}}, \qquad (5.3.13)$$

where

$$d^2 = \frac{q+b}{p-b}. \qquad (5.3.14)$$

We have now arrived at a standard form and can make use of the result (3.2.3) to show that

$$\alpha^{-1/2}\theta = -\sqrt{\frac{2}{p+q}} \, \mathrm{cn}^{-1} x, \qquad (5.3.15)$$

the modulus being given by

$$k^2 = \frac{p-b}{p+q}. \qquad (5.3.16)$$

Thus,

$$x = \mathrm{cn}\,\gamma\theta, \qquad (5.3.17)$$

where

$$\gamma = \sqrt{\{(p+q)/2\alpha\}}. \qquad (5.3.18)$$

The polar equation of the orbit now follows in the form

$$r = \frac{1 - \mathrm{cn}\,\gamma\theta}{q + p\,\mathrm{cn}\,\gamma\theta}, \qquad (5.3.19)$$

where $p < q$. We deduce that, as θ increases from 0, the trajectory spirals outward from the center, the mass being at its maximum distance $2/(q-p) = 1/a$ from O when $\theta = 2K/\gamma$. Thereafter, the orbit spirals inward and reaches the center again when $\theta = 4K/\gamma$. As before, negative values of θ yield the mirror image trajectory, which is identical with the original.

In case (iii), $f(u) = (u - u_1)(u - u_2)(u - u_3)$, where $u_1 < 0 < u_2 < u_3$. Hence, either $0 \leqslant u \leqslant u_2$ or $u \geqslant u_3$ to make $f(u) \geqslant 0$, and there are two types of orbit. Integrating equation (5.3.2), we get

$$\alpha^{-1/2}\theta = \int \frac{du}{\sqrt{\{(u-u_1)(u-u_2)(u-u_3)\}}}. \tag{5.3.20}$$

Changing the variable by the transformation

$$u = u_1 + 1/x^2 \qquad (x > 0), \tag{5.3.21}$$

we reduce the integral to standard form, thus

$$\alpha^{-1/2}\theta = -\frac{2}{\sqrt{\{(u_2-u_1)(u_3-u_1)\}}} \int \frac{dx}{\sqrt{\{(a^2-x^2)(b^2-x^2)\}}}, \tag{5.3.22}$$

where

$$a^2 = 1/(u_2 - u_1), \qquad b^2 = 1/(u_3 - u_1). \tag{5.3.23}$$

Clearly $a > b$ and, if $u \geqslant u_3$, then $x \leqslant b$ and the result (3.1.7) may be applied to give

$$\alpha^{-1/2}\theta = -\frac{2}{\sqrt{(u_3-u_1)}} \mathrm{sn}^{-1}\{\sqrt{(u_3-u_1)}x\}, \tag{5.3.24}$$

with modulus k, where

$$k^2 = \frac{u_2 - u_1}{u_3 - u_1}. \tag{5.3.25}$$

Then, the orbit is found to have equation

$$\frac{1}{r} = u = u_1 + (u_3 - u_1)\mathrm{ns}^2\gamma\theta, \tag{5.3.26}$$

where

$$\gamma = \tfrac{1}{2}\sqrt{\{(u_3 - u_1)/\alpha\}}. \tag{5.3.27}$$

If, however, $0 \leqslant u \leqslant u_2$, then $x \geqslant a$ and the standard form (3.2.9) is used to yield

$$\alpha^{-1/2}\theta = \frac{2}{\sqrt{(u_3-u_1)}} \mathrm{ns}^{-1}\{\sqrt{(u_2-u_1)}x\}, \tag{5.3.28}$$

whence

$$\frac{1}{r} = u = u_1 + (u_2 - u_1)\mathrm{sn}^2\gamma\theta. \tag{5.3.29}$$

The constants k and γ take the same values as before.

If the orbit is governed by the equation (5.3.26), as θ increases from 0, the trajectory spirals outward from the center, achieving maximum distance $1/u_3$ when $\theta = K/\gamma$; it then spirals back into the pole, arriving there when $\theta = 2K/\gamma$.

For orbital equation (5.3.29), $\gamma\theta$ must equal or exceed $\omega = \mathrm{sn}^{-1}\sqrt{\{-u_1/(u_2-u_1)\}}$ to give positive values for u and r. When θ has this limiting value ω, r is infinite and further increase in θ causes r to decrease to a minimum of $1/u_2$ when $\theta = K/\gamma$. If θ is increased again, r approaches infinity as $\theta \to (2K - \omega)/\gamma$. Thus, the trajectory first approaches the center of attraction from infinity and later recedes again to an infinite distance.

Finally, in case (v),

$$f(u) = (u + a)\{(u - b)^2 + c^2\}, \qquad (5.3.30)$$

where a, b, c are all positive; all positive values of u are now admissible. The analysis proceeds as for case (i), the signs of a and b being reversed. Thus

$$p = \sqrt{\{(a + b)^2 + c^2\}} + a, \qquad q = \sqrt{\{(a + b)^2 + c^2\}} - a, \qquad (5.3.31)$$

and $p > q$. The orbital equation is as given at (5.3.19) and, as θ increases from 0, r increases from 0 and the orbit spirals outward from the center. However, the equation cn $\gamma\theta = -q/p$ now has a real root θ for which r becomes infinite and the trajectory does not return to the center of attraction. Negative values of θ provide a mirror image orbit along which the mass can fall into the center of attraction from an infinite distance.

It should be noted that the only closed orbit is the circular one and that this is unstable.

5.4. Orbits under a μ/r^5 Law of Attraction

If μ/r^5 is the attraction per unit mass, the particle's potential energy in the field is $-\mu/4r^4$ and the equations of energy and angular momentum are

$$\tfrac{1}{2}(\dot{r}^2 + r^2\dot{\theta}^2) - \frac{\mu}{4r^4} = E, \qquad r^2\dot{\theta} = h. \qquad (5.4.1)$$

These equations lead to the equation

$$\alpha\left(\frac{du}{d\theta}\right)^2 = u^4 - \alpha u^2 + \beta = g(u^2) \qquad (5.4.2)$$

determining the orbits, where

$$\alpha = 2h^2/\mu > 0, \qquad \beta = 4E/\mu. \qquad (5.4.3)$$

$g(v)$ $(v = u^2)$ is a quadratic and its zeros distinguish five cases (graph $v^2 - \alpha v$): (i) $\beta < 0$, both zeros v_1, v_2 are real and $v_1 < 0, v_2 > \alpha$. $v \geqslant v_2$ for g to be positive. (ii) $\beta = 0$, zeros are 0 and α. $v \geqslant \alpha$ on the orbit. (iii) $0 < \beta < \tfrac{1}{4}\alpha^2$, both zeros are real and satisfy $0 < v_1 < v_2 < \alpha$. $v \leqslant v_1$ or $v \geqslant v_2$ on the orbit. (iv) $\beta = \tfrac{1}{4}\alpha^2$, coincident zeros at $v = \tfrac{1}{2}\alpha$; all positive values of v are admissible. (v) $\beta > \tfrac{1}{4}\alpha^2$, zeros are complex with positive real parts; all positive values of v are admissible.

As before, we first treat cases (ii) and (iv), whose analyses require only elementary functions. We shall continue to suppose θ increases with t $(h > 0)$.

If $\beta = 0$, integration of (5.4.2) leads to

$$\alpha^{-1/2}\theta = \int \frac{du}{u\sqrt{(u^2 - \alpha)}} = \alpha^{-1/2} \cos^{-1} \alpha^{1/2} r. \tag{5.4.4}$$

(Put $u = 1/r$.) Clearly $u > \alpha^{1/2}$ and $r < \alpha^{-1/2}$. The polar equation of the orbit is therefore

$$r = \alpha^{-1/2} \cos \theta, \tag{5.4.5}$$

which is a circle through the center of attraction. Thus, the particle first recedes from O along one semicircle and then falls back into O along the remaining semicircle (a striking result!).

If $\beta = \tfrac{1}{4}\alpha^2$, we find that

$$\alpha^{-1/2}\theta = \int \frac{du}{\tfrac{1}{2}\alpha - u^2} = \frac{1}{\sqrt{(2\alpha)}} \ln \left| \frac{\sqrt{(\tfrac{1}{2}\alpha)} + u}{\sqrt{(\tfrac{1}{2}\alpha)} - u} \right|. \tag{5.4.6}$$

Solving for u, we derive

$$r = 1/u = \sqrt{(2/\alpha)} \tanh(\theta/\sqrt{2}) \quad \text{or} \quad \sqrt{(2/\alpha)} \coth(\theta/\sqrt{2}) \tag{5.4.7}$$

as possible equations for the orbit. The first equation corresponds to a trajectory which spirals outward from O, approaching the circle $r = \sqrt{(2/\alpha)}$ asymptotically. The alternative orbit spirals inward from infinity and approaches the same circle asymptotically.

In case (i),

$$\alpha \left(\frac{du}{d\theta} \right)^2 = (u^2 - v_1)(u^2 - v_2), \tag{5.4.8}$$

where

$$v_1 = \tfrac{1}{2}\alpha - \sqrt{(\tfrac{1}{4}\alpha^2 - \beta)} < 0, \qquad v_2 = \tfrac{1}{2}\alpha + \sqrt{(\tfrac{1}{4}\alpha^2 - \beta)} > \alpha. \tag{5.4.9}$$

We must have $u \geqslant \sqrt{v_2}$. Integration leads to the orbital equation

$$\alpha^{-1/2}\theta = (v_2 - v_1)^{-1/2} \mathrm{nc}^{-1}(u/\sqrt{v_2}), \tag{5.4.10}$$

using the standard integral (3.2.12), the modulus being given by

$$k^2 = -\frac{v_1}{v_2 - v_1}. \tag{5.4.11}$$

We deduce that

$$r = \frac{1}{\sqrt{v_2}} \mathrm{cn}\, \gamma\theta, \tag{5.4.12}$$

where

$$\gamma^2 = (v_2 - v_1)/\alpha. \tag{5.4.13}$$

Thus, as θ increases from $-K/\gamma$ to K/γ, the particle spirals out from the center of attraction to a maximum distance $1/\sqrt{v_2}$ and then falls back into the center along the mirror image spiral.

In case (iii), $0 < v_1 < v_2 < \alpha$ and either $u \leqslant \sqrt{v_1}$ or $u \geqslant \sqrt{v_2}$. In the former case, we write

$$\alpha\left(\frac{du}{d\theta}\right)^2 = (v_1 - u^2)(v_2 - u^2) \tag{5.4.14}$$

and then apply the standard integral (3.1.7) to give

$$r = \frac{1}{\sqrt{v_1}} \, \text{ns} \, \gamma\theta, \tag{5.4.15}$$

with

$$k^2 = v_1/v_2 \quad \text{and} \quad \gamma^2 = v_2/\alpha. \tag{5.4.16}$$

Thus, as θ increases from 0 to $2K/\gamma$, the particle approaches the center of attraction from infinity to a minimum distance $1/\sqrt{v_1}$ and then recedes again to infinity.

If, however, $u \geqslant v_2$, then we write

$$\alpha\left(\frac{du}{d\theta}\right)^2 = (u^2 - v_1)(u^2 - v_2) \tag{5.4.17}$$

and apply the standard integral (3.2.9). This shows that

$$r = \frac{1}{\sqrt{v_2}} \, \text{sn} \, \gamma\theta, \tag{5.4.18}$$

k and γ being given as before.

Hence, as θ increases from 0 to $2K/\gamma$, the particle leaves the center of attraction and recedes from it to a maximum distance $1/\sqrt{v_2}$; thereafter, it falls back into the pole O. This is similar to case (i).

Finally, in case (v), for convenience in later calculations, we shall take the zeros of $g(v)$ to be

$$v_1, v_2 = \tfrac{1}{2}a^2 - b^2 \pm ia\sqrt{(b^2 - \tfrac{1}{4}a^2)}, \tag{5.4.19}$$

where

$$a^2 = \alpha + 2\sqrt{\beta}, \qquad b^2 = \sqrt{\beta}, \tag{5.4.20}$$

taking a and b positive. Note that $b^2 > a^2/4$. Thus

$$\alpha\left(\frac{du}{d\theta}\right)^2 = (u^2 + b^2 - \tfrac{1}{2}a^2)^2 + a^2(b^2 - \tfrac{1}{4}a^2). \tag{5.4.21}$$

Transforming by

$$u = b\frac{1+t}{1-t}, \tag{5.4.22}$$

the last equation becomes

$$\alpha\left(\frac{dt}{d\theta}\right)^2 = (b^2 - \tfrac{1}{4}a^2)(t^2 + p^2)(t^2 + q^2), \tag{5.4.23}$$

where

$$p^2 = \frac{2b - a}{2b + a}, \qquad q^2 = \frac{2b + a}{2b - a}. \tag{5.4.24}$$

We shall take p, q to be positive and, clearly, $p < q$. Integration, using the standard form (3.2.14), now yields

$$\alpha^{-1/2}\theta = (b^2 - \tfrac{1}{4}a^2)^{-1/2} \int \frac{dt}{\sqrt{\{(t^2 + p^2)(t^2 + q^2)\}}},$$

$$= \frac{2}{a + 2b} \operatorname{sc}^{-1}(t/p), \tag{5.4.25}$$

where the modulus is given by

$$k = \frac{\sqrt{(8ab)}}{a + 2b}. \tag{5.4.26}$$

The polar equation of the orbit now follows in the form

$$r = \frac{1}{b} \cdot \frac{\operatorname{cn} \gamma\theta - p \operatorname{sn} \gamma\theta}{\operatorname{cn} \gamma\theta + p \operatorname{sn} \gamma\theta}, \tag{5.4.27}$$

where

$$\gamma = (a + 2b)/(2\alpha^{1/2}). \tag{5.4.28}$$

Consideration of equation (5.4.27) reveals that $r = \infty$ when $\theta = -\omega$ and $r = 0$ when $\theta = \omega$, where

$$\omega = \frac{1}{\gamma} \operatorname{sn}^{-1} \sqrt{\frac{a + 2b}{4b}}. \tag{5.4.29}$$

(Since $a/b < 2$, ω can be found in the interval $(0, K)$.) We conclude that the particle spirals into the center of attraction from infinity as θ increases from $-\omega$ to ω.

5.5. Relativistic Planetary Orbits

According to the general theory of relativity, a planet falling freely in the gravitational field of a spherically symmetric sun behaves as if it were governed by the Newtonian laws and were attracted to the sun by a non-Newtonian gravitational force of

$$\mu\left(\frac{1}{r^2} + \frac{3h^2}{c^2 r^4}\right) \tag{5.5.1}$$

per unit mass, where h is the angular momentum per unit mass of the planet about the center of the sun and c is the velocity of light (see, e.g., D.F. Lawden, *An Introduction to Tensor Calculus, Relativity and Cosmology*, 3rd ed., p. 148, Wiley, 1982; it is assumed both r and t are measured according to Schwarzschild's convention). Thus, its equations of motion are

$$\tfrac{1}{2}(\dot{r}^2 + r^2\dot{\theta}^2) - \mu\left(\frac{1}{r} + \frac{h^2}{c^2r^3}\right) = E, \qquad r^2\dot{\theta} = h. \tag{5.5.2}$$

As in the previous sections, putting $u = 1/r$, we now arrive at the equation

$$\left(\frac{du}{d\theta}\right)^2 = \frac{2\mu}{h^2}u - u^2 + \frac{2\mu}{c^2}u^3 + \frac{2E}{h^2} \tag{5.5.3}$$

determining the orbit.

For all planets in the solar system, the term $2\mu u^3/c^2$ is always very small by comparison with the remaining terms in equation (5.5.3) and it will be convenient to introduce a small dimensionless parameter to assimilate this feature into the analysis. Thus, we define a dimensionless variable v by the equation

$$u = \mu v/h^2 \tag{5.5.4}$$

and then write equation (5.5.3) in the form

$$\left(\frac{dv}{d\theta}\right)^2 = 2v - v^2 + \alpha v^3 - \beta = f(v), \tag{5.5.5}$$

where

$$\alpha = 2(\mu/ch)^2, \qquad \beta = -2Eh^2/\mu^2. \tag{5.5.6}$$

The circumstance that a planet's energy is insufficient to permit its escape from the sun's field requires that $\beta > 0$. Also $\beta \leqslant 1$, for otherwise $(dv/d\theta)^2 < 0$ in the absence of the relativistic term. α is very small and positive for all planets in the solar system, taking its largest value of 5.09×10^{-8} for Mercury.

By graphing the function $2v - v^2 + \alpha v^3$, it is easy to establish that the zeros of $f(v)$ (equation (5.5.5)) are all real and satisfy the inequalities $0 < v_1 < 1 < v_2 < 2 < v_3$, v_3 being very large. Thus

$$f(v) = \alpha(v - v_1)(v_2 - v)(v_3 - v) \tag{5.5.7}$$

and, since $f(v) \geqslant 0$, v must lie in the interval (v_1, v_2) ($v \geqslant v_3$ is excluded since this would lead to $v \to \infty$ as $\theta \to \infty$; i.e., the planet would fall into the sun).

α being small, the zeros of $f(v)$ can be expanded in series of ascending powers of α, thus (provided $e \neq 0$):

$$\left.\begin{aligned} v_1 &= 1 - e - \frac{\alpha}{2e}(1 - e)^3 + O(\alpha^2), \\[2mm] v_2 &= 1 + e + \frac{\alpha}{2e}(1 + e)^3 + O(\alpha^2), \\[2mm] v_3 &= \frac{1}{\alpha} - 2 + O(\alpha), \end{aligned}\right\} \tag{5.5.8}$$

where

$$e^2 = 1 - \beta. \tag{5.5.9}$$

(e is the eccentricity of the classical orbit (see below).)

Integrating equation (5.5.5), we deduce that

$$\alpha^{1/2}\theta = \int \frac{dv}{\sqrt{(v-v_1)(v_2-v)(v_3-v)}}. \tag{5.5.10}$$

Changing the variable in the elliptic integral by $v = v_1 + 1/t^2$, we bring it to standard form, thus:

$$\alpha^{1/2}\theta = -\frac{2}{\sqrt{\{(v_2-v_1)(v_3-v_1)\}}}\int \frac{dt}{\sqrt{\{(t^2-a^2)(t^2-b^2)\}}}, \tag{5.5.11}$$

where

$$a^2 = 1/(v_2-v_1), \qquad b^2 = 1/(v_3-v_1). \tag{5.5.12}$$

Use of (3.2.9) now yields the result

$$\alpha^{1/2}\theta = \frac{1}{\sqrt{(v_3-v_1)}}\,\mathrm{ns}^{-1}\{t\sqrt{(v_2-v_1)}\}, \tag{5.5.13}$$

with modulus given by

$$k^2 = \frac{v_2-v_1}{v_3-v_1}. \tag{5.5.14}$$

Thus,

$$v = v_1 + (v_2-v_1)\,\mathrm{sn}^2\{\tfrac{1}{2}\sqrt{\alpha(v_3-v_1)}\theta\} \tag{5.5.15}$$

is the equation of the orbit.

Substituting the expansions (5.5.8), we now calculate that

$$\frac{1}{r} = \frac{\mu}{h^2}(A + B\,\mathrm{sn}^2\eta\theta), \tag{5.5.16}$$

where

$$\left.\begin{aligned}
A &= 1 - e - \frac{\alpha}{2e}(1-e)^3 + O(\alpha^2),\\
B &= 2e + \alpha(3e + 1/e) + O(\alpha^2),\\
\eta &= \tfrac{1}{2} - \tfrac{1}{4}(3-e)\alpha + O(\alpha^2).
\end{aligned}\right\} \tag{5.5.17}$$

The modulus is determined by

$$k^2 = 2e\alpha + O(\alpha^2). \tag{5.5.18}$$

If $\alpha = 0$, then $A = 1 - e$, $B = 2e$, $\eta = \tfrac{1}{2}$, $k = 0$, and the orbital equation reduces to

$$\frac{l}{r} = 1 - e\cos\theta, \tag{5.5.19}$$

where $l = h^2/\mu$. This represents the classical elliptical orbit with semi-latus rectum l and eccentricity e.

On the relativistic orbit given by equation (5.5.16), perihelion occurs when $\theta = K/\eta$ and, on the next occasion, when $\theta = 3K/\eta$. Thus, θ increases by $2K/\eta$ between two passages through perihelion, instead of the increase of 2π expected from the classical theory. The advance of perihelion per revolution is accordingly

$$\frac{2K}{\eta} - 2\pi = \frac{\pi(1 + \frac{1}{4}k^2 + \cdots)}{\frac{1}{2} - \frac{1}{4}(3 - e)\alpha + \cdots} - 2\pi = 3\pi\alpha \qquad (5.5.20)$$

(using equation (3.8.5)). For Mercury, $\alpha = 5.09 \times 10^{-8}$ and its period is 88 days. Thus, the advance of perihelion per century predicted by the theory is 43"; this is exactly the residual advance remaining to be explained at the time the new theory was proposed by Einstein.

5.6. Whirling Chain

In this section, we consider a uniform length l of rope or chain, whose ends are fixed at points O and A and which is set rotating about the axis OA with constant angular velocity ω. Gravity will be neglected and it will be assumed that the chain always lies in a plane through the axis of rotation. A skipping rope which is being rapidly whirled by hand about a horizontal axis approximates this physical situation.

We shall take O to be the origin of axes Ox, Oy, the x-axis lying along OA, and the y-axis lying in the plane of the chain at some instant t (Fig. 5.2). Consider the motion of an element $ds = PQ$ of the chain, where P and Q have coordinates (x, y), $(x + dx, y + dy)$ respectively. The forces acting on this element are the tensions T and $T + dT$ at its ends P and Q, and their lines of action are the tangents to the chain at these points; let these tangents make angles ψ, $\psi + d\psi$ respectively with the x-axis. Resolving the forces tangentially and normally, we obtain components $(dT, Td\psi)$ respectively (to the first order in the differentials). The element moves around a circle of radius y with angular velocity ω and its acceleration is accordingly $\omega^2 y$ directed in the negative sense parallel to the y-axis. We can now write down the tangential and normal

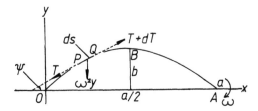

Figure 5.2. Whirling rope.

components of the equation of motion thus:

$$dT = -\sigma \, ds\omega^2 y \sin\psi, \qquad T \, d\psi = -\sigma \, ds\omega^2 y \cos\psi, \qquad (5.6.1)$$

where σ is the mass per unit length of the chain.

Dividing these equations, we find that

$$dT/T = \tan\psi \cdot d\psi, \qquad (5.6.2)$$

which integrates to give the equation

$$T = T_0 \sec\psi, \qquad (5.6.3)$$

T_0 being the tension at the point B where $\psi = 0$.

Substituting for T in the second equation (5.6.1), we now deduce that

$$\lambda \, d\psi/ds = -y \cos^2\psi, \qquad (5.6.4)$$

where

$$\lambda = T_0/\sigma\omega^2. \qquad (5.6.5)$$

But $dy/ds = \sin\psi$ and it therefore follows that

$$\lambda \tan\psi \sec\psi \, d\psi = -y \, dy. \qquad (5.6.6)$$

This equation integrates to

$$\lambda(\sec\psi - 1) = \tfrac{1}{2}(b^2 - y^2), \qquad (5.6.7)$$

where $y = b$, $\psi = 0$ at B.

Thus,

$$dy/dx = \tan\psi = \sqrt{(\sec^2\psi - 1)} = \frac{1}{2\lambda}\sqrt{\{(b^2 - y^2)(b^2 + 4\lambda - y^2)\}} \qquad (5.6.8)$$

and, after integration from $x = y = 0$, this leads to the equation

$$x = 2\lambda \int_0^y \frac{dy}{\sqrt{\{(b^2 - y^2)(c^2 - y^2)\}}}, \qquad (5.6.9)$$

where

$$c^2 = b^2 + 4\lambda. \qquad (5.6.10)$$

Reference to the standard form (3.1.7) now shows that

$$x = \frac{2\lambda}{c} \, \mathrm{sn}^{-1}(y/b), \qquad (5.6.11)$$

the modulus being given by

$$k^2 = b^2/c^2 = (1 + 4\lambda/b^2)^{-1}. \qquad (5.6.12)$$

We conclude that the equation of the chain is

$$y = b \, \mathrm{sn}(cx/2\lambda). \qquad (5.6.13)$$

Supposing the end A to lie at the point $x = a$ on the x-axis, we must have

$y = 0$ at $x = a$. Clearly, therefore, it is necessary that

$$ac/2\lambda = 2K \tag{5.6.14}$$

and the equation of the chain can be written

$$y = b\,\text{sn}(2Kx/a). \tag{5.6.15}$$

By eliminating λ and c between equations (5.6.10), (5.6.12), and (5.6.14), we arrive at the equation

$$\frac{ak}{b(1 - k^2)} = K. \tag{5.6.16}$$

Since K is a known function of k, this equation determines k and K when a, b are given. λ can then be found from equation (5.6.12). For example, from Table B (p. 278), if $k = 0.5$, then $K = 1.6858$ and, thus, $a/b = 2.53$ and $\lambda = \frac{3}{4}b^2$.

Instead of b being specified, the total length l of the chain may be given. This can be related to the other parameters, thus:

$$
\begin{aligned}
l &= \int_0^a \sqrt{[1 + (dy/dx)^2]}\,dx \\
&= \int_0^a \sqrt{[1 + (2bK/a)^2\,\text{cn}^2(2Kx/a)\,\text{dn}^2(2Kx/a)]}\,dx \\
&= \int_0^a \sqrt{[1 + (4k^2/k'^4)\,\text{cn}^2(2Kx/a)\,\text{dn}^2(2Kx/a)]}\,dx \\
&= \int_0^a [(2/k'^2)\,\text{dn}^2(2Kx/a) - 1]\,dx \\
&= \frac{a}{k'^2K}\int_0^{2K} \text{dn}^2 u\,du - a \\
&= \frac{2aE}{k'^2K} - a \tag{5.6.17}
\end{aligned}
$$

using equation (3.5.4). With $k = 0.5$, $K = 1.6858$ as before, we read from the table $E = 1.4675$ and, hence, $l = 1.321a$.

5.7. Body Rotating Freely about a Fixed Point

Consider a rigid body which is smoothly supported in light gimbals so that it is free to rotate in any manner about one of its points O, not necessarily its center of mass. Let $O123$ be a set of rectangular axes through O, fixed in the body, and forming principal axes of inertia for the body at the point. We shall suppose that no forces act upon the body, except the forces of reaction at the bearings whose resultant will be assumed directed through O. In particular, gravity will be neglected unless O is the center of mass, when the weight can be compounded with the reactions of the bearings. Alternatively, the body may be regarded as moving in empty space under no forces; in these circumstances, if O is the mass center, an inertial frame $Oxyz$ can be constructed, relative to which the rotation of the body will be measured (e.g., the rotation of the earth can be so treated, if we neglect the attraction of the sun and moon). If $\boldsymbol{\omega}$ is the angular velocity of the body and of the axes $O123$ relative to an inertial frame $Oxyz$, let $(\omega_1, \omega_2, \omega_3)$ be its components in the rotating frame $O123$. Then, if (A, B, C) are the principal moments of inertia of the body about the axes $O1, O2, O3$ respectively, the angular momentum $\mathbf{h}$ of the body has components given by

$$\mathbf{h} = (A\omega_1, B\omega_2, C\omega_3) \tag{5.7.1}$$

in the frame $O123$. We shall assume $A > B > C$.

Choosing the inertial frame $Oxyz$ to be instantaneously coincident with $O123$ at the instant t under consideration, if $d\mathbf{h}/dt$ represents the rate of change in the angular momentum relative to the inertial frame and $\partial\mathbf{h}/\partial t$ denotes the rate of change in the moving frame, we have

$$d\mathbf{h}/dt = \partial\mathbf{h}/\partial t + \boldsymbol{\omega} \times \mathbf{h}. \tag{5.7.2}$$

But $d\mathbf{h}/dt = 0$, since the forces acting on the body have no moment about O. Equations (5.7.1), (5.7.2) accordingly lead to Euler's equations of motion, viz.

$$\left.\begin{array}{l} A\dot{\omega}_1 + (C - B)\omega_2\omega_3 = 0, \\ B\dot{\omega}_2 + (A - C)\omega_3\omega_1 = 0, \\ C\dot{\omega}_3 + (B - A)\omega_1\omega_2 = 0. \end{array}\right\} \tag{5.7.3}$$

Two first integrals of these equations are easily derived. Thus, multiplying the equations in succession by ω_1, ω_2, ω_3, and adding, we find that

$$A\omega_1\dot{\omega}_1 + B\omega_2\dot{\omega}_2 + C\omega_3\dot{\omega}_3 = 0, \tag{5.7.4}$$

which integrates immediately to give

$$A\omega_1^2 + B\omega_2^2 + C\omega_3^2 = T \quad \text{(constant)}. \tag{5.7.5}$$

T is twice the kinetic of the body and this is the energy integral, therefore. Further, multiplication in succession by $A\omega_1$, $B\omega_2$, $C\omega_3$, and addition, leads in the same way to the integral

$$A^2\omega_1^2 + B^2\omega_2^2 + C^2\omega_3^2 = H^2 \quad \text{(constant)}. \tag{5.7.6}$$

It is evident from equation (5.7.1) that H is the magnitude of the angular momentum and this integral accordingly follows from the principle that $\mathbf{h}$ is a constant vector relative to the inertial frame.

It is next easily verified that

$$\omega_1 = P \operatorname{cn} p(t - t_0), \qquad \omega_2 = -Q \operatorname{sn} p(t - t_0), \qquad \omega_3 = R \operatorname{dn} p(t - t_0) \tag{5.7.7}$$

can be made to satisfy the Euler equations at all times t, provided P, Q, R and p, and the modulus k are chosen so that

$$APp = (B - C)QR, \qquad BQp = (A - C)PR, \qquad CRk^2p = (A - B)PQ. \tag{5.7.8}$$

The integrals (5.7.5) and (5.7.6) are also necessarily satisfied by the solution (5.7.7) and yield the further equations

$$AP^2 + CR^2 = T, \qquad A^2P^2 + C^2R^2 = H^2. \tag{5.7.9}$$

(Note: Either substitute from (5.7.7) and use (5.7.8), or substitute from (5.7.7) at $t = t_0$.)

We now have five equations (5.7.8), (5.7.9) to determine the five unknowns P, Q, R, p, k in terms of T and H. Thus, the equations (5.7.7) provide a solution to the Euler equations, involving three arbitrary constants T, H, and t_0, indicating that this solution is a general one. Solving (5.7.8) and (5.7.9), we calculate that

$$P^2 = \frac{H^2 - CT}{A(A - C)}, \quad Q^2 = \frac{H^2 - CT}{B(B - C)}, \quad R^2 = \frac{AT - H^2}{C(A - C)}, \tag{5.7.10}$$

$$p^2 = \frac{1}{ABC}(B - C)(AT - H^2), \quad k^2 = \frac{(A - B)(H^2 - CT)}{(B - C)(AT - H^2)}. \tag{5.7.11}$$

The conditions for P, Q, etc. to be real are $A > H^2/T > C$; it easily follows from (5.7.5) and (5.7.6) that these conditions are always satisfied. As usual, we shall require that $k^2 < 1$; the condition for this is $H^2/T < B$. Thus, the solution (5.7.7) is acceptable only when

$$A > B > H^2/T > C. \tag{5.7.12}$$

If $H^2/T = B$, then $k = 1$ and the solution we have found reduces to

$$\omega_1 = P \operatorname{sech} p(t - t_0), \qquad \omega_2 = - Q \tanh p(t - t_0),$$

$$\omega_3 = R \operatorname{sech} p(t - t_0), \tag{5.7.13}$$

where

$$P^2 = \frac{(B - C)T}{A(A - C)}, \qquad Q^2 = T/B, \qquad R^2 = \frac{(A - B)T}{C(A - C)}, \tag{5.7.14}$$

$$p^2 = \frac{1}{ABC}(A - B)(B - C)T. \tag{5.7.15}$$

In this case, as t increases from $-\infty$ to $+\infty$, the body moves from a state in which it is rotating about $O2$ with angular velocity Q to a state in which it is rotating about this axis with angular velocity $-Q$. Evidently, it is theoretically possible for the body to rotate indefinitely about the $O2$ axis with constant velocity, but this motion is seen to be unstable.

If $H^2/T = C$, then $P = Q = k = 0$ and the solution is

$$\omega_1 = \omega_2 = 0, \qquad \omega_3 = T/C, \tag{5.7.16}$$

i.e., the body rotates with constant angular velocity about its axis of minimum moment. This motion is stable, since any small departure will be governed by the equations (5.7.7) and will result in small oscillations of the three components ω_1, ω_2, ω_3.

In case

$$A > H^2/T > B > C, \tag{5.7.17}$$

it is necessary to assume a solution

$$\omega_1 = P \operatorname{dn} p(t - t_0), \qquad \omega_2 = -Q \operatorname{sn} p(t - t_0), \qquad \omega_3 = R \operatorname{cn} p(t - t_0). \tag{5.7.18}$$

It is left for the reader to verify that

$$P^2 = \frac{H^2 - CT}{A(A - C)}, \qquad Q^2 = \frac{AT - H^2}{B(A - B)}, \qquad R^2 = \frac{AT - H^2}{C(A - C)}, \tag{5.7.19}$$

$$p^2 = \frac{1}{ABC}(A - B)(H^2 - CT), \qquad k^2 = \frac{(B - C)(AT - H^2)}{(A - B)(H^2 - CT)}. \tag{5.7.20}$$

The extreme cases $H^2 = AT$, $H^2 = BT$ are treated as before.

Equations (5.7.11) show that, in the special case $A = B$, the modulus vanishes and only trigonometric functions appear in the solution. In this case, equations (5.7.7) show that ω_1 and ω_2 oscillate sinusoidally with the same amplitude, but their phases are in quadrature; ω_3 is constant, as is evident from the third Euler equation. Similar remarks apply to the case $B = C$—this time equations (5.7.18) are applicable and ω_1 is constant.

The physical interpretation of our results is not straightforward and will now be studied: The instantaneous axis of rotation is the line through O parallel to ω. Thus, if x_i $(i = 1, 2, 3)$ are coordinates with respect to the frame 0123, this axis has parametric equations

$$x_1 = \lambda \omega_1, \qquad x_2 = \lambda \omega_2, \qquad x_3 = \lambda \omega_3. \tag{5.7.21}$$

The equation of the surface traced out by this line as the motion proceeds is obtained by eliminating $\omega_1, \omega_2, \omega_3, \lambda$ between the equations (5.7.5), (5.7.6), and (5.7.21); it is found to have equation

$$A(AT - H^2)x_1^2 + B(BT - H^2)x_2^2 + C(CT - H^2)x_3^2 = 0. \tag{5.7.22}$$

If $B > H^2/T > C$, this represents an elliptical cone with its vertex at O and its axis along 03 (axis of minimum moment). If $A > H^2/T > B$, the surface is again an elliptical cone, but the axis is now along 01 (axis of maximum moment). The surface is called the *polhode* and is fixed in the body.

In the intermediate case $H^2 = BT$, the polhode comprises a pair of planes having equation

$$A(A - B)x_1^2 - C(B - C)x_3^2 = 0. \tag{5.7.23}$$

These planes pass through the x_2-axis and are equally inclined to the $x_1 x_2$-

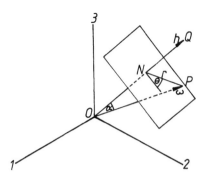

Figure 5.3. The herpolhode.

plane. In the special cases $A = B$ or $B = C$, the polhode is a right circular cone with its axis along Ox_3 or Ox_1 respectively.

The conical locus of the axis of rotation in space (i.e., as seen from the stationary frame $Oxyz$) is called the *herpolhode*. Its equation is much more difficult to establish (see below). Since, at any instant, the body is rotating about the instantaneous axis which lies in both the polhode and the herpolhode, the body's motion can be reproduced by rolling the former on the latter. This is Poinsot's geometrical method of generating the motion.

In Fig. 5.3, OP represents the vector $\boldsymbol{\omega}$ and OQ the vector $\mathbf{h}$. Taking their scalar product, using equations (5.7.1) and (5.7.5), we have

$$\omega H \cos \alpha = \boldsymbol{\omega} \cdot \mathbf{h} = A\omega_1^2 + B\omega_2^2 + C\omega_3^2 = T, \tag{5.7.24}$$

α being the angle PON. Hence, if PN is perpendicular to OQ, it follows that

$$ON = \omega \cos \alpha = T/H. \tag{5.7.25}$$

But $\mathbf{h}$ is constant relative to stationary axes and the direction of OQ in space never changes therefore. We conclude that N is a fixed point in space (i.e., relative to the frame $Oxyz$) and that P moves in the invariant plane through N perpendicular to OQ. The section of the herpolhode by this plane is the locus of P and, hence, by calculating this path, we can fix the herpolhode.

Taking N as pole, let (r, θ) be the polar coordinates of P in the invariant plane. Then

$$
\begin{aligned}
r^2 = OP^2 - ON^2 &= \omega^2 - T^2/H^2 = \omega_1^2 + \omega_2^2 + \omega_3^2 - T^2/H^2 \\
&= \frac{(A-D)(D-C)T}{ACD} - \frac{(A-B)(D-C)T}{ABC} \operatorname{sn}^2 p(t-t_0), \quad \text{if } B > D > C, \\
&= \frac{(A-D)(D-C)T}{ACD} - \frac{(B-C)(A-D)T}{ABC} \operatorname{sn}^2 p(t-t_0), \quad \text{if } A > D > B,
\end{aligned}
\tag{5.7.26}
$$

where $D = H^2/T$ and we have used equations (5.7.7) and (5.7.18).

To find θ, we first note that P lies on the axis of rotation and its velocities with respect to the fixed and moving frames must accordingly be the same. If $\mathbf{v}$ denotes this velocity, we can obtain an equation determining θ by calculating the scalar triple product $\mathbf{h} \cdot \boldsymbol{\omega} \times \mathbf{v}$ in both frames and equating the results.

In the fixed frame, taking axes along OQ and through O parallel and perpendicular to the instantaneous position of NP, the relevant vectors have the components shown:

$$\mathbf{h} = (H, 0, 0), \qquad \boldsymbol{\omega} = \mathbf{OP} = (T/H, r, 0), \qquad \mathbf{v} = (0, \dot{r}, r\dot{\theta}). \qquad (5.7.27)$$

Hence $[\mathbf{h}, \boldsymbol{\omega}, \mathbf{v}] = Hr^2\dot{\theta}$.

In the moving frame 0123, $\mathbf{v} = \dot{\boldsymbol{\omega}}$ and the components are

$$\mathbf{h} = (A\omega_1, B\omega_2, C\omega_3), \qquad \boldsymbol{\omega} = (\omega_1, \omega_2, \omega_3), \qquad \mathbf{v} = (\dot{\omega}_1, \dot{\omega}_2, \dot{\omega}_3). \qquad (5.7.28)$$

Calculating the triple product and equating it to the previous value, we get

$$Hr^2\dot{\theta} = A\omega_1(\omega_2\dot{\omega}_3 - \omega_3\dot{\omega}_2) + B\omega_2(\omega_3\dot{\omega}_1 - \omega_1\dot{\omega}_3) + C\omega_3(\omega_1\dot{\omega}_2 - \omega_2\dot{\omega}_1).$$

$$(5.7.29)$$

To reduce the right-hand member of this equation, we first solve

$$\omega_1^2 + \omega_2^2 + \omega_3^2 = r^2 + T^2/H^2 \qquad (5.7.30)$$

together with equations (5.7.5) and (5.7.6), for ω_1^2, ω_2^2, ω_3^2, to give

$$\omega_1^2 = \frac{BC(r^2 + \alpha)}{(A - B)(A - C)}, \qquad \omega_2^2 = \frac{CA(r^2 + \beta)}{(B - C)(B - A)}, \qquad \omega_3^2 = \frac{AB(r^2 + \gamma)}{(C - A)(C - B)},$$

$$(5.7.31)$$

where

$$\alpha = \frac{(B - D)(C - D)}{BCD} T, \qquad \beta = \frac{(C - D)(A - D)}{CAD} T, \qquad \gamma = \frac{(A - D)(B - D)}{ABD} T.$$

$$(5.7.32)$$

Then, using Euler's equations,

$$\begin{aligned}
A\omega_1(\omega_2\dot{\omega}_3 - \omega_3\dot{\omega}_2) &= A\omega_1^2\left[\frac{1}{C}(A - B)\omega_2^2 + \frac{1}{B}(A - C)\omega_3^2\right] \\
&= A\omega_1^2\left[\frac{A}{C - B}(r^2 + \beta) + \frac{A}{B - C}(r^2 + \gamma)\right] \\
&= \frac{A^2\omega_1^2}{B - C}(\gamma - \beta) \\
&= \frac{A(A - D)(r^2 + \alpha)T}{(A - C)(A - B)}.
\end{aligned} \qquad (5.7.33)$$

Similar expressions for the remaining groups of terms on the right-hand side of equation (5.7.29) can be written down by cyclic permutation of A, B, C and α, β, γ.

Thus, equation (5.7.29) can be expressed in the form

$$Hr^2\dot\theta = \frac{T}{(B-C)(C-A)(A-B)}[A(D-A)(B-C)(r^2+\alpha)$$

$$+ B(D-B)(C-A)(r^2+\beta) + C(D-C)(A-B)(r^2+\gamma)]$$

$$= Tr^2 + \frac{T}{ABCD}(A-D)(B-D)(C-D), \tag{5.7.34}$$

whence

$$\dot\theta = \frac{T}{H} + \frac{T^3}{ABCH^3}(A-D)(B-D)(C-D)r^{-2}. \tag{5.7.35}$$

Substitution for r^2 from the relevant equation (5.7.26) now establishes $\dot\theta$ as a function of t. This can be expressed in the form

$$\dot\theta = \lambda - \mu\frac{v\,sn^2u}{1-v\,sn^2u}, \tag{5.7.36}$$

where

$$\lambda = TD/HB, \qquad \mu = T(B-D)/HB, \qquad u = p(t-t_0), \tag{5.7.37}$$

and

$$v = \frac{D(A-B)}{B(A-D)} \quad \text{if } B>D>C, \left.\begin{array}{l}\\[2em]\\[2em]\end{array}\right\} \tag{5.7.38}$$

$$= \frac{D(B-C)}{B(D-C)} \quad \text{if } A>D>B.$$

We now define a parameter a by the equation

$$sn^2a = v/k^2 = \frac{D(B-C)}{B(D-C)} \quad \text{if } B>D>C, \left.\begin{array}{l}\\[2em]\\[2em]\end{array}\right\} \tag{5.7.39}$$

$$= \frac{D(A-B)}{B(A-D)} \quad \text{if } A>D>B.$$

It follows immediately that

$$cn^2a = -\frac{C(B-D)}{B(D-C)} \quad \text{if } B>D>C, \left.\begin{array}{l}\\[2em]\\[2em]\end{array}\right\} \tag{5.7.40}$$

$$= -\frac{A(D-B)}{B(A-D)} \quad \text{if } A>D>B,$$

$$\mathrm{dn}^2 a = \frac{A(B-D)}{B(A-D)} \quad \text{if } B > D > C,$$

$$= \frac{C(D-B)}{B(D-C)} \quad \text{if } A > D > B.$$

$$(5.7.41)$$

Since $\mathrm{cn}^2 a$ is negative, a must be complex. Suppose we define a new parameter b by the equation

$$a = K + ib. \tag{5.7.42}$$

Then b can be shown to be real: For, using equations (2.2.17)–(2.2.19), (2.6.12), we can show that

$$\mathrm{sn}(a, k) = \mathrm{nd}(b, k'), \qquad \mathrm{cn}(a, k) = -ik' \,\mathrm{sd}(b, k'), \qquad \mathrm{dn}(a, k) = k' \,\mathrm{cd}(b, k').$$

$$(5.7.43)$$

Thus, referring to equations (5.7.39)–(5.7.41), it now follows that, if $A > B > D > C$,

$$\mathrm{sn}^2(b, k') = \frac{C(A-D)}{D(A-C)}, \qquad \mathrm{cn}^2(b, k') = \frac{A(D-C)}{D(A-C)}, \qquad \mathrm{dn}^2(b, k') = \frac{B(D-C)}{D(B-C)},$$

$$(5.7.44)$$

and if $A > D > B > C$, then

$$\mathrm{sn}^2(b, k') = \frac{A(D-C)}{D(A-C)}, \qquad \mathrm{cn}^2(b, k') = \frac{C(A-D)}{D(A-C)}, \qquad \mathrm{dn}^2(b, k') = \frac{B(A-D)}{D(A-B)}.$$

$$(5.7.45)$$

The right-hand members of all these equations are positive and less than unity and accordingly can be used to define a real parameter b in the interval $(0, K')$. For reference, we note that

$$k'^2 = \frac{(A-C)(B-D)}{(B-C)(A-D)} \quad \text{if } B > D > C,$$

$$= \frac{(A-C)(D-B)}{(A-B)(D-C)} \quad \text{if } A > D > B.$$

$$(5.7.46)$$

Equation (5.7.36) can now be written as

$$\dot{\theta} = \frac{H}{B} - \frac{ipk^2 \,\mathrm{sn}\, a \,\mathrm{cn}\, a \,\mathrm{dn}\, a \,\mathrm{sn}^2 u}{1 - k^2 \,\mathrm{sn}^2 a \,\mathrm{sn}^2 u} \quad \text{if } B > D > C,$$

$$= \frac{H}{B} + \frac{ipk^2 \,\mathrm{sn}\, a \,\mathrm{cn}\, a \,\mathrm{dn}\, a \,\mathrm{sn}^2 u}{1 - k^2 \,\mathrm{sn}^2 a \,\mathrm{sn}^2 u} \quad \text{if } A > D > B.$$

$$(5.7.47)$$

(Note: $\mathrm{sn}\, a$, $i\,\mathrm{cn}\, a$, and $\mathrm{dn}\, a$ are all positive by equations (5.7.43).) Integration over the range (t_0, t) then yields the result

$$\theta = \frac{H}{B}(t - t_0) \pm i\Pi(u, a, k), \qquad (5.7.48)$$

the upper sign being taken if $A > D > B$ and the lower sign if $B > D > C$ (we have chosen $\theta = 0$ at $t = t_0$). Equations (5.7.26) and (5.7.48) determine the herpolhode.

By expressing $\Pi(u, a, k)$ in terms of theta functions, it is possible to verify that its value is always purely imaginary and, hence, that equation (5.7.48) yields a real value for θ. We make use of the result (3.7.8). Thus,

$$\Theta(u - a) = \Theta(u - K - ib) = \theta_4\left[\frac{\pi}{2K}(u - ib) - \tfrac{1}{2}\pi\right] = \theta_3\{\pi(u - ib)/2K\}$$

$$= X + iY, \qquad (5.7.49)$$

where

$$X = \sum_{-\infty}^{\infty} q^{n^2} e^{n\pi b/K} \cos(n\pi u/K), \qquad Y = \sum_{-\infty}^{\infty} q^{n^2} e^{n\pi b/K} \sin(n\pi u/K), \quad (5.7.50)$$

and $q = e^{i\pi\tau} = e^{-\pi K'/K}$ (using equation (2.2.3)). Similarly, we show that

$$\Theta(u + a) = X - iY. \qquad (5.7.51)$$

It now follows that

$$\frac{1}{2}\ln\frac{\Theta(u - a)}{\Theta(u + a)} = i\tan^{-1}(Y/X). \qquad (5.7.52)$$

Further, using equation (3.6.1), we calculate that

$$Z(a) = Z(K + ib) = \frac{1}{\theta_3^2(0)}\frac{\theta_4'(\tfrac{1}{2}\pi + i\pi b/2K)}{\theta_4(\tfrac{1}{2}\pi + i\pi b/2K)} = \frac{1}{\theta_3^2(0)}\frac{\theta_3'(i\pi b/2K)}{\theta_3(i\pi b/2K)}$$

$$= -i\zeta, \qquad (5.7.53)$$

where

$$\zeta = \frac{2}{(\sum q^{n^2})^2}\frac{\sum n q^{n^2} e^{n\pi b/K}}{\sum q^{n^2} e^{n\pi b/K}}. \qquad (5.7.54)$$

Equation (3.7.8) now yields the result

$$\theta = \frac{H}{B}(t - t_0) \pm \{\tan^{-1}(Y/X) - \zeta u\}, \qquad (5.7.55)$$

the plus sign being taken when $B > D > C$ and the minus sign when $A > D > B$. As t increases, $\tan^{-1}(Y/X)$ oscillates about a zero mean and the average rate of increase of θ is accordingly given by

$$\dot{\theta}_{\text{mean}} = H/B \mp \zeta p. \qquad (5.7.56)$$

This result establishes the mean rate of precession of the instantaneous axis about the invariant axis.

5.8. Current Flow in a Rectangular Conducting Plate

The steady flow of electric charge in a conducting plate can be described by a complex potential $w = \phi + i\psi$, where ϕ is the scalar potential of the electric field responsible for the flow and ψ is the stream function. If σ is the conductivity per unit area and AB is any arc drawn on the plate, then the current flow across AB is $\sigma(\psi_B - \psi_A)$ (in the sense for which A is to the left and B is to the right). It follows that the curves $\psi = $ constant are lines of flow. ϕ and ψ are both harmonic functions related by the Cauchy-Riemann equations $\partial\phi/\partial x = \partial\psi/\partial y, \partial\phi/\partial y = - \partial\psi/\partial x$, where (x, y) are rectangular Cartesian coordinates in the plane of the plate. Hence, if $z = x + iy, w(z)$ is an analytic function of z, whose real and imaginary parts are ϕ and ψ respectively. The equipotentials $\phi = $ const. and the lines of flow $\psi = $ const. intersect orthogonally.

Consider the particular case

$$w = \phi + i\psi = \ln(\operatorname{cn} z). \tag{5.8.1}$$

We will show that this is the complex potential for current flowing in a rectangular plate $OABC$, whose vertices have the following coordinates: $O(0,0), A(K,0), B(K,K'), C(0,K')$. (We assume $0 < k < 1$.)

On OA, as x increases from 0 to $K, \operatorname{cn} z$ is real and decreases from 1 to 0. Thus, w is real and decreases from 0 to $-\infty$. Hence, $\psi = 0$ along this segment, which must accordingly be a line of flow for the current described by the complex potential (5.8.1).

On $OC, z = iy$ and $\operatorname{cn}(z, k) = \operatorname{nc}(y, k')$ (equations (2.6.12)). But K' is a quarter-period for the elliptic functions of modulus k' and it follows that $\operatorname{cn} z$ is real on the segment and increases from 1 to $+\infty$ as y increases from 0 to K'. We conclude that $\psi = 0$ on OC, which must therefore be a line of flow also.

On $AB, z = K + iy$ and so $\operatorname{cn}(z, k) = - k' \operatorname{sd}(iy, k) = - ik' \operatorname{sd}(y, k')$ (refer to equations (2.2.18) and (2.6.12)). Hence, as y increases from 0 to $K', \operatorname{cn} z$ changes from 0 to $- ik'/k$ through purely imaginary values. Thus, $\psi = - \frac{1}{2}\pi$ and this segment is also a line of flow, therefore.

Finally, on $BC, z = x + iK'$ and $\operatorname{cn} z = - (i/k)\operatorname{ds} x$ (equation (2.2.18)). Thus, as x increases from 0 to $K, \operatorname{cn} z$ is purely imaginary and changes from $-i\infty$ to $- ik'/k$. Hence, $\psi = - \frac{1}{2}\pi$ on BC and it is a line of flow.

The boundary conditions on ψ for flow in the plate being satisfied, it follows from the general theory that the complex potential describes a possible flow.

w is finite everywhere on the rectangle $OABC$, except at A and C, where ϕ tends to $-\infty$ and $+\infty$ respectively. These are the points at which we station the electrodes, therefore. Suppose we take these to be circular quadrants of small radius δ and infinite conductivity. Then, on the edge of the electrode $A, z = K + \delta e^{i\theta}$ ($\frac{1}{2}\pi \leqslant \theta \leqslant \pi$), and the complex potential takes the form

$$\ln\{\operatorname{cn}(K + \delta e^{i\theta})\} = \ln\{-k'\operatorname{sd}(\delta e^{i\theta})\} = \ln(-k'\delta e^{i\theta})$$
$$= \ln(k'\delta) + i(\theta - \pi), \tag{5.8.2}$$

to the first order in δ. Thus, $\phi = \ln(k'\delta)$, $\psi = \theta - \pi$, showing that the potential of the electrode has the constant value $\ln(k'\delta)$ and that the current taken out of the plate by the electrode is $\frac{1}{2}\pi\sigma$. Similarly, on the edge of the electrode $C, z = iK' + \delta e^{i\theta}$ ($-\frac{1}{2}\pi \leqslant \theta \leqslant 0$) and the complex potential is

$$\ln\{\operatorname{cn}(iK' + \delta e^{i\theta})\} = \ln\left\{\frac{1}{ik}\operatorname{ds}(\delta e^{i\theta})\right\} = -\ln(ik\delta e^{i\theta})$$
$$= -\ln(k\delta) - i(\theta + \tfrac{1}{2}\pi), \tag{5.8.3}$$

to $O(\delta)$. It follows that this electrode has potential $-\ln(k\delta)$ and it feeds a current $\frac{1}{2}\pi\sigma$ into the plate.

If the dimensions of the plate are to be a and b, we first calculate a modulus k such that $K/K' = a/b$ and then write $a = \alpha K$, $b = \alpha K'$. If I is the current entering at C and leaving at A, the appropriate complex potential is then given by

$$w = \frac{2I}{\pi\sigma}\ln\{\operatorname{cn}(z/\alpha)\}. \tag{5.8.4}$$

The potentials of the electrodes are found to be

$$\phi_A = \frac{2I}{\pi\sigma}\ln(k'\delta/\alpha), \qquad \phi_C = -\frac{2I}{\pi\sigma}\ln(k\delta/\alpha). \tag{5.8.5}$$

Dividing the potential difference between the electrodes by the current, we find that the effective overall resistance of the plate is

$$R = \frac{4}{\pi\sigma}\ln\frac{\alpha}{\delta\sqrt{(kk')}}. \tag{5.8.6}$$

For a square plate, $K = K'$ and so $k = k' = 1/\sqrt{2}$. The effective resistance is accordingly $(4/\pi\sigma)\ln(\alpha\sqrt{2}/\delta)$.

To calculate the potential distribution over the plate, we can proceed thus:

$$\phi = \mathcal{R}\ln(\operatorname{cn} z) = \ln|\operatorname{cn} z| = \tfrac{1}{2}\ln|\operatorname{cn} z|^2$$
$$= \tfrac{1}{2}\ln(\operatorname{cn} z \ \operatorname{cn} z^*)$$
$$= \tfrac{1}{2}\ln\frac{\operatorname{cn} 2x \operatorname{dn} 2iy + \operatorname{cn} 2iy \operatorname{dn} 2x}{\operatorname{dn} 2x + \operatorname{dn} 2iy}, \tag{5.8.7}$$

by use of the identity (2.4.25). This expression can now be reduced to a purely real form by use of the results (2.6.12). Thus,

$$\phi = \tfrac{1}{2}\ln\frac{\operatorname{cn}2x\,\operatorname{dc}2y + \operatorname{nc}2y\,\operatorname{dn}2x}{\operatorname{dn}2x + \operatorname{dc}2y}$$

$$= \tfrac{1}{2}\ln\frac{\operatorname{cn}2x\,\operatorname{dn}2y + \operatorname{dn}2x}{\operatorname{dn}2x\,\operatorname{cn}2y + \operatorname{dn}2y}, \qquad (5.8.8)$$

where it is understood that all functions of x are to be taken to modulus k and all functions of y to modulus k'.

The stream function determining the lines of flow $\psi = $ constant is easily found using the formula (2.4.2),

$$\psi = \mathscr{I}\ln(\operatorname{cn}z) = \arg(\operatorname{cn}z),$$

$$= \arg\frac{\operatorname{cn}x\,\operatorname{cn}iy - \operatorname{sn}x\,\operatorname{sn}iy\,\operatorname{dn}x\,\operatorname{dn}iy}{1 - k^2\,\operatorname{sn}^2x\,\operatorname{sn}^2iy},$$

$$= \arg\frac{\operatorname{cn}x\,\operatorname{nc}y - i\,\operatorname{sn}x\,\operatorname{sc}y\,\operatorname{dn}x\,\operatorname{dc}y}{1 + k^2\,\operatorname{sn}^2x\,\operatorname{sc}^2y},$$

$$= -\tan^{-1}(\operatorname{sc}x\,\operatorname{sc}y\,\operatorname{dn}x\,\operatorname{dn}y). \qquad (5.8.9)$$

Thus, the family of lines of flow has equation

$$\operatorname{sc}x\,\operatorname{sc}y\,\operatorname{dn}x\,\operatorname{dn}y = \text{constant}, \qquad (5.8.10)$$

remembering that functions of y are to modulus k'.

An alternative physical interpretation of this complex potential is that ψ represents the steady-state temperature of a rectangular heat-conducting plate when its edges OA, OC are maintained at zero temperature and its edges AB, BC are kept at temperature $-\tfrac{1}{2}\pi$. Since, for steady conditions, the temperature ψ is known to be harmonic and there is only one such function satisfying given boundary conditions, the formula (5.8.9) must give the temperature at (x, y) for this problem (it is assumed there is no loss of heat from the plate except at its edges). The curves $\phi = $ const. are then the lines of heat flow.

5.9. Parallel Plate Capacitor

The idea of a complex potential, introduced in the previous section, can be employed to solve many problems in electrostatics. Thus, if $w = w(z) = \phi + i\psi$, where $z = x + iy$ and $w(z)$ is an analytic function, then $\phi(x, y)$ and $\psi(x, y)$ both satisfy Laplace's equation $\nabla^2 V = 0$, and either may be treated as the potential function of a two-dimensional electrostatic field in the plane of rectangular Cartesian coordinates x, y. Having identified a pair of equipotentials, e.g., $\phi = \phi_1, \phi = \phi_2$, the field can be generated by setting up a pair of infinitely long conducting cylinders, with generators perpendicular to the xy-plane and having these equipotentials for their right sections, and then raising these to the potentials ϕ_1 and ϕ_2 respectively.

It follows from electrostatic theory that, if $\phi = $ constant represents a conductor, then the electric charge on a conducting segment over which ψ increases from ψ_1 to ψ_2 is $\varepsilon_0(\psi_2 - \psi_1)$, where ε_0 is the permittivity of free space ($= 8.85 \times 10^{-12}$ in SI units); it is understood that this gives the charge per unit length of the infinite cylinder and that the sense along the arc in which the change in ψ is measured is such that the metallic substance of the conductor lies to the right along the direction of travel. If, instead, ψ is taken as the potential function, then the formula for the charge is $\varepsilon_0(\phi_2 - \phi_1)$, where the change in ϕ is calculated by traversing the arc with the conductor to the left.

As an example, suppose we take

$$z = cZ(w, k), \qquad (5.9.1)$$

where Z is Jacobi's zeta function, c is real, and $0 < k < 1$ (this defines w inversely as an analytic function of z). Putting $w = \phi + i\psi$ and separating out real and imaginary parts (use the identities (3.6.8) and (2.6.12)), it will be found that

$$x = cZ(\phi) + \frac{ck^2 \operatorname{sn}\phi \operatorname{cn}\phi \operatorname{dn}\phi \operatorname{sn}^2\psi}{1 - \operatorname{dn}^2\phi \operatorname{sn}^2\psi}, \qquad (5.9.2)$$

$$y = -cZ(\psi) - \frac{c\pi}{2KK'}\psi + \frac{c \operatorname{sn}\psi \operatorname{cn}\psi \operatorname{dn}\psi \operatorname{dn}^2\phi}{1 - \operatorname{dn}^2\phi \operatorname{sn}^2\psi}, \qquad (5.9.3)$$

where all functions of ϕ are taken with modulus k and all functions of ψ with modulus k' (K and K' are the usual functions of k).

We shall now choose ψ to be the potential function and assume conductors occupy the equipotentials $\psi = 0, \psi = 2K'$. Putting $\psi = 0$ in equations (5.9.2), (5.9.3), these reduce to

$$x = cZ(\phi), \qquad y = 0. \qquad (5.9.4)$$

These are parametric equations for a segment of the x-axis. $Z(\phi)$ is periodic with period $2K$ and behaves roughly like $\sin(\pi\phi/K)$ multiplied by a factor which increases with k. Hence, as ϕ increases from 0 to $2K, x$ first increases from 0 to a positive maximum x_1 ($\phi = \phi_1$), then decreases to 0 ($\phi = K$), then further decreases to a negative minimum $-x_1$ ($\phi = 2K - \phi_1$), and finally increases to 0 again ($\phi = 2K$). Thus, the conductor at zero potential is a plane plate of width $2x_1$. For the same variation of ϕ, $\operatorname{dn}\phi$ first decreases from 1 to k' at $\phi = K$, and then increases back to 1. Thus, had ψ been very small and positive instead of zero, y would first decrease to a minimum when $\phi = K$ and then return to its initial value at $\phi = 2K$. This shows that the *upper* surface of the conducting plate is associated with the values $0 < \phi < \phi_1, 2K - \phi_1 < \phi < 2K$, and the *lower* surface with the values $\phi_1 < \phi < 2K - \phi_1$ (see Fig. 5.4).

To calculate x_1, referring to equation (3.6.1), we note that

$$\frac{dx}{d\phi} = c(\operatorname{dn}^2\phi - E/K). \qquad (5.9.5)$$

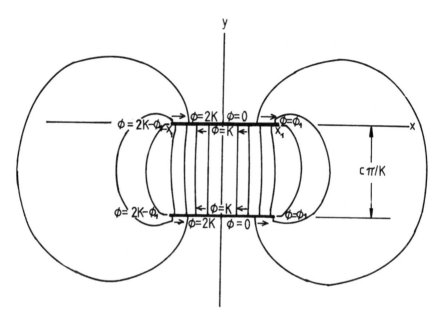

Figure 5.4. Parallel plate capacitor.

Thus, $dx/d\phi = 0$ when $\phi = \phi_1$, where $dn^2\phi_1 = E/K$ or

$$\text{sn } \phi_1 = \frac{1}{k}\sqrt{(1 - E/K)}. \tag{5.9.6}$$

Defining α such that

$$\sin \alpha = \frac{1}{k}\sqrt{(1 - E/K)}, \tag{5.9.7}$$

it follows from equation (3.1.8) that

$$\phi_1 = F(\alpha, k). \tag{5.9.8}$$

Then,

$$x_1 = cZ(\phi_1, k) = c\{E(\phi_1, k) - E\phi_1/K\}$$

$$= \frac{c}{K}\{KD(\alpha, k) - EF(\alpha, k)\}, \tag{5.9.9}$$

transforming to Legendre's form of the integral of the second kind by equation (3.4.27). If, therefore, a is the width of the conducting plate,

$$a = \frac{2c}{K}\{KD(\alpha, k) - EF(\alpha, k)\}. \tag{5.9.10}$$

Putting $\psi = 2K'$ in equations (5.9.2), (5.9.3), the parametric equations of the second conductor at potential $2K'$ are found to be

$$x = cZ(\phi), \qquad y = -c\pi/K. \tag{5.9.11}$$

(Note: $\mathrm{sn}(2K', k') = 0$.) These equations define a plane plate, also of width a, in the plane $y = -c\pi/K$ (see Fig. 5.4). As ϕ increases from 0 to $2K$, the two sides of the plate are described as explained earlier (in this case, it is necessary to suppose $\psi = 2K' - \varepsilon$, where ε is small; the multiplier of $\mathrm{dn}^2\phi$ in the last term of equation (5.9.3) is then negative and it is the *upper* side of the plate which is traversed as ϕ increases from ϕ_1 to $2K - \phi_1$).

The two plates constitute a parallel plate capacitor, the plates of which are separated by an air gap of width d, where

$$d = c\pi/K. \tag{5.9.12}$$

Dividing equations (5.9.10) and (5.9.12), we obtain the result

$$\frac{a}{d} = \frac{2}{\pi}\{KD(\alpha, k) - EF(\alpha, k)\}. \tag{5.9.13}$$

Thus, given the dimensions of the capacitor, equations (5.9.7) and (5.9.13) can be used to determine k. Then, c follows from equation (5.9.12).

The capacitance of the device can be found by calculating the charge on either plate. Thus, proceeding around the two surfaces of the lower plate, keeping the substance of the plate to the left, ϕ increases from 0 to $2K$ and the total charge must be $2\varepsilon_0 K$. The charge on the upper (earthed) plate is, similarly, found to be $-2\varepsilon_0 K$. The potential difference between the plates is $2K'$. Thus, the capacitance C (per unit length) is determined to be

$$C = \varepsilon_0 K/K'. \tag{5.9.14}$$

The charge on the inside of the lower plate is $2\varepsilon_0(K - \phi_1)$ and on the outside is $2\varepsilon_0\phi_1$. These charges are in the ratio

$$\frac{K}{\phi_1} - 1 = \frac{K}{F(\alpha, k)} - 1. \tag{5.9.15}$$

The equipotentials $\psi = \text{constant}$ and lines of force $\phi = \text{constant}$ can be plotted using equations (5.9.2) and (5.9.3). A set of lines of force for the case $q = 0.3$ is given in Fig. 5.4.

EXERCISES

1. Using the notation of Fig. 5.1, let M be the midpoint of PN. Prove that the acceleration of the bob has magnitude $2\omega^2 MD$ and that its direction makes an angle θ with MD. If R is the tension in the string, show that $R/m\omega^2 = 3ND + OD$.

2. A uniform beam of length $2a$ rests on level ground in a vertical plane against a vertical wall, making an angle α with the wall. It slips down the wall from rest. If the friction between the beam and wall and between beam and ground is neglected,

show that the angle θ it makes with the wall a time t after it begins to move is given by

$$\sin \tfrac{1}{2}\theta = \sin \tfrac{1}{2}\alpha \, \mathrm{nd}\left(\sqrt{\frac{3g}{4a}}\,t\right),$$

the modulus being $k = \cos \tfrac{1}{2}\alpha$.

3. A particle having unit mass oscillates on the line joining two centers of force, it being repelled from each by a force μ/r^2, where r is its distance from a center. Initially, the particle is at rest a distance kc ($k < 1$) from the point O midway between the centers, where $2c$ is the distance between the centers. If x is the particle's distance from O a time t after passing through this point, prove that

$$t = \left[\frac{c^3(1 - k^2)}{4\mu}\right]^{1/2} D(\phi, k),$$

where $\sin \phi = x/kc$. Deduce that the period of oscillation is

$$2\left[\frac{c^3(1 - k^2)}{\mu}\right]^{1/2} E(k).$$

4. A particle having unit mass moves in a plane under the action of an attractive central force of magnitude μr^5, r being its distance from the center of attraction O. Its total energy is $2\mu a^6$ and its angular momentum about O is $\mu^{1/2} a^4$. Show that its orbit lies between two concentric circles and that, if the time t is measured from an instant when the particle lies on the outer circle, its polar coordinates with respect to O are given by

$$r^2 = a^2 \frac{\alpha + 3\mathrm{cn}(\beta t, k)}{\alpha - \mathrm{cn}(\beta t, k)},$$

$$\theta = \left(\frac{3}{2}\right)^{1/4}\left[\frac{\alpha^2}{3(\alpha^2 - 9)}\Lambda\left(\beta t, \frac{3i}{\sqrt{(\alpha^2 - 9)}}, k\right)\right.$$

$$\left.- \sqrt{\frac{2}{\alpha^2 - 9}}\tan^{-1}\left(\frac{\alpha\,\mathrm{sd}(\beta t, k)}{\sqrt{\{2(\alpha^2 - 9)\}}}\right)\right] - \frac{1}{3}\mu^{1/2}a^2 t,$$

where $\alpha^2 = 9 + 4\sqrt{6}$, $\beta = 4(2/3)^{1/4}\mu^{1/2}a^2$, $k^2 = (8 - 3\sqrt{6})/16$. (Note: Use the transformation $r^2 = a^2(3s + 1)/(1 - s)$ to bring the elliptic integral to canonical form.)

5. A particle having unit mass moves in a plane under the action of an attractive central force μr^3, r being its distance from the center of attraction O. When its distance from O is $a\sqrt{2}$, it is projected with velocity $a^2\sqrt{(7\mu/2)}$ at right angles to the radius from O. Show that it moves between two circles having radii $a\sqrt{2}$ and $a\sqrt{(2\sqrt{2} - 1)}$ and that, if time is measured from an instant when the particle lies on the inner circle, its polar coordinates at time t afterward are given by

$$r^2 = a^2 \frac{\sqrt{2}\,\mathrm{dn}(\beta t) + 3}{\sqrt{2}\,\mathrm{dn}(\beta t) + 1},$$

$$\theta = a\sqrt{(7\mu)}\,t - \frac{3}{\sqrt{7}}\Lambda(\beta t, i/\sqrt{7}, 1/\sqrt{2}) + \frac{1}{2}\tan^{-1}\left(2\sqrt{\frac{2}{7}}\,\mathrm{sc}(\beta t)\right),$$

the modulus being $1/\sqrt{2}$ and $\beta = 2a\mu^{1/2}$. (Note: The transformation $r^2 = a^2(s + 3)/(s + 1)$ brings the elliptic integral to canonical form.)

6. A particle having unit mass moves in a plane under the action of a constant attractive force μ toward a fixed point O. It is projected with velocity $\sqrt{(3\mu a)}/2$ at right angles to the radius from O when its distance from this point is a. Show that it moves between two circles whose radii are a and $(3 + \sqrt{105})a/16$ and that, when its distance from O is r, the time t which has elapsed since its projection is given by

$$t = 2\sqrt{\frac{a}{21\mu}}\left[(1 - 2k'^2/k^2)(x - K) + \frac{2}{k^2}\{E(x) - E\} - \frac{2\,\mathrm{sn}\,x\,\mathrm{cn}\,x}{\mathrm{dn}\,x + k'}\right],$$

where the modulus $k = 4/\sqrt{21}$ and

$$\mathrm{dn}\,x = k'\frac{3a - 2r}{2r - a}.$$

(Note: The change of variable $r = (a/2)(s + 3)/(s + 1)$ reduces the elliptic integral.)

7. A particle whose mass is unity moves in a plane under an attraction μ/r^7 toward the pole O of coordinates (r, θ). Initially it is at a distance a from the pole and it is projected with velocity $(1/a^3)\sqrt{(13\mu/12)}$ at right angles to the radius vector. Show that, thereafter, it must steadily recede from the pole and that its orbit (with appropriate choice of the line $\theta = 0$) has polar equation

$$r^2 = \frac{2}{3}a^2\frac{2 + \mathrm{sn}\,\alpha\theta}{1 + \mathrm{sn}\,\alpha\theta},$$

where $\alpha^2 = 20/13$ and $k^2 = 2/5$. (Note: The change of variable $r^2 = (2/3)a^2(s + 2)/(s + 1)$ brings the integral to canonical form.) Deduce that the asymptote along which the particle escapes to infinity makes an angle $2K/\alpha$ with the initial direction of the radius vector.

8. A bow is constructed from a uniform straight wand by flexing its two ends toward one another and connecting them by a string. s is arc length measured along the bow from its midpoint and ψ is the angle made by the tangent to the bow with the string. Taking the center of the string as origin, x-axis along the string, and y-axis through the center of the wand, by assuming the curvature of the bow at any point is proportional to the bending moment of the applied forces, obtain the equation $d\psi/ds = \omega^2 y$. Deduce that

$$\frac{d^2\psi}{ds^2} + \omega^2 \sin\psi = 0$$

and, by treating this as the equation of motion of a simple pendulum and referring to section 5.1, prove that

$$\sin\tfrac{1}{2}\psi = k\,\mathrm{sn}(\omega s, k),$$

where $k = \sin\tfrac{1}{2}\alpha$ and α is the angle made by the wand with the string at its ends. Deduce, further, that parametric equations for the wand are

$$x = \frac{2}{\omega}E(\omega s, k) - s, \qquad y = \frac{2k}{\omega}\mathrm{cn}(\omega s, k),$$

and that the length of the wand is $2K/\omega$.

9. Show that the complex potential $w = \ln(\operatorname{sn} z)$ describes the electric current which flows in the rectangular plate $OABC$ of section 5.8, when a current $\frac{1}{2}\pi\sigma$ enters by a small quadrant electrode at C and leaves by a similar electrode at O. Deduce that the equivalent resistance between the electrodes is $-(4/\pi\sigma)\ln(\delta\sqrt{k})$, where δ is the electrodes' radius. Show, also, that the lines of flow have equation

$$\operatorname{cs} x \operatorname{dn} x \operatorname{sn} y \operatorname{cd} y = \text{constant},$$

it being understood that functions of y are to modulus k'.

10. Show that the transformation $x + iy = c\operatorname{sn}(\phi + i\psi, k)$ solves the problem of the electric field between a charged conducting plane plate of breadth b lying in the plane $y = 0$ a distance c from an infinite earthed plate occupying the plane $x = 0$. (Hint: Consider the equipotentials $\phi = 0, K$, and take $k = c/(b + c)$.) Show, also, that the capacitance of the charged plate is $2\varepsilon_0 K'/K$ per unit length.

11. Show that the transformation $x + iy = c\operatorname{dn}(\phi + i\psi, k)$ solves the problem of the electric field between a charged plane plate and a pair of earthed semi-infinite plates lying in the plane of the first plate, at equal distances from its two sides (take ϕ to be the potential). If $c = b + d$ and $k' = d/(b + d)$, show that the width of the insulated plate is $2d$ and that the gaps between this plate and the earthed plates are both of length b. Show that the capacitance of the insulated plate is $4\varepsilon_0 K'/K$ (per unit length).

12. A bead P is free to move on a smooth circular wire of radius a, which is constrained to rotate about one of its points O with angular velocity ω in its own plane, which is horizontal. If ϕ is the angle between OP and the diameter OD through O, obtain the equation of motion

$$\ddot{\phi} + \omega^2 \sin\phi \cos\phi = 0.$$

If the speed of the bead relative to the wire as it passes through D at $t = 0$ is $2\omega b$, prove that

$$\sin\phi = \operatorname{sn}(\omega bt/a, a/b) \quad \text{if } a < b,$$
$$\sin\phi = (b/a)\operatorname{sn}(\omega t, b/a) \quad \text{if } a > b.$$

Describe the two types of motion.

CHAPTER 6

Weierstrass's Elliptic Function

6.1. Jacobi's Functions with Specified Periods

Given the complex parameter τ (with positive imaginary part), the quarter-periods K and iK' of the Jacobian elliptic functions are determined by the equations (2.2.7) and (2.2.8); according to equation (2.2.3), τ is precisely the ratio iK'/K of these quarter-periods. Thus it is possible to construct a set of Jacobi functions having a common pair of arbitrary periods $2\omega_1, 2\omega_3$ in the following manner: First, choose the notation so that $\omega_3/\omega_1 = \tau$ has positive imaginary part (if this ratio is real, the elliptic functions cannot be defined (see section 8.1)). Secondly, construct Jacobian elliptic functions, sn u, etc., from theta functions with parameter τ. Thirdly, transform from the variable u to a new variable v by the equation $u = 2Kv/\omega_1$. Clearly, the resulting functions of v will have periods $2\omega_1$ and $2\omega_3$. As proved in section 2.8, these functions will have exactly two simple poles in each of their primitive period parallelograms.

There are many other elliptic functions having periods $2\omega_1, 2\omega_3$, and, in this chapter, we shall provide an alternative construction of such a function, which is due to Weierstrass. Weierstrass's elliptic function having primitive periods $2\omega_1, 2\omega_3$ is denoted by $\mathscr{P}(u, \omega_1, \omega_3)$ (read as "pea of yew, etc.") and possesses only one singularity inside each cell, viz. a double pole. It is analytically more elementary than the Jacobi functions, but is not so useful in regard to applications.

In the same way that it is very convenient to approach the Jacobi functions via the theta functions, we shall first define the sigma functions and then derive Weierstrass's function from these.

6.2. The Sigma Functions

With $\tau = \omega_3/\omega_1$ and $z = \pi u/2\omega_1$, we define

$$\sigma(u, \omega_1, \omega_3) = A e^{au^2} \theta_1(z|\tau), \tag{6.2.1}$$

where the complex parameters A, a are yet to be fixed. The parameters ω_1, ω_3 will often be left understood, the sigma function being written $\sigma(u)$.

Note that $\sigma(0) = 0$. We choose A so that $\sigma'(0) = 1$; this requires that

$$A = 2\omega_1/\{\pi\theta_1'(0)\}. \tag{6.2.2}$$

Since $\theta_1(z)$ is odd, so also is $\sigma(u)$ and we conclude that $\sigma''(0) = 0$. a is next chosen so that the third-order derivative also vanishes for $u = 0$. Hence, we take

$$a = -\frac{\pi^2}{24\omega_1^2} \cdot \frac{\theta_1'''(0)}{\theta_1'(0)}. \tag{6.2.3}$$

If ω_1 is real and ω_3 is imaginary, then $\sigma(u)$ is real for real u.

σ, like θ_1, is an integral function, whose zeros (all simple) lie at the points

$$u = \frac{2\omega_1}{\pi}(m\pi + n\pi\tau) = 2m\omega_1 + 2n\omega_2 \tag{6.2.4}$$

(see (1.3.10)).

The quasi-periodicity of θ_1 induces the same property in $\sigma(u)$. Thus, the identities (1.3.2) imply that

$$\sigma(u + 2\omega_1) = -\exp\{4a\omega_1(u + \omega_1)\}\sigma(u), \tag{6.2.5}$$

$$\sigma(u + 2\omega_3) = -\exp\{(4a\omega_3 - i\pi/\omega_1)(u + \omega_3)\}\sigma(u). \tag{6.2.6}$$

A more symmetric notation can now be achieved by writing

$$a = \eta_1/2\omega_1, \quad \text{i.e.,} \quad \eta_1 = -\frac{\pi^2}{12\omega_1} \cdot \frac{\theta_1'''(0)}{\theta_1'(0)}, \tag{6.2.7}$$

and defining η_3 so that

$$\eta_1\omega_3 - \eta_3\omega_1 = \tfrac{1}{2}i\pi. \tag{6.2.8}$$

The equations (6.2.1), (6.2.5), and (6.2.6) then reduce to the forms

$$\sigma(u) = \frac{2\omega_1}{\pi\theta_1'(0)} e^{\eta_1 u^2/2\omega_1}\theta_1(z), \tag{6.2.9}$$

$$\sigma(u + 2\omega_1) = -e^{2\eta_1(u + \omega_1)}\sigma(u), \tag{6.2.10}$$

$$\sigma(u + 2\omega_3) = -e^{2\eta_3(u + \omega_3)}\sigma(u). \tag{6.2.11}$$

Replacing u by $-u$ in these equations, we also calculate that

$$\sigma(u - 2\omega_1) = -e^{-2\eta_1(u - \omega_1)}\sigma(u), \tag{6.2.12}$$

$$\sigma(u - 2\omega_3) = -e^{-2\eta_3(u - \omega_3)}\sigma(u). \tag{6.2.13}$$

Clearly, $2\omega_1 + 2\omega_3$ will also be a quasi-period of $\sigma(u)$, as also will $-(2\omega_1 + 2\omega_3)$. It is convenient to write $2\omega_2 = -(2\omega_1 + 2\omega_3)$ for this quasi-period and to associate with it the parameter $\eta_2 = -(\eta_1 + \eta_3)$. Thus,

$$\omega_1 + \omega_2 + \omega_3 = 0, \tag{6.2.14}$$

$$\eta_1 + \eta_2 + \eta_3 = 0. \tag{6.2.15}$$

The symmetrical relations

$$\eta_3\omega_2 - \eta_2\omega_3 = \eta_2\omega_1 - \eta_1\omega_2 = \tfrac{1}{2}\pi i \tag{6.2.16}$$

follow immediately from equation (6.2.8). It is now easily verified, using equations (6.2.8), (6.2.12), and (6.2.13), that

$$\sigma(u + 2\omega_2) = -e^{2\eta_2(u + \omega_2)}\sigma(u), \tag{6.2.17}$$

which completes the set of these identities.

It is customary to define three other sigma functions $\sigma_\alpha(u)$ ($\alpha = 1, 2, 3$) by the equations

$$\sigma_\alpha(u) = e^{-\eta_\alpha u}\frac{\sigma(u + \omega_\alpha)}{\sigma(\omega_\alpha)}. \tag{6.2.18}$$

It is easy to see that these functions are closely related to the remaining theta functions. Thus

$$\sigma_1(u) = \exp\left[-\eta_1 u + \frac{\eta_1}{2\omega_1}(u + \omega_1)^2 - \frac{\eta_1}{2\omega_1}\omega_1^2\right]\theta_1(z + \tfrac{1}{2}\pi)/\theta_1(\tfrac{1}{2}\pi),$$

$$= \exp(\eta_1 u^2/2\omega_1)\theta_2(z)/\theta_2(0), \tag{6.2.19}$$

having used the identity (1.3.7). Very similarly, it may be shown that

$$\sigma_2(u) = \exp(\eta_1 u^2/2\omega_1)\theta_3(z)/\theta_3(0), \tag{6.2.20}$$

$$\sigma_3(u) = \exp(\eta_1 u^2/2\omega_1)\theta_4(z)/\theta_4(0). \tag{6.2.21}$$

The sigma functions $\sigma_\alpha(u)$ are also quasi-periodic with periods $2\omega_1, 2\omega_3$. It is left for the reader to prove that

$$\sigma_\alpha(u + 2\omega_\beta) = \pm e^{2\eta_\beta(u + \omega_\beta)}\sigma_\alpha(u), \tag{6.2.22}$$

the plus sign being taken if $\alpha \neq \beta$ and the minus sign if $\alpha = \beta$.

It follows from the definition (6.2.18) that $\sigma_\alpha(0) = 1$ and that the sigma functions are regular for all finite u. Also,

$$\sigma_1(-u) = e^{\eta_1 u}\sigma(-u + \omega_1)/\sigma(\omega_1) = -e^{\eta_1 u}\sigma(u - \omega_1)/\sigma(\omega_1)$$

$$= e^{\eta_1 u}e^{-2\eta_1 u}\sigma(u + \omega_1)/\sigma(\omega_1) = \sigma_1(u), \tag{6.2.23}$$

having used the identity (6.2.12). Thus, $\sigma_1(u)$ is even, as are $\sigma_2(u)$ and $\sigma_3(u)$. Hence $\sigma_\alpha'(0) = 0$.

Note that, for any complex nonzero multiplier λ,

$$\sigma(\lambda u, \lambda\omega_1, \lambda\omega_3) = \lambda\sigma(u, \omega_1, \omega_3), \tag{6.2.24}$$

$$\sigma_\alpha(\lambda u, \lambda\omega_1, \lambda\omega_3) = \sigma_\alpha(u, \omega_1, \omega_3). \tag{6.2.25}$$

In particular, taking $\lambda = 1/\omega_1$,

$$\omega_1^{-1}\sigma(u, \omega_1, \omega_3) = \sigma(u/\omega_1, 1, \tau), \qquad \sigma_\alpha(u, \omega_1, \omega_3) = \sigma_\alpha(u/\omega_1, 1, \tau). \tag{6.2.26}$$

Thus, the functions $\omega_1^{-1}\sigma(u, \omega_1, \omega_3)$ and $\sigma_\alpha(u, \omega_1, \omega_3)$ can be tabulated against u/ω_1 and τ. This has been done for real values of u/ω_1 and positive imaginary values of τ in Table F (pp. 298–301), where $u/\omega_1 = 0(0.1)2$, $\kappa = -i\tau = 0.2(0.2)2$. To extend the table to values of u/ω_1 outside this range, we make use of their quasi-periodicity.

6.3. Alternative Definitions for Jacobi's Elliptic Functions

Taking

$$\omega_1 = \tfrac{1}{2}\pi\theta_3^2(0) = K, \qquad \omega_3 = iK' = \tfrac{1}{2}\pi\tau\theta_3^2(0) = \tau K, \tag{6.3.1}$$

where K, iK' are the quarter-periods of the Jacobi functions, it follows that

$$z = \frac{\pi u}{2\omega_1} = \frac{u}{\theta_3^2(0)}, \tag{6.3.2}$$

in agreement with the definition of z adopted in section 2.1. Thus, referring to equation (2.1.1),

$$\text{sn}(u, k) = \frac{\theta_3(0)}{\theta_2(0)} \cdot \frac{\theta_1(z)}{\theta_4(z)} = \frac{\pi\theta_3(0)\theta_1'(0)}{2\omega_1\theta_2(0)\theta_4(0)} \cdot \frac{\sigma(u)}{\sigma_3(u)} = \frac{\sigma(u)}{\sigma_3(u)}, \tag{6.3.3}$$

after making use of the results (1.5.11), (6.2.9), and (6.2.21).

Similarly, we can show from equations (2.1.2), (2.1.3) that

$$\text{cn}(u, k) = \frac{\sigma_1(u)}{\sigma_3(u)}, \qquad \text{dn}(u, k) = \frac{\sigma_2(u)}{\sigma_3(u)}. \tag{6.3.4}$$

The periodicity of the Jacobi functions can now be established from the quasi-periodicity of the sigma functions.

6.4. Identities Relating Sigma Functions

By substituting from equations (6.2.9), (6.2.19)–(6.2.21) for theta functions appearing in the identities established in section 1.4, a large number of identities involving the sigma functions can be derived.

For example, equations (1.4.16), (1.4.23), and (1.4.30) lead to the identities

$$\sigma(u+v)\sigma(u-v) = \sigma^2(u)\sigma_\alpha^2(v) - \sigma^2(v)\sigma_\alpha^2(u), \qquad (6.4.1)$$

for $\alpha = 1, 2, 3$.

Equations (1.4.19), (1.4.25), (1.4.31) yield the results

$$\left.\begin{aligned}
\sigma_3(u+v)\sigma_3(u-v)\theta_4^4(0) &= \sigma_2^2(u)\sigma_2^2(v)\theta_3^4(0) - \sigma_1^2(u)\sigma_1^2(v)\theta_2^4(0), \\
\sigma_2(u+v)\sigma_2(u-v)\theta_3^4(0) &= \sigma_1^2(u)\sigma_1^2(v)\theta_2^4(0) + \sigma_3^2(u)\sigma_3^2(v)\theta_4^4(0), \\
\sigma_1(u+v)\sigma_1(u-v)\theta_2^4(0) &= \sigma_2^2(u)\sigma_2^2(v)\theta_3^4(0) - \sigma_3^2(u)\sigma_3^2(v)\theta_4^4(0).
\end{aligned}\right\} \qquad (6.4.2)$$

Putting $v = 0$, these all reduce to the same identity, viz.

$$\sigma_2^2(u)\theta_3^4(0) = \sigma_1^2(u)\theta_2^4(0) + \sigma_3^2(u)\theta_4^4(0). \qquad (6.4.3)$$

With $u = 0$ also, this last identity becomes equation (1.4.53).

The sigma function forms of the identities (1.4.38), (1.4.40), (1.4.48) are

$$\left.\begin{aligned}
\sigma(u+v)\sigma_3(u-v) &= \sigma(u)\sigma_3(u)\sigma_1(v)\sigma_2(v) + \sigma(v)\sigma_3(v)\sigma_1(u)\sigma_2(u), \\
\sigma(u+v)\sigma_2(u-v) &= \sigma(u)\sigma_2(u)\sigma_1(v)\sigma_3(v) + \sigma(v)\sigma_2(v)\sigma_1(u)\sigma_3(u), \\
\sigma(u+v)\sigma_1(u-v) &= \sigma(u)\sigma_1(u)\sigma_2(v)\sigma_3(v) + \sigma(v)\sigma_1(v)\sigma_2(u)\sigma_3(u).
\end{aligned}\right\} \qquad (6.4.4)$$

A double application of the identity (6.4.1) leads to the result

$$\begin{aligned}
\sigma(u+v)\sigma(u-v)\sigma(x+y)\sigma(x-y) = {} &\sigma^2(u)\sigma^2(x)\sigma_\alpha^2(v)\sigma_\alpha^2(y) - \sigma^2(v)\sigma^2(x)\sigma_\alpha^2(u)\sigma_\alpha^2(y) \\
&+ \sigma^2(v)\sigma^2(y)\sigma_\alpha^2(u)\sigma_\alpha^2(x) - \sigma^2(u)\sigma^2(y)\sigma_\alpha^2(v)\sigma_\alpha^2(x).
\end{aligned}$$
$$(6.4.5)$$

The important identity

$$\begin{aligned}
&\sigma(u+v)\sigma(u-v)\sigma(x+y)\sigma(x-y) + \sigma(v+x)\sigma(v-x)\sigma(u+y)\sigma(u-y) \\
&+ \sigma(x+u)\sigma(x-u)\sigma(v+y)\sigma(v-y) = 0
\end{aligned} \qquad (6.4.6)$$

now follows by application of the previous identity to each term.

Expressing the identities (1.4.39), (1.4.41), and (1.4.47) in terms of sigma functions, we prove that

$$\left.\begin{aligned}
\sigma_1(u+v)\sigma_2(u-v) &= \sigma_1(u)\sigma_2(u)\sigma_1(v)\sigma_2(v) - \frac{\pi^2\theta_4^4(0)}{4\omega_1^2}\sigma(u)\sigma_3(u)\sigma(v)\sigma_3(v), \\
\sigma_1(u+v)\sigma_3(u-v) &= \sigma_1(u)\sigma_3(u)\sigma_1(v)\sigma_3(v) - \frac{\pi^2\theta_3^4(0)}{4\omega_1^2}\sigma(u)\sigma_2(u)\sigma(v)\sigma_2(v), \\
\sigma_2(u+v)\sigma_3(u-v) &= \sigma_2(u)\sigma_3(u)\sigma_2(v)\sigma_3(v) - \frac{\pi^2\theta_2^4(0)}{4\omega_1^2}\sigma(u)\sigma_1(u)\sigma(v)\sigma_1(v),
\end{aligned}\right\} \qquad (6.4.7)$$

after making use of the result (1.5.11).

Further identities, which we shall find useful, can be derived from ones already established, by simple algebraic manipulation. Thus, multiplying the second and last of the set (6.4.7) by $\sigma_2(u)\sigma_2(v)$ and $\sigma_1(u)\sigma_1(v)$ respectively, and subtracting, we find that

$$[\sigma_1(u)\sigma_1(v)\sigma_2(u+v) - \sigma_2(u)\sigma_2(v)\sigma_1(u+v)]\sigma_3(u-v)$$

$$= \frac{\pi^2}{4\omega_1^2}\sigma(u)\sigma(v)[\theta_3^4(0)\sigma_2^2(u)\sigma_2^2(v) - \theta_2^4(0)\sigma_1^2(u)\sigma_1^2(v)]$$

$$= \frac{\pi^2}{4\omega_1^2}\sigma(u)\sigma(v)\sigma_3(u+v)\sigma_3(u-v)\theta_4^4(0), \tag{6.4.8}$$

having used the first of the identities (6.4.2) in the last step. Division through by $\sigma_3(u-v)$ now shows that

$$\sigma_1(u)\sigma_1(v)\sigma_2(u+v) - \sigma_2(u)\sigma_2(v)\sigma_1(u+v) = \frac{\pi^2\theta_4^4(0)}{4\omega_1^2}\sigma(u)\sigma(v)\sigma_3(u+v).$$

$$\tag{6.4.9}$$

Using a similar procedure, the reader may prove the identities

$$\sigma_2(u)\sigma_2(v)\sigma_3(u+v) - \sigma_3(u)\sigma_3(v)\sigma_2(u+v) = \frac{\pi^2\theta_2^4(0)}{4\omega_1^2}\sigma(u)\sigma(v)\sigma_1(u+v),$$

$$\tag{6.4.10}$$

$$\sigma_1(u)\sigma_1(v)\sigma_3(u+v) - \sigma_3(u)\sigma_3(v)\sigma_1(u+v) = \frac{\pi^2\theta_3^4(0)}{4\omega_1^2}\sigma(u)\sigma(v)\sigma_2(u+v).$$

$$\tag{6.4.11}$$

6.5. Sigma Functions as Infinite Products

Having already expressed the theta functions as infinite products at (1.6.23)–(1.6.26), it is straightforward to obtain similar expressions for the sigma functions. To do this, we need an expression for $\theta_1'(0)$ as an infinite product; this follows from equations (1.6.16) and (1.6.22) in the form

$$\theta_1'(0) = 2q^{1/4}\prod_{n=1}^{\infty}(1-q^{2n})^3. \tag{6.5.1}$$

From equations (6.2.9) and (1.6.23), it now follows that

$$\sigma(u) = \frac{2\omega_1}{\pi}\exp(\eta_1 u^2/2\omega_1)\sin\frac{\pi u}{2\omega_1}\prod_{n=1}^{\infty}(1-q^{2n})^{-2}\left(1-2q^{2n}\cos\frac{\pi u}{\omega_1}+q^{4n}\right). \tag{6.5.2}$$

Also, substituting from equations (1.6.24)–(1.6.26) into equations (6.2.19)–(6.2.21), we obtain the results

$$\sigma_1(u) = \exp(\eta_1 u^2/2\omega_1)\cos\frac{\pi u}{2\omega_1}\prod_{n=1}^{\infty}(1+q^{2n})^{-2}\left(1+2q^{2n}\cos\frac{\pi u}{\omega_1}+q^{4n}\right), \tag{6.5.3}$$

$$\sigma_2(u) = \exp(\eta_1 u^2/2\omega_1) \prod_{n=1}^{\infty} (1 + q^{2n-1})^{-2} \left(1 + 2q^{2n-1} \cos \frac{\pi u}{\omega_1} + q^{4n-2}\right),$$

$$(6.5.4)$$

$$\sigma_3(u) = \exp(\eta_1 u^2/2\omega_1) \prod_{n=1}^{\infty} (1 - q^{2n-1})^{-2} \left(1 - 2q^{2n-1} \cos \frac{\pi u}{\omega_1} + q^{4n-2}\right).$$

$$(6.5.5)$$

6.6. Weierstrass's Elliptic Function

We first define Weierstrass's zeta function by the equation

$$\zeta(u) = \sigma'(u)/\sigma(u) = \frac{d}{du} \ln \sigma(u). \tag{6.6.1}$$

In section 6.2 we proved the quasi-periodicity of $\sigma(u)$ in the form

$$\sigma(u + 2\omega_\alpha) = - e^{2\eta_\alpha(u + \omega_\alpha)} \sigma(u), \tag{6.6.2}$$

for $\alpha = 1, 2, 3$. Differentiation yields the further result

$$\sigma'(u + 2\omega_\alpha) = - e^{2\eta_\alpha(u + \omega_\alpha)} \{2\eta_\alpha \sigma(u) + \sigma'(u)\}. \tag{6.6.3}$$

Hence, dividing the last two equations,

$$\zeta(u + 2\omega_\alpha) = \zeta(u) + 2\eta_\alpha. \tag{6.6.4}$$

Clearly, the zeta function is doubly periodic with periods $2\omega_1, 2\omega_3$, to within the addition of a constant. Since $\sigma(u)$ is odd, $\sigma'(u)$ is even and $\zeta(u)$ is therefore odd. Putting $u = -\omega_\alpha$ in equation (6.6.4), we conclude that

$$\zeta(\omega_\alpha) = \eta_\alpha. \tag{6.6.5}$$

A further differentiation of equation (6.6.4) will eliminate the constant and so yield a true elliptic function. We define Weierstrass's elliptic function $\mathscr{P}(u)$ thus:

$$\mathscr{P}(u) = - \zeta'(u) = - \frac{d^2}{du^2} \{\ln \sigma(u)\} \tag{6.6.6}$$

and then equation (6.6.4) shows that

$$\mathscr{P}(u + 2\omega_\alpha) = \mathscr{P}(u). \tag{6.6.7}$$

$\mathscr{P}(u)$ is an even function.

Both these functions can be expressed in terms of the theta function $\theta_1(z|\tau)$. Thus, substitution from equation (6.2.9) into equation (6.6.1) leads to the result

$$\zeta(u) = \frac{\eta_1}{\omega_1} u + \frac{\pi}{2\omega_1} \frac{d}{dz} \{\ln \theta_1(z)\}, \tag{6.6.8}$$

where $z = \pi u/2\omega_1$. It now follows that

$$\mathscr{P}(u) = \left(\frac{\pi}{2\omega_1}\right)^2 \left[\frac{1}{3}\frac{\theta_1'''(0)}{\theta_1'(0)} - \frac{d^2}{dz^2}\{\ln\theta_1(z)\}\right], \qquad (6.6.9)$$

using equation (6.2.7).

Since $\theta_1(z)$ is odd, it is expressible in a Maclaurin series containing odd powers of z only, thus:

$$\theta_1(z) = z\theta_1'(0) + \tfrac{1}{6}z^3\theta_1'''(0) + \cdots. \qquad (6.6.10)$$

The series must converge for all values of z, since $\theta_1(z)$ has no singularities. Differentiating, we get

$$\theta_1'(z) = \theta_1'(0) + \tfrac{1}{2}z^2\theta_1'''(0) + \cdots. \qquad (6.6.11)$$

It now follows that, for sufficiently small z,

$$\theta_1'(z)/\theta_1(z) = \frac{1}{z} + \frac{\theta_1'''(0)}{3\theta_1'(0)}z + O(z^3) \qquad (6.6.12)$$

and, hence, by equation (6.6.9),

$$\mathscr{P}(u) = \frac{1}{u^2} + O(u^2). \qquad (6.6.13)$$

This last result demonstrates that $\mathscr{P}(u)$ possesses a double pole at the origin with zero residue. The periodicity of $\mathscr{P}(u)$ requires that this double pole should be repeated at all the points congruent to the origin, viz. where

$$u = 2m\omega_1 + 2n\omega_3. \qquad (6.6.14)$$

Elsewhere, $\mathscr{P}(u)$ is regular.

By differentiating equation (6.2.24) with respect to u, we find that

$$\sigma'(\lambda u, \lambda\omega_1, \lambda\omega_3) = \sigma'(u, \omega_1, \omega_3). \qquad (6.6.15)$$

It follows from equation (6.6.1) that

$$\zeta(\lambda u, \lambda\omega_1, \lambda\omega_3) = \lambda^{-1}\zeta(u, \omega_1, \omega_3). \qquad (6.6.16)$$

Then, differentiating the last equation with respect to u, we find that

$$\mathscr{P}(\lambda u, \lambda\omega_1, \lambda\omega_3) = \lambda^{-2}\mathscr{P}(u, \omega_1, \omega_3). \qquad (6.6.17)$$

In particular, if $\lambda = 1/\omega_1$, we deduce that

$$\omega_1\zeta(u, \omega_1, \omega_3) = \zeta(u/\omega_1, 1, \tau), \qquad (6.6.18)$$

$$\omega_1^2\mathscr{P}(u, \omega_1, \omega_3) = \mathscr{P}(u/\omega_1, 1, \tau). \qquad (6.6.19)$$

These indicate that $\omega_1\zeta(u, \omega_1, \omega_3)$ and $\omega_1^2\mathscr{P}(u, \omega_1, \omega_3)$ can be tabulated against u/ω_1 and τ. This has been done for the zeta function in Table G (p. 302) with $u/\omega_1 = 0.1(0.1)1.9$ and $\kappa = -i\tau = 0.2(0.2)2$. The corresponding table for the Weierstrass function is Table H (p. 303) with $u/\omega_1 = 0.02(0.02)1$,

$\kappa = -i\tau = 0.1(0.1)2$. Both functions take real values in these cases. The periodicity of these functions and their oddness or evenness permit an immediate extension of these tables to all real values of u.

Supposing ω_1 to be real, as $\tau \to +i\infty$, $q \to 0$ and we can substitute from equation (2.1.14) into equation (6.6.9) to derive the limiting form of $\mathscr{P}(u)$ as $2\omega_3$ becomes infinite along the positive imaginary axis. We calculate that

$$\omega_1^2 \mathscr{P}(u, \omega_1, +i\infty) = \tfrac{1}{4}\pi^2(\operatorname{cosec}^2 z - \tfrac{1}{3}) \tag{6.6.20}$$

where $z = \pi u/2\omega_1$.

For any values of ω_1, ω_3, we shall denoted $\mathscr{P}(\omega_\alpha)$ by e_α. Putting $u = \omega_1$ in equation (6.6.20), then $z = \tfrac{1}{2}\pi$ and then $\omega_1^2 e_1 = \pi^2/6$. Putting $u = \omega_2$ or ω_3, the imaginary parts of u and z become infinite and, hence, $\operatorname{cosec} z = 0$; we conclude that $\omega_1^2 e_2 = \omega_1^2 e_3 = -\pi^2/12$ for this limiting case.

6.7. Differential Equation Satisfied by $\mathscr{P}(u)$

Differentiating the identity (6.4.6) partially with respect to x, we obtain the equation

$$\sigma(u+v)\sigma(u-v)\{\sigma'(x+y)\sigma(x-y) + \sigma(x+y)\sigma'(x-y)\}$$
$$+ \sigma(u+y)\sigma(u-y)\{\sigma'(v+x)\sigma(v-x) - \sigma(v+x)\sigma'(v-x)\}$$
$$+ \sigma(v+y)\sigma(v-y)\{\sigma'(x+u)\sigma(x-u) + \sigma(x+u)\sigma'(x-u)\} = 0. \tag{6.7.1}$$

Putting $y = x$ and recalling that $\sigma(0) = 0$, $\sigma'(0) = 1$, the last equation reduces to

$$\sigma(u+v)\sigma(u-v)\sigma(2x) + \sigma(v+x)\sigma(v-x)\{\sigma'(x+u)\sigma(x-u) + \sigma(x+u)\sigma'(x-u)\}$$
$$+ \sigma(u+x)\sigma(u-x)\{\sigma'(v+x)\sigma(v-x) - \sigma(v+x)\sigma'(v-x)\} = 0. \tag{6.7.2}$$

Dividing through by $\sigma(2x)\sigma(u+x)\sigma(u-x)\sigma(v+x)\sigma(v-x)$ (assumed non-vanishing), the last identity can be written in the form

$$\frac{\sigma(u+v)\sigma(u-v)}{\sigma(u+x)\sigma(u-x)\sigma(v+x)\sigma(v-x)}$$
$$= \frac{1}{\sigma(2x)}\{\zeta(u+x) - \zeta(u-x) - \zeta(v+x) + \zeta(v-x)\}. \tag{6.7.3}$$

We now let $x \to 0$ on both sides of this identity, noting that, by Taylor's theorem,

$$\left.\begin{array}{r}
\zeta(u+x) = \zeta(u) - x\mathscr{P}(u) + O(x^2), \\
\zeta(u-x) = \zeta(u) + x\mathscr{P}(u) + O(x^2), \\
\zeta(v+x) = \zeta(v) - x\mathscr{P}(v) + O(x^2), \\
\zeta(v-x) = \zeta(v) + x\mathscr{P}(v) + O(x^2), \\
\sigma(2x) = 2x + O(x^5).
\end{array}\right\} \tag{6.7.4}$$

(Note: The last equation follows since $\sigma(u)$ is odd, $\sigma'(0) = 1$, and $\sigma'''(0) = 0$.)
Thus, in the limit, the identity becomes

$$\frac{\sigma(u+v)\sigma(u-v)}{\sigma^2(u)\sigma^2(v)} = \mathscr{P}(v) - \mathscr{P}(u). \tag{6.7.5}$$

Now, in the limit as $v \to u$,

$$\frac{\mathscr{P}(v) - \mathscr{P}(u)}{v - u} \to \mathscr{P}'(u), \tag{6.7.6}$$

$$\frac{\sigma(v-u)}{v-u} \to \sigma'(0) = 1. \tag{6.7.7}$$

Hence, dividing both sides of (6.7.5) by $(v - u)$ and letting $v \to u$, we find in the limit that

$$\mathscr{P}'(u) = -\sigma(2u)/\sigma^4(u). \tag{6.7.8}$$

Putting $v = \omega_1$ in equation (6.7.5) and making double use of the identity (6.2.18), we next prove that

$$\mathscr{P}(u) - \mathscr{P}(\omega_1) = \sigma_1^2(u)/\sigma^2(u). \tag{6.7.9}$$

Thus, writing $\mathscr{P}(\omega_1) = e_1$, we have found that

$$\{\mathscr{P}(u) - e_1\}^{1/2} = +\sigma_1(u)/\sigma(u). \tag{6.7.10}$$

(Note: $\{\mathscr{P}(u) - e_1\}^{1/2}$ will always be taken with the positive sign shown.)
Clearly, the square root of $\mathscr{P}(u) - e_1$ has no branch point—like $\sqrt{z^2}$ (and unlike $\sqrt{z}$), it separates into an independent pair of single-valued analytic functions.

Similarly, writing $\mathscr{P}(\omega_2) = e_2$, $\mathscr{P}(\omega_3) = e_3$, we can prove that

$$\{\mathscr{P}(u) - e_2\}^{1/2} = \sigma_2(u)/\sigma(u), \tag{6.7.11}$$

$$\{\mathscr{P}(u) - e_3\}^{1/2} = \sigma_3(u)/\sigma(u). \tag{6.7.12}$$

Multiplication of the results (6.7.10)–(6.7.12) now yields

$$\sqrt{\{(\mathscr{P} - e_1)(\mathscr{P} - e_2)(\mathscr{P} - e_3)\}} = \sigma_1(u)\sigma_2(u)\sigma_3(u)/\sigma^3(u), \tag{6.7.13}$$

where the sign of the root is to be chosen appropriately. Putting $v = u$ in any one of the identities (6.4.4), we deduce that

$$\sigma(2u) = 2\sigma(u)\sigma_1(u)\sigma_2(u)\sigma_3(u). \tag{6.7.14}$$

It now follows that

$$\sqrt{\{(\mathscr{P} - e_1)(\mathscr{P} - e_2)(\mathscr{P} - e_3)\}} = \sigma(2u)/2\sigma^4(u). \tag{6.7.15}$$

Referring to equation (6.7.8), we see that we have proved that $\mathscr{P}(u)$ satisfies the first-order differential equation

$$\mathscr{P}'(u) = -2\sqrt{\{(\mathscr{P}(u) - e_1)(\mathscr{P}(u) - e_2)(\mathscr{P}(u) - e_3)\}}. \tag{6.7.16}$$

$\mathscr{P}'(u)$ is an odd elliptic function with periods $2\omega_1$ and $2\omega_3$. Hence,

$$\mathscr{P}'(\omega_1) = -\mathscr{P}'(-\omega_1) = -\mathscr{P}'(-\omega_1 + 2\omega_1) = -\mathscr{P}'(\omega_1). \qquad (6.7.17)$$

Thus $\mathscr{P}'(u)$ vanishes at $u = \omega_1$. Similarly, it vanishes at $u = \omega_2$, $u = \omega_3$. These results also follow from equation (6.7.8) or (6.7.16). We have accordingly identified e_1, e_2, e_3 as stationary values of $\mathscr{P}(u)$.

Squaring the equation (6.7.16), we find

$$\{\mathscr{P}'(u)\}^2 = 4\mathscr{P}^3(u) - g_1\mathscr{P}^2(u) - g_2\mathscr{P}(u) - g_3, \qquad (6.7.18)$$

where

$$g_1 = 4(e_1 + e_2 + e_3), \qquad g_2 = -4(e_2 e_3 + e_3 e_1 + e_1 e_2), \qquad g_3 = 4e_1 e_2 e_3. \qquad (6.7.19)$$

Now, equation (6.6.13) shows that $\mathscr{P}(u) - 1/u^2$ is regular and $O(u^2)$ in a neighborhood of $u = 0$ and so has a Taylor expansion

$$\mathscr{P}(u) - \frac{1}{u^2} = au^2 + bu^4 + \cdots \qquad (6.7.20)$$

(there are no odd powers, since $\mathscr{P}(u)$ is even). Thus,

$$\mathscr{P}'(u) = -\frac{2}{u^3} + 2au + 4bu^3 + \cdots \qquad (6.7.21)$$

and, substituting in equation (6.7.18), we obtain

$$\frac{4}{u^6}\{1 - 2au^4 - 4bu^6 + \cdots\} = \frac{4}{u^6}\{1 + 3au^4 + 3bu^6 + \cdots\}$$
$$- \frac{g_1}{u^4}\{1 + 2au^4 + 2bu^6 + \cdots\}$$
$$- \frac{g_2}{u^2}\{1 + au^4 + bu^6 + \cdots\} - g_3. \qquad (6.7.22)$$

Equating coefficients of like powers of u on the two sides of this equation, we deduce that

$$g_1 = 0, \qquad a = \frac{1}{20}g_2, \qquad b = \frac{1}{28}g_3. \qquad (6.7.23)$$

Thus, we have proved that

$$e_1 + e_2 + e_3 = \mathscr{P}(\omega_1) + \mathscr{P}(\omega_2) + \mathscr{P}(\omega_3) = 0, \qquad (6.7.24)$$

$$\mathscr{P}(u) = \frac{1}{u^2} + \frac{1}{20}g_2 u^2 + \frac{1}{28}g_3 u^4 + O(u^6), \qquad (6.7.25)$$

and

$$\{\mathscr{P}'(u)\}^2 = 4\mathscr{P}^3(u) - g_2\mathscr{P}(u) - g_3. \qquad (6.7.26)$$

e_1, e_2, e_3 are therefore the roots of the cubic equation

$$4s^3 - g_2 s - g_3 = 0. \tag{6.7.27}$$

g_2, g_3 are called the *invariants* of $\mathscr{P}$.

The vanishing of $e_1 + e_2 + e_3$ can also be proved directly, by calculation of $\mathscr{P}(\omega_\alpha)$ for $\alpha = 1, 2, 3$.

Thus, equation (6.6.9) shows that

$$\mathscr{P}(u) = \left(\frac{\pi}{2\omega_1}\right)^2 \left[\frac{1}{3}\frac{\theta_1'''(0)}{\theta_1'(0)} - \frac{\theta_1(z)\theta_1''(z) - \theta_1'^2(z)}{\theta_1^2(z)}\right]. \tag{6.7.28}$$

Putting $u = \omega_1$, $z = \frac{1}{2}\pi$ and referring to equations (1.3.6)–(1.3.9), we calculate that

$$e_1 = \left(\frac{\pi}{2\omega_1}\right)^2 \left[\frac{1}{3}\frac{\theta_1'''(0)}{\theta_1'(0)} - \frac{\theta_2''(0)}{\theta_2(0)}\right]. \tag{6.7.29}$$

Similarly, putting $u = \omega_2 = -\omega_1 - \omega_3$, $z = -\frac{1}{2}\pi - \frac{1}{2}\pi\tau$, we find that

$$e_2 = \left(\frac{\pi}{2\omega_1}\right)^2 \left[\frac{1}{3}\frac{\theta_1'''(0)}{\theta_1'(0)} - \frac{\theta_3''(0)}{\theta_3(0)}\right] \tag{6.7.30}$$

and putting $u = \omega_3$, $z = \frac{1}{2}\pi\tau$, we get

$$e_3 = \left(\frac{\pi}{2\omega_1}\right)^2 \left[\frac{1}{3}\frac{\theta_1'''(0)}{\theta_1'(0)} - \frac{\theta_4''(0)}{\theta_4(0)}\right]. \tag{6.7.31}$$

Addition of these expressions for e_1, e_2, e_3 now leads to the required result by virtue of the theta function identity

$$\theta_1'''(0)/\theta_1'(0) = \theta_2''(0)/\theta_2(0) + \theta_3''(0)/\theta_3(0) + \theta_4''(0)/\theta_4(0) \tag{6.7.32}$$

(see Exercises 11, 12, 13 in Chapter 1 for an outline proof of this result). Equations (6.7.29)–(6.7.31) show that the e_α are all real when ω_1 is real and ω_3 is purely imaginary.

When evaluating elliptic integrals by means of the Weierstrass function, it is more convenient to regard it as being dependent upon the invariants g_2, g_3 instead of the half-periods ω_1, ω_3. To distinguish the two points of view, we shall adopt the notation

$$\mathscr{P}(u, \omega_1, \omega_3) = \mathscr{P}(u|g_2, g_3). \tag{6.7.33}$$

Noting from equations (6.7.29)–(6.7.31) that

$$e_\alpha(\lambda\omega_1, \lambda\omega_3) = \lambda^{-2} e_\alpha(\omega_1, \omega_3), \tag{6.7.34}$$

we deduce that

$$g_2(\lambda\omega_1, \lambda\omega_3) = \lambda^{-4} g_2(\omega_1, \omega_3), \qquad g_3(\lambda\omega_1, \lambda\omega_3) = \lambda^{-6} g_3(\omega_1, \omega_3). \tag{6.7.35}$$

Thus, putting $\lambda = 1/\omega_1$, we find that

$$\omega_1^2 e_\alpha(\omega_1, \omega_3) = e_\alpha(1, \tau), \qquad \omega_1^4 g_2(\omega_1, \omega_3) = g_2(1, \tau), \qquad \omega_1^6 g_3(\omega_1, \omega_3) = g_3(1, \tau). \tag{6.7.36}$$

These results show that $\omega_1^2 e_\alpha, \omega_1^4 g_2, \omega_1^6 g_3$ can be tabulated as functions of τ. Table K (p. 315) lists these quantities for the cases $\tau = i\kappa, \frac{1}{2}(1 + i\nu)$, when the invariants are real ($\kappa = 0.1(0.1)2, \nu = 0.2(0.2)4$).

Equations (6.6.17) and (6.7.35) together imply that

$$\mathcal{P}(\lambda u | \lambda^{-4} g_2, \lambda^{-6} g_3) = \lambda^{-2} \mathcal{P}(u | g_2, g_3). \tag{6.7.37}$$

The limiting values of the e_α as $\kappa = -i\tau \to +\infty$ ($q \to 0$) are easily found by substitution from equations (2.1.14)–(2.1.17) into equations (6.7.29)–(6.7.31). We find, as at the end of the previous section, that

$$\omega_1^2 e_1 = \pi^2/6, \qquad \omega_1^2 e_2 = \omega_1^2 e_3 = -\pi^2/12, \tag{6.7.38}$$

and then it follows that

$$\omega_1^4 g_2 = \pi^4/12, \qquad \omega_1^6 g_3 = \pi^6/216. \tag{6.7.39}$$

6.8. Addition Theorem for $\mathcal{P}(u)$

Taking the logarithmic derivative of both sides of the identity (6.7.5) with respect to u, we obtain the equation

$$\frac{\sigma'(u+v)}{\sigma(u+v)} + \frac{\sigma'(u-v)}{\sigma(u-v)} - 2\frac{\sigma'(u)}{\sigma(u)} = \frac{\mathcal{P}'(u)}{\mathcal{P}(u) - \mathcal{P}(v)}, \tag{6.8.1}$$

or

$$\zeta(u+v) + \zeta(u-v) - 2\zeta(u) = \frac{\mathcal{P}'(u)}{\mathcal{P}(u) - \mathcal{P}(v)}. \tag{6.8.2}$$

Exchanging u and v, we also have

$$\zeta(u+v) - \zeta(u-v) - 2\zeta(v) = \frac{\mathcal{P}'(v)}{\mathcal{P}(v) - \mathcal{P}(u)}. \tag{6.8.3}$$

Addition of the last two equations now gives

$$\zeta(u+v) = \zeta(u) + \zeta(v) + \frac{1}{2}\frac{\mathcal{P}'(u) - \mathcal{P}'(v)}{\mathcal{P}(u) - \mathcal{P}(v)}. \tag{6.8.4}$$

Differentiating the last equation with respect to u, we find

$$\mathcal{P}(u+v) = \mathcal{P}(u) - \frac{1}{2}\frac{\mathcal{P}''(u)}{\mathcal{P}(u) - \mathcal{P}(v)} + \frac{1}{2}\frac{\{\mathcal{P}'(u) - \mathcal{P}'(v)\}\mathcal{P}'(u)}{\{\mathcal{P}(u) - \mathcal{P}(v)\}^2}. \tag{6.8.5}$$

Exchanging u and v in the last result gives

$$\mathcal{P}(u+v) = \mathcal{P}(v) + \frac{1}{2}\frac{\mathcal{P}''(v)}{\mathcal{P}(u) - \mathcal{P}(v)} - \frac{1}{2}\frac{\{\mathcal{P}'(u) - \mathcal{P}'(v)\}\mathcal{P}'(v)}{\{\mathcal{P}(u) - \mathcal{P}(v)\}^2}. \tag{6.8.6}$$

Adding the last two equations, we obtain

$$2\mathcal{P}(u+v) = \mathcal{P}(u) + \mathcal{P}(v) - \frac{1}{2}\frac{\mathcal{P}''(u) - \mathcal{P}''(v)}{\mathcal{P}(u) - \mathcal{P}(v)} + \frac{1}{2}\left[\frac{\mathcal{P}'(u) - \mathcal{P}'(v)}{\mathcal{P}(u) - \mathcal{P}(v)}\right]^2. \tag{6.8.7}$$

Differentiating equation (6.7.26), we find that

$$\mathscr{P}''(u) = 6\mathscr{P}^2(u) - \tfrac{1}{2}g_2. \tag{6.8.8}$$

It follows that

$$\mathscr{P}''(u) - \mathscr{P}''(v) = 6\{\mathscr{P}^2(u) - \mathscr{P}^2(v)\}. \tag{6.8.9}$$

Thus, the identity (6.8.7) can be reduced to the form

$$\mathscr{P}(u+v) = \frac{1}{4}\left[\frac{\mathscr{P}'(u) - \mathscr{P}'(v)}{\mathscr{P}(u) - \mathscr{P}(v)}\right]^2 - \mathscr{P}(u) - \mathscr{P}(v). \tag{6.8.10}$$

Since the derivatives can be expressed algebraically in terms of the Weierstrass function by equation (6.7.26), this is a true addition theorem for $\mathscr{P}(u)$.

As an application of this result, put $v = \omega_1$. Then, $\mathscr{P}'(\omega_1) = 0$, $\mathscr{P}(\omega_1) = e_1$, and we obtain

$$\mathscr{P}(u + \omega_1) = \frac{\{\mathscr{P}'(u)\}^2}{4\{\mathscr{P}(u) - e_1\}^2} - \mathscr{P}(u) - e_1$$

$$= \frac{\{\mathscr{P}(u) - e_2\}\{\mathscr{P}(u) - e_3\}}{\mathscr{P}(u) - e_1} - \mathscr{P}(u) - e_1$$

$$= \frac{e_1^2 + e_2 e_3 - (e_2 + e_3)\mathscr{P}(u)}{\mathscr{P}(u) - e_1}$$

$$= e_1 + \frac{(e_1 - e_2)(e_1 - e_3)}{\mathscr{P}(u) - e_1}, \tag{6.8.11}$$

since $e_1 + e_2 + e_3 = 0$. Similarly, we can prove

$$\mathscr{P}(u + \omega_2) = e_2 + \frac{(e_2 - e_3)(e_2 - e_1)}{\mathscr{P}(u) - e_2}, \tag{6.8.12}$$

$$\mathscr{P}(u + \omega_3) = e_3 + \frac{(e_3 - e_1)(e_3 - e_2)}{\mathscr{P}(u) - e_3}. \tag{6.8.13}$$

6.9. Relationship between Jacobi's and Weierstrass's Functions

We have proved at (6.7.10)–(6.7.12) that

$$\{\mathscr{P}(u) - e_\alpha\}^{1/2} = \sigma_\alpha(u)/\sigma(u), \qquad \alpha = 1, 2, 3. \tag{6.9.1}$$

With $\alpha = 3$ and $u = \omega_1$, this identity leads to the result

$$(e_1 - e_3)^{1/2} = \frac{\pi}{2\omega_1} \frac{\theta_1'(0)\theta_4(\tfrac{1}{2}\pi)}{\theta_1(\tfrac{1}{2}\pi)\theta_4(0)} = \frac{\pi}{2\omega_1}\theta_3^2(0), \tag{6.9.2}$$

where we have used equations (6.2.9) and (6.2.21). Similarly, taking $\alpha = 3$, $u = \omega_2$ and then $\alpha = 2$, $u = \omega_1$, we obtain the results

$$(e_2 - e_3)^{1/2} = -\frac{\pi}{2\omega_1} \theta_2^2(0), \qquad (e_1 - e_2)^{1/2} = \frac{\pi}{2\omega_1} \theta_4^2(0). \qquad (6.9.3)$$

We conclude from these results that, if ω_1 is real and ω_3 is imaginary, then $\theta_2(0)$, $\theta_3(0)$, $\theta_4(0)$ are real and, hence, that

$$e_1 > e_2 > e_3. \qquad (6.9.4)$$

Expressions for the stationary values e_α in terms of $\theta_2(0)$, $\theta_3(0)$, and $\theta_4(0)$ now follow immediately from equations (6.9.2), (6.9.3) (see Exercise 4).

If we now take

$$\omega_1 = K, \qquad \omega_3 = iK' \qquad (6.9.5)$$

as at equations (6.3.1), then equations (6.3.3) and (6.3.4) are valid, and using equations (6.9.1), we calculate that

$$\operatorname{sn} u = [\mathscr{P}(u) - e_3]^{-1/2}, \qquad (6.9.6)$$

$$\operatorname{cn} u = \left[\frac{\mathscr{P}(u) - e_1}{\mathscr{P}(u) - e_3}\right]^{1/2}, \qquad (6.9.7)$$

$$\operatorname{dn} u = \left[\frac{\mathscr{P}(u) - e_2}{\mathscr{P}(u) - e_3}\right]^{1/2}. \qquad (6.9.8)$$

Observe that $2\omega_1$ and $2\omega_3$ are the periods of $\mathscr{P}(u)$, but are only the half-periods of the set of Jacobi functions. This discrepancy is allowed for by the sign ambiguity associated with the square roots occurring in the equations (6.9.6)–(6.9.8). Thus, $\operatorname{sn}(u + 2\omega_1) = \operatorname{sn}(u + 2K) = -\operatorname{sn} u$, the difference in sign between $\operatorname{sn}(u + 2\omega_1)$ and $\operatorname{sn} u$ being covered in equation (6.9.6) by the two values of the right-hand member.

Since $\omega_1 = K = \frac{1}{2}\pi\theta_3^2(0)$, equations (6.9.2) and (6.9.3) now reduce to the results

$$(e_1 - e_3)^{1/2} = 1, \qquad (e_2 - e_3)^{1/2} = -k, \qquad (e_1 - e_2)^{1/2} = k', \qquad (6.9.9)$$

expressing the Jacobian modulus and complementary modulus in terms of the e_α. Alternatively, since $e_1 + e_2 + e_3 = 0$, we can solve for the e_α in terms of modulus to obtain the results

$$e_1 = \tfrac{1}{3}(2 - k^2), \qquad e_2 = \tfrac{1}{3}(2k^2 - 1), \qquad e_3 = -\tfrac{1}{3}(1 + k^2). \qquad (6.9.10)$$

Combining equations (6.8.11)–(6.8.13) with equations (6.9.6)–(6.9.8) and using the last set of results, we can also show that

$$\left.\begin{aligned}
\mathscr{P}(u) &= -\tfrac{1}{3}(1 + k^2) + \operatorname{ns}^2 u, \\
\mathscr{P}(u + \omega_1) &= \tfrac{1}{3}(2 - k^2) + k'^2 \operatorname{sc}^2 u, \\
\mathscr{P}(u + \omega_2) &= \tfrac{1}{3}(2k^2 - 1) - k^2 k'^2 \operatorname{sd}^2 u, \\
\mathscr{P}(u + \omega_3) &= -\tfrac{1}{3}(1 + k^2) + k^2 \operatorname{sn}^2 u.
\end{aligned}\right\} \qquad (6.9.11)$$

6.10. Jacobi's Transformation of the Weierstrass Function

The half-periods ω_1, ω_3 have been ordered to ensure that $\tau = \omega_3/\omega_1$ has a positive imaginary part. However, $-\omega_3, \omega_1$ are also half-periods of $\mathscr{P}$ and their ratio, taken in that order, is $\tau' = -\omega_1/\omega_3 = -1/\tau$, which will also have positive imaginary part. This suggests that we study the doubly periodic function $\mathscr{P}(u, -\omega_3, \omega_1)$ and its relationship to $\mathscr{P}(u, \omega_1, \omega_3)$. Period parallelograms with a vertex at $u = 0$ have been drawn in Fig. 6.1 for these two functions. Thus, the lattices of period parallelograms for the two functions are identical and our expectation, therefore, is that the functions are the same.

Both these functions have a double pole at $u = 0$, with principal part $1/u^2$, and their only other poles are at points congruent to this point. Thus, their difference $\mathscr{P}(u, \omega_1, \omega_3) - \mathscr{P}(u, -\omega_3, \omega_1)$ is a doubly periodic function having no singularities and it is proved in section 8.2 that such a function is constant. Thus, the two functions can differ by a constant at most. In fact, in the neighborhood of $u = 0$, equation (6.6.13) shows that the difference is $O(u^2)$ and only a zero constant has this property. We conclude that

$$\mathscr{P}(u, -\omega_3, \omega_1) = \mathscr{P}(u, \omega_1, \omega_3). \tag{6.10.1}$$

However, we note that $\tau\tau' = -1$ and, hence, $\mathscr{P}(u, -\omega_3, \omega_1)$ is obtained from $\mathscr{P}(u, \omega_1, \omega_3)$ by a Jacobi transformation (section 1.7) and the identity (6.10.1) can therefore be established by application of this transformation. This we proceed to do:

Equation (6.2.1) shows that

$$\sigma(u, -\omega_3, \omega_1) = A' e^{a'u^2} \theta_1(z'|\tau'), \tag{6.10.2}$$

where

$$z' = -\pi u/2\omega_3 = -\pi u/(2\tau\omega_1) = -z/\tau, \tag{6.10.3}$$

and the constants A', a' are given by equations (6.2.2), (6.2.3) (suitably

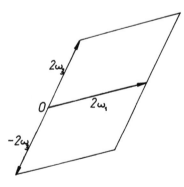

Figure 6.1. Period parallelograms for $\mathscr{P}(u, \omega_1, \omega_3)$ and $\mathscr{P}(u, -\omega_3, \omega_1)$.

amended), in the forms

$$A' = -\frac{2\omega_3}{\pi\theta'_1(0|\tau')}, \qquad a' = -\frac{\pi^2}{24\omega_3^2}\cdot\frac{\theta'''_1(0|\tau')}{\theta'_1(0|\tau')}. \tag{6.10.4}$$

Differentiating the identity (1.7.14) first once and then three times with respect to z and putting $z = 0$ on each occasion, we next show that

$$\theta'_1(0|\tau') = -i(-i\tau)^{1/2}\tau\theta'_1(0|\tau), \tag{6.10.5}$$

$$\theta'''_1(0|\tau') = -i\tau^2(-i\tau)^{1/2}\left[\frac{6i}{\pi}\theta'_1(0|\tau) + \tau\theta'''_1(0|\tau)\right]. \tag{6.10.6}$$

It now follows that

$$A' = -i(-i\tau)^{-1/2}A, \qquad a' = a - \frac{i\pi}{4\tau\omega_1^2}. \tag{6.10.7}$$

Also, replacing z by $z' = -z/\tau$ in equation (1.7.14), we find that

$$\theta_1(z'|\tau') = i(-i\tau)^{1/2}e^{iz^2/\pi\tau}\theta_1(z|\tau). \tag{6.10.8}$$

Substituting from equations (6.10.7) and (6.10.8) into equation (6.10.2), we conclude that

$$\sigma(u, -\omega_3, \omega_1) = Ae^{au^2}\theta_1(z|\tau) = \sigma(u, \omega_1, \omega_3). \tag{6.10.9}$$

This is Jacobi's transformation for the sigma function.

Taking logarithms of both sides of this identity and differentiating with respect to u, we now prove that

$$\zeta(u, -\omega_3, \omega_1) = \zeta(u, \omega_1, \omega_3) \tag{6.10.10}$$

and a further differentiation leads to the result (6.10.1).

The identities (6.6.17) and (6.10.1) together now give

$$\mathscr{P}(u, -i\omega_3, i\omega_1) = i^{-2}\mathscr{P}(-iu, -\omega_3, \omega_1) = -\mathscr{P}(-iu, \omega_1, \omega_3) = -\mathscr{P}(iu, \omega_1, \omega_3). \tag{6.10.11}$$

Let us consider in more detail the elliptic function $\mathscr{P}(u, -i\omega_3, i\omega_1)$. If e'_1, e'_2, e'_3 represent its stationary values at $u = -i\omega_3, i\omega_3 - i\omega_1$, and $i\omega_1$ respectively, then

$$\left.\begin{array}{l}
e'_1 = \mathscr{P}(-i\omega_3, -i\omega_3, i\omega_1) = -\mathscr{P}(\omega_3, \omega_1, \omega_3) = -e_3, \\[2pt]
e'_2 = \mathscr{P}(i\omega_3 - i\omega_1, -i\omega_3, i\omega_1) = -\mathscr{P}(\omega_1 - \omega_3, \omega_1, \omega_3), \\[2pt]
\quad = -\mathscr{P}(-\omega_1 - \omega_3, \omega_1, \omega_3) = -e_2, \\[2pt]
e'_3 = \mathscr{P}(i\omega_1, -i\omega_3, i\omega_1) = -\mathscr{P}(-\omega_1, \omega_1, \omega_3) = -\mathscr{P}(\omega_1, \omega_1, \omega_3) = -e_1.
\end{array}\right\} \tag{6.10.12}$$

Thus, if g_2, g_3 are the invariants of $\mathscr{P}(u, \omega_1, \omega_3)$, then the invariants of $\mathscr{P}(u, -i\omega_3, i\omega_1)$ are

$$g'_2 = g_2, \qquad g'_3 = -g_3, \tag{6.10.13}$$

and we can therefore write the identity (6.10.11) in the form

$$\mathscr{P}(u|g_2, -g_3) = -\mathscr{P}(iu|g_2, g_3). \tag{6.10.14}$$

The limiting form of $\mathscr{P}(u, \omega_1, \omega_3)$ when ω_1 is positive and $\omega_3 \to +i\infty$ $(q \to 0)$ has been calculated at (6.6.20). Using this result and the identity (6.10.11), we can now obtain the limiting form of $\mathscr{P}(u, \omega_1, \omega_3)$ when ω_3 is positive imaginary, as $\omega_1 \to +\infty$ $(q \to 1)$ thus:

$$\mathscr{P}(u, \infty, \omega_3) = \lim_{\omega_1 \to \infty} \mathscr{P}(u, \omega_1, \omega_3) = - \lim_{\omega_1 \to \infty} \mathscr{P}(iu, -i\omega_3, i\omega_1)$$

$$= -\left(\frac{\pi}{2i\omega_3}\right)^2 \left[\operatorname{cosec}^2\left(-\frac{\pi u}{2\omega_3}\right) - \frac{1}{3}\right]$$

$$= \left(\frac{\pi}{2\omega_3}\right)^2 \left[\operatorname{cosec}^2\left(\frac{\pi u}{2\omega_3}\right) - \frac{1}{3}\right]. \tag{6.10.15}$$

6.11. Weierstrass Function with Real and Purely Imaginary Periods

If ω_1 is real and ω_3 is purely imaginary, the period parallelograms of $\mathscr{P}$ are rectangles, each of which can be divided into four smaller rectangles by the half-period lines as indicated in Fig. 6.2. The values taken by the function over the four quarters of a period rectangle are closely related. For consider the central cell $ABCD$. If $\mathscr{P}$ denotes the function's value at a point u in the

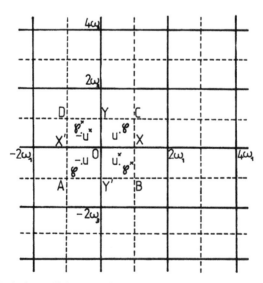

Figure 6.2. Period parallelograms for $\mathscr{P}$ with ω_1 real and ω_3 purely imaginary.

quarter lying in the first quadrant of the complex plane, since the function is even it will take the same value at the point $-u$ in the third quadrant. Also, since the function takes real values on the real axis, its value at the conjugate point u^* in the fourth quadrant will be $\mathscr{P}^*$. It then follows that the function's value at the point $-u^*$ in the second quadrant is also $\mathscr{P}^*$. To summarize, the function's values over the first half-period rectangle are repeated over the third half-period rectangle; conjugate values are taken over the remaining two half-period rectangles.

On the boundary of a half-period rectangle, $\mathscr{P}(u)$ is real. For consider the rectangle $OXCY$. For values of u in the interval $(0, \omega_1)$, $\sigma(2u)$ is positive and hence, by equation (6.7.8), $\mathscr{P}'(u)$ is negative. Thus, as u increases from 0 to $\omega_1, \mathscr{P}(u)$ decreases monotonically from $+\infty$, at its pole O, to e_1 at X.

On OY, u is purely imaginary. Putting $u = iv$ (v real), the identity (6.10.11) shows that $\mathscr{P}(iv, \omega_1, \omega_3) = -\mathscr{P}(v, -i\omega_3, i\omega_1)$. But $-i\omega_3$ is real and positive, and $i\omega_1$ is purely imaginary, so that $\mathscr{P}(v, -i\omega_3, i\omega_1)$ is real. Further, as v increases from 0 to $-i\omega_3$, $\mathscr{P}(v, -i\omega_3, i\omega_1)$ must decrease from $+\infty$ to $e_1' = -e_3$ (equations (6.10.12)). Thus, as u varies from 0 at O to ω_3 at $Y, \mathscr{P}(u, \omega_1, \omega_3)$ increases from $-\infty$ to e_3.

On $XC, u = \omega_1 + iv$ (v real). As v increases from 0 to $-i\omega_3$, we have just proved that $\mathscr{P}(iv, \omega_1, \omega_3)$ increases from $-\infty$ to e_3. It accordingly follows from the identity (6.8.11) that $\mathscr{P}(u, \omega_1, \omega_3)$ decreases from e_1 to e_2 as u varies from ω_1 at X to $\omega_1 + \omega_3$ at C.

On $YC, u = \omega_3 + w$ (w real). As w increases from 0 to ω_1, we have already proved that $\mathscr{P}(w, \omega_1, \omega_3)$ decreases from $+\infty$ to e_1. Hence, using the identity (6.8.13), we deduce that $\mathscr{P}(u, \omega_1, \omega_3)$ increases from e_3 to e_2 as u varies from ω_3 at Y to $\omega_1 + \omega_3$ at C (see Table I, p. 307).

To summarize, if u is taken around the rectangle $OXCY, \mathscr{P}(u)$ first decreases from $+\infty$ to e_1 along OX, then decreases from e_1 to e_2 along XC, then decreases from e_2 to e_3 along CY, and finally decreases from e_3 to $-\infty$ along YO. The same sequence of values is assumed by $\mathscr{P}$ for a circuit $OX'DYO$ of the rectangle in the second quadrant, or for circuits $OX'AY'O$ $OXBY'O$ of the rectangles in the other two quadrants. Any other half-period rectangle can be treated by appeal to congruency.

6.12. Weierstrass's Evaluation of Elliptic Integrals of the First Kind

In this section, it will be assumed that the half-periods ω_1, ω_3 are real and imaginary respectively. We shall write

$$\mathscr{P}(u + \omega_\alpha) = \mathscr{P}_\alpha(u), \qquad \alpha = 1, 2, 3. \tag{6.12.1}$$

Then, as shown in the last section, as u increases from 0 to ω_1, $\mathscr{P}(u)$ takes real values, decreasing monotonically from $+\infty$ to e_1. It follows from the identities (6.8.11)–(6.8.13) that, for the same range of increase of u, the functions $\mathscr{P}_\alpha(u)$ take real values, $\mathscr{P}_1(u)$ increasing from e_1 to $+\infty$, $\mathscr{P}_2(u)$ decreasing from e_2 to e_3, and $\mathscr{P}_3(u)$ increasing from e_3 to e_2.

Since the functions $\mathscr{P}(u), \mathscr{P}_\alpha(u)$ are continuous and monotonic in the interval $(0, \omega_1)$, their inverse functions $\mathscr{P}^{-1}(t), \mathscr{P}_\alpha^{-1}(t)$ all have unique values in the range $(0, \omega_1)$, for $t \geqslant e_1$ in the cases of $\mathscr{P}^{-1}(t), \mathscr{P}_1^{-1}(t)$ and for $e_3 \leqslant t \leqslant e_2$ in the cases of $\mathscr{P}_2^{-1}(t), \mathscr{P}_3^{-1}(t)$. Such a choice of values for the inverse functions will be assumed throughout this section, t being restricted always to the appropriate range.

If $u = \mathscr{P}^{-1}(t)$, then $t = \mathscr{P}(u)$ and, hence, by equation (6.7.16)

$$\frac{dt}{du} = -2\sqrt{\{(t - e_1)(t - e_2)(t - e_3)\}}. \tag{6.12.2}$$

Integrating this equation from (u, t) to $u = 0$, $t = +\infty$, we prove that

$$u = \mathscr{P}^{-1}(t) = \frac{1}{2}\int_t^\infty \{(s - e_1)(s - e_2)(s - e_3)\}^{-1/2}\,ds, \qquad t \geqslant e_1. \tag{6.12.3}$$

If $t = e_1$, then $u = \omega_1$ and, thus,

$$\omega_1 = \frac{1}{2}\int_{e_1}^\infty \{(s - e_1)(s - e_2)(s - e_3)\}^{-1/2}\,ds \tag{6.12.4}$$

is a formula for the real half-period ω_1 in terms of the e_α.

If $u = \mathscr{P}_1^{-1}(t)$, then $t = \mathscr{P}_1(u) = \mathscr{P}(u + \omega_1)$ and we have

$$\frac{dt}{du} = 2\sqrt{\{(t - e_1)(t - e_2)(t - e_3)\}}, \tag{6.12.5}$$

where the positive root must be appropriate, since t increases with u. Integrating the last equation from $u = 0$, $t = e_1$ to (u, t), we show that

$$\mathscr{P}_1^{-1}(t) = \frac{1}{2}\int_{e_1}^t \{(s - e_1)(s - e_2)(s - e_3)\}^{-1/2}\,ds, \qquad t \geqslant e_1. \tag{6.12.6}$$

But, equation (6.8.11) can be written in the form

$$t = e_1 + \frac{(e_1 - e_2)(e_1 - e_3)}{\mathscr{P}(u) - e_1} \tag{6.12.7}$$

and it follows that

$$\mathscr{P}(u) = e_1 + \frac{(e_1 - e_2)(e_1 - e_3)}{t - e_1}. \tag{6.12.8}$$

Thus,

$$\mathscr{P}_1^{-1}(t) = u = \mathscr{P}^{-1}\left[e_1 + \frac{(e_1 - e_2)(e_1 - e_3)}{t - e_1}\right] \tag{6.12.9}$$

and, hence,

$$\mathscr{P}^{-1}\left[e_1 + \frac{(e_1 - e_2)(e_1 - e_3)}{t - e_1}\right] = \frac{1}{2}\int_{e_1}^{t} \{(s - e_1)(s - e_2)(s - e_3)\}^{-1/2}\,ds,$$

$$(6.12.10)$$

provided $t \geqslant e_1$.

Clearly, the results (6.12.3) and (6.12.10) imply (using (6.12.4)) that

$$\mathscr{P}^{-1}(t) + \mathscr{P}^{-1}\left[e_1 + \frac{(e_1 - e_2)(e_1 - e_3)}{t - e_1}\right] = \omega_1. \qquad (6.12.11)$$

Now suppose $u = \mathscr{P}_2^{-1}(t)$, so that $t = \mathscr{P}_2(u) = \mathscr{P}(u + \omega_2)$ and, thus, $e_3 \leqslant t \leqslant e_2$. Using equation (6.7.16), we have

$$\frac{dt}{du} = -2\sqrt{\{(e_1 - t)(e_2 - t)(t - e_3)\}}, \qquad (6.12.12)$$

the negative root being taken, since t decreases with u for $0 \leqslant u \leqslant \omega_1$. Integration from (u, t) to $u = 0$, $t = e_2$ now leads to the result

$$u = \mathscr{P}_2^{-1}(t) = \frac{1}{2}\int_{t}^{e_2} \{(e_1 - s)(e_2 - s)(s - e_3)\}^{-1/2}\,ds, \qquad e_3 \leqslant t \leqslant e_2.$$

$$(6.12.13)$$

Equation (6.8.12) requires that

$$t = e_2 + \frac{(e_2 - e_3)(e_2 - e_1)}{\mathscr{P}(u) - e_2}, \qquad (6.12.14)$$

from which it follows that

$$\mathscr{P}(u) = e_2 + \frac{(e_2 - e_3)(e_2 - e_1)}{t - e_2}. \qquad (6.12.15)$$

Hence,

$$\mathscr{P}^{-1}\left[e_2 + \frac{(e_2 - e_3)(e_2 - e_1)}{t - e_2}\right] = u = \frac{1}{2}\int_{t}^{e_2} \{(e_1 - s)(e_2 - s)(s - e_3)\}^{-1/2}\,ds,$$

$$(6.12.16)$$

where $e_3 \leqslant t \leqslant e_2$.

Similarly, by putting $u = \mathscr{P}_3^{-1}(t)$ and using equation (6.8.13), we prove that

$$\mathscr{P}^{-1}\left[e_3 + \frac{(e_3 - e_1)(e_3 - e_2)}{t - e_3}\right] = \frac{1}{2}\int_{e_3}^{t} \{(e_1 - s)(e_2 - s)(s - e_3)\}^{-1/2}\,ds,$$

$$(6.12.17)$$

provided $e_3 \leqslant t \leqslant e_2$.

Putting $t = e_3$ in (6.12.16) or $t = e_2$ in (6.12.17), the following alternative expression for the real half-period ω_1 is obtained:

$$\omega_1 = \frac{1}{2}\int_{e_3}^{e_2} \{(e_1 - s)(e_2 - s)(s - e_3)\}^{-1/2}\,ds. \qquad (6.12.18)$$

Equations (6.12.16), (6.12.17) together now imply that

$$\mathscr{P}^{-1}\left[e_2+\frac{(e_2-e_3)(e_2-e_1)}{t-e_2}\right]+\mathscr{P}^{-1}\left[e_3+\frac{(e_3-e_1)(e_3-e_2)}{t-e_3}\right]=\omega_1.$$

(6.12.19)

Now suppose we apply the results just found to the function $\mathscr{P}(u|g_2,-g_3)$ instead of to the function $\mathscr{P}(u|g_2,g_3)$. Then, by equations (6.10.12), we must replace $e_1>e_2>e_3$ by $-e_3>-e_2>-e_1$ respectively. Thus, equation (6.12.3) is reinterpreted to read

$$\mathscr{P}^{-1}(t|g_2,-g_3)=\frac{1}{2}\int_t^\infty\{(s+e_3)(s+e_2)(s+e_1)\}^{-1/2}\,ds,\qquad t\geqslant-e_3.$$

(6.12.20)

Changing the variable of integration by putting $s=-r$ and then replacing t by $-t$ everywhere, we obtain

$$\mathscr{P}^{-1}(-t|g_2,-g_3)=\frac{1}{2}\int_{-\infty}^t\{(e_1-r)(e_2-r)(e_3-r)\}^{-1/2}\,dr,\qquad t\leqslant e_3.$$

(6.12.21)

The same mode of transformation applied to the result (6.12.10) gives the new result

$$\mathscr{P}^{-1}\left[-e_3-\frac{(e_3-e_1)(e_3-e_2)}{t-e_3}\bigg|g_2,-g_3\right]=\frac{1}{2}\int_t^{e_3}\{(e_1-r)(e_2-r)(e_3-r)\}^{-1/2}\,dr,$$

(6.12.22)

provided $t\leqslant e_3$.

Applying the transformation to the results (6.12.16) and (6.12.17), we also obtain the identities

$$\mathscr{P}^{-1}\left[-e_2-\frac{(e_2-e_1)(e_2-e_3)}{t-e_2}\bigg|g_2,-g_3\right]=\frac{1}{2}\int_{e_2}^t\{(e_1-r)(r-e_2)(r-e_3)\}^{-1/2}\,dr,$$

(6.12.23)

$$\mathscr{P}^{-1}\left[-e_1-\frac{(e_1-e_2)(e_1-e_3)}{t-e_1}\bigg|g_2,-g_3\right]=\frac{1}{2}\int_t^{e_1}\{(e_1-r)(r-e_2)(r-e_3)\}^{-1/2}\,dr,$$

(6.12.24)

both valid for $e_2\leqslant t\leqslant e_1$.

For convenience, we now collect together the complete set of elliptic integrals which have been evaluated:

$$\mathscr{P}^{-1}(t|g_2,g_3)=\frac{1}{2}\int_t^\infty\{(s-e_1)(s-e_2)(s-e_3)\}^{-1/2}\,ds,\qquad t\geqslant e_1,$$

$$\mathscr{P}^{-1}\left[e_1+\frac{(e_1-e_2)(e_1-e_3)}{t-e_1}\bigg|g_2,g_3\right]=\frac{1}{2}\int_{e_1}^t\{(s-e_1)(s-e_2)(s-e_3)\}^{-1/2}\,ds,$$

$$t\geqslant e_1,$$

$$\wp^{-1}\left[-e_1-\frac{(e_1-e_2)(e_1-e_3)}{t-e_1}\bigg|g_2,-g_3\right]=\frac{1}{2}\int_t^{e_1}\{(e_1-s)(s-e_2)(s-e_3)\}^{-1/2}\,ds.$$
$$e_2\leqslant t\leqslant e_1,$$

$$\wp^{-1}\left[-e_2-\frac{(e_2-e_1)(e_2-e_3)}{t-e_2}\bigg|g_2,-g_3\right]=\frac{1}{2}\int_{e_2}^{t}\{(e_1-s)(s-e_2)(s-e_3)\}^{-1/2}\,ds,$$
$$e_2\leqslant t\leqslant e_1,$$

$$\wp^{-1}\left[e_2+\frac{(e_2-e_1)(e_2-e_3)}{t-e_2}\bigg|g_2,g_3\right]=\frac{1}{2}\int_t^{e_2}\{(e_1-s)(e_2-s)(s-e_3)\}^{-1/2}\,ds,$$
$$e_3\leqslant t\leqslant e_2,$$

$$\wp^{-1}\left[e_3+\frac{(e_3-e_1)(e_3-e_2)}{t-e_3}\bigg|g_2,g_3\right]=\frac{1}{2}\int_{e_3}^{t}\{(e_1-s)(e_2-s)(s-e_3)\}^{-1/2}\,ds,$$
$$e_3\leqslant t\leqslant e_2,$$

$$\wp^{-1}\left[-e_3-\frac{(e_3-e_1)(e_3-e_2)}{t-e_3}\bigg|g_2,g_3\right]=\frac{1}{2}\int_t^{e_3}\{(e_1-s)(e_2-s)(e_3-s)\}^{-1/2}\,ds,$$
$$t\leqslant e_3,$$

$$\wp^{-1}(-t|g_2,-g_3)=\frac{1}{2}\int_{-\infty}^{t}\{(e_1-s)(e_2-s)(e_3-s)\}^{-1/2}\,ds,\qquad t\leqslant e_3.$$

$$(6.12.25)$$

Application of the result (6.12.4) to the function $\wp(u|g_2,-g_3)$ yields its real half-period, viz.

$$-i\omega_3=\frac{1}{2}\int_{-e_3}^{\infty}\{(s+e_1)(s+e_2)(s+e_3)\}^{-1/2}\,ds.\qquad(6.12.26)$$

Changing the integration variable, this can be reinterpreted as a formula for the imaginary period of $\wp(u,\omega_1,\omega_3)$, thus:

$$\omega_3=\frac{1}{2}i\int_{-\infty}^{e_3}\{(e_1-r)(e_2-r)(e_3-r)\}^{-1/2}\,dr.\qquad(6.12.27)$$

An alternative formula for ω_3 can be derived similarly from (6.12.18); it is found to be

$$\omega_3=\frac{1}{2}i\int_{e_2}^{e_1}\{(e_1-r)(r-e_2)(r-e_3)\}^{-1/2}\,dr.\qquad(6.12.28)$$

The last two results enable us to link together another two pairs of the formulae (6.12.25), thus:

$$i\wp^{-1}\left[-e_1-\frac{(e_1-e_2)(e_1-e_3)}{t-e_1}\bigg|g_2,-g_3\right]$$
$$+i\wp^{-1}\left[-e_2-\frac{(e_2-e_1)(e_2-e_3)}{t-e_2}\bigg|g_2,-g_3\right]=\omega_3,\qquad(6.12.29)$$
$$i\wp^{-1}[-t|g_2,-g_3]+i\wp^{-1}\left[-e_3-\frac{(e_3-e_1)(e_3-e_2)}{t-e_3}\bigg|g_2,-g_3\right]=\omega_3.$$

Any integral of the first kind can now be reduced to one of the eight forms (6.12.25).

For example, transforming the variable of integration in (3.1.7) by $t^2 = r$, this integral takes the form

$$\frac{1}{2}\int_0^{x^2} \{r(a^2 - r)(b^2 - r)\}^{-1/2}\,dr, \qquad 0 \leqslant x \leqslant b < a. \qquad (6.12.30)$$

The zeros of the cubic under the square root do not sum to zero, so we make a further transformation $r = s + \frac{1}{3}(a^2 + b^2)$ to give

$$\frac{1}{2}\int_{e_3}^{\alpha} \{(e_1 - s)(e_2 - s)(s - e_3)\}^{-1/2}\,ds, \qquad (6.12.31)$$

where

$$e_1 = \tfrac{1}{3}(2a^2 - b^2), \qquad e_2 = \tfrac{1}{3}(2b^2 - a^2), \qquad e_3 = -\tfrac{1}{3}(a^2 + b^2), \qquad (6.12.32)$$
$$\alpha = x^2 - \tfrac{1}{3}(a^2 + b^2).$$

The integral is now in the sixth of the canonical forms (6.12.25) and can be evaluated thus:

$$\int_0^x \{(a^2 - t^2)(b^2 - t^2)\}^{-1/2}\,dt = \mathscr{P}^{-1}\left[\frac{a^2 b^2}{x^2} - \tfrac{1}{3}(a^2 + b^2)|g_2, g_3\right], \qquad (6.12.33)$$

where

$$g_2 = \tfrac{4}{3}(a^4 - a^2 b^2 + b^4), \qquad g_3 = \tfrac{4}{27}(2a^2 - b^2)(a^2 + b^2)(a^2 - 2b^2). \qquad (6.12.34)$$

6.13. Weierstrass's Method for Integrals of the Second Kind

Twelve canonical forms for elliptic integrals of the second kind have been listed in section 3.4 ((3.4.2)–(3.4.13)). These can all be brought to a Weierstrass canonical form by transformations $t^2 = r$, $r = s + \text{constant}$, as explained at the end of the previous section. The integrals so obtained will be identical with the eight listed at (6.12.25), but with an additional factor $(s + \alpha)$ (α constant) being present in the integrand. The results (6.12.25) suggest further transformations by which each of these forms can be evaluated in terms of Weierstrass's zeta function $\zeta(u)$.

Thus, to evaluate

$$\frac{1}{2}\int_t^T (s + \alpha)\{(s - e_1)(s - e_2)(s - e_3)\}^{-1/2}\,ds, \qquad (6.13.1)$$

where $T \geqslant t \geqslant e_1$ (N.B. $T = \infty$ yields a divergent integral), we split it into the integral

$$\frac{1}{2} \int_t^T s\{(s - e_1)(s - e_2)(s - e_3)\}^{-1/2} \, ds \qquad (6.13.2)$$

and an integral of the first kind. We then put $s = \mathscr{P}(u)$ and transform the last integral to

$$-\int_v^V \mathscr{P}(u) \, du = \int_v^V \zeta'(u) \, du = \zeta(V) - \zeta(v), \qquad (6.13.3)$$

where $t = \mathscr{P}(v)$, $T = \mathscr{P}(V)$.

In the case of the integral of the second kind corresponding to the fourth member of the list (6.12.25), viz.

$$\frac{1}{2} \int_{e_2}^t s\{(e_1 - s)(s - e_2)(s - e_3)\}^{-1/2} \, ds, \qquad e_2 \leqslant t \leqslant e_1, \qquad (6.13.4)$$

we are presented with a more difficult evaluation problem. We first use the transformation

$$r = -e_2 - \frac{(e_2 - e_1)(e_2 - e_3)}{s - e_2}, \qquad (6.13.5)$$

suggested by the result for the integral of the first kind, to bring the integral (6.13.4) to the form

$$\frac{1}{2} \int_p^\infty \left[e_2 - \frac{(e_2 - e_1)(e_2 - e_3)}{r + e_2} \right] \{(r + e_1)(r + e_2)(r + e_3)\}^{-1/2} \, dr, \qquad (6.13.6)$$

where $r = p$ when $s = t$, and $p \geqslant -e_3$.

Changing the notation by $e_1 = -e_3', e_2 = -e_2', e_3 = -e_1'$, the last integral can be written

$$\frac{1}{2} \int_p^\infty \left[-e_2' - \frac{(e_2' - e_1')(e_2' - e_3')}{r - e_2'} \right] \{(r - e_1')(r - e_2')(r - e_3')\}^{-1/2} \, dr, \qquad (6.13.7)$$

where $p \geqslant e_1'$ and it remains true that $e_1' + e_2' + e_3' = 0$ and $e_1' > e_2' > e_3'$. Having made this change of notation, we shall now discard the primes, it being understood that the functions $\mathscr{P}(u), \zeta(u)$, introduced later, are to be calculated with the new values e_1', e_2', e_3' (i.e., new invariants $g_2' = g_2, g_3' = -g_3$).

We next change the integration variable in (6.13.7) to u by the transformation $r = \mathscr{P}(u)$ to bring the integral to the form

$$\int_0^v \left[-e_2 - \frac{(e_2 - e_1)(e_2 - e_3)}{\mathscr{P}(u) - e_2} \right] du, \qquad (6.13.8)$$

where $v = \mathscr{P}^{-1}(p)$. Then, using the identity (6.8.12), this immediately reduces to

$$-\int_0^v \mathscr{P}(u + \omega_2)\, du = \int_0^v \zeta'(u + \omega_2)\, du,$$

$$= \zeta(v + \omega_2) - \zeta(\omega_2),$$

$$= \zeta(v) + \frac{1}{2} \frac{\mathscr{P}'(v)}{\mathscr{P}(v) - e_2}, \qquad (6.13.9)$$

the final step being justified by the identity (6.8.4).

The identity (6.7.16) permits us to write (6.13.9) in its final form

$$\zeta(v) - \sqrt{\left[\frac{\{\mathscr{P}(v) - e_1\}\{\mathscr{P}(v) - e_3\}}{\mathscr{P}(v) - e_2}\right]} = \zeta(v) - \sqrt{\left[\frac{(p - e_1)(p - e_3)}{p - e_2}\right]}.$$

$$(6.13.10)$$

The remaining integrals of the second kind associated with the forms (6.12.25) can be evaluated by similar methods.

6.14. Weierstrass's Integral of the Third Kind

Using the transformation $t^2 = s + \text{constant}$, we bring the integral of the third kind (3.3.37) to the Weierstrassian form

$$\frac{1}{2} \int_t^\infty (s + \alpha)^{-1} \{(s - e_1)(s - e_2)(s - e_3)\}^{-1/2}\, ds, \qquad (6.14.1)$$

where $t \geqslant e_1$, or to one of the eight similar forms corresponding to the various types of the first-kind integrals (6.12.25).

Further transformations may be needed, as explained in the previous section, if the integral does not immediately assume the form (6.14.1). Thus, an integral of the type

$$\frac{1}{2} \int_{e_2}^t (s + \alpha)^{-1} \{(e_1 - s)(s - e_2)(s - e_3)\}^{-1/2}\, ds, \qquad e_2 \leqslant t \leqslant e_1, \quad (6.14.2)$$

can be brought to the form

$$\frac{1}{2} \int_p^\infty \frac{r - e_2}{\beta r + \gamma} \{(r - e_1)(r - e_2)(r - e_3)\}^{-1/2}\, dr, \qquad (6.14.3)$$

where $p > e_1$, by the transformation (6.13.5) and a change of notation in respect of the e_α. Any integral of the third kind can be brought to the form

$$\frac{1}{2} \int_p^\infty \frac{r + \varepsilon}{\beta r + \gamma} \{(r - e_1)(r - e_2)(r - e_3)\}^{-1/2}\, dr, \qquad (6.14.4)$$

where the multiplier $(r + \varepsilon)/(\beta r + \gamma)$ increases or decreases monotonically as r increases from p to ∞.

We now put $r = \mathscr{P}(u)$ to bring the integral to the form

$$\int_0^v \frac{\mathscr{P}(u) + \varepsilon}{\beta\mathscr{P}(u) + \gamma} du = v/\beta + A \int_0^v \frac{du}{\mathscr{P}(u) + B}, \qquad (6.14.5)$$

where $A = (\beta\varepsilon - \gamma)/\beta^2$ and $B = \gamma/\beta$. Then, choosing w (in general complex) such that $\mathscr{P}(w) = -B$, we can complete the integration by evaluating

$$\int \frac{\mathscr{P}'(w)}{\mathscr{P}(u) - \mathscr{P}(w)} du = \int [\zeta(u - w) - \zeta(u + w) + 2\zeta(w)] du$$

$$= \ln \sigma(u - w) - \ln \sigma(u + w) + 2u\zeta(w)$$

$$= \ln \frac{\sigma(u - w)}{\sigma(u + w)} + 2u\zeta(w), \qquad (6.14.6)$$

after making use of the identities (6.6.1) and (6.8.3).

This result should be compared with the analogous Jacobi result (3.7.8).

6.15. Further Conditions for $\mathscr{P}(u)$ to Be Real

We have shown that, provided the modulus k is real and lies in the interval $(0, 1)$, the Jacobi functions $\operatorname{sn} u$ and $\operatorname{dn} u$ have each one real and one purely imaginary period. Primitive periods for $\operatorname{cn} u$, however, are $4K$ and $2K(1 + iv)$, where $v = K'/K > 0$. All these functions take real values for real values of u. It has also been shown that $\mathscr{P}(u, \omega_1, \omega_3)$ takes real values for real u, whenever ω_1 is real and ω_3 is purely imaginary. Comparison with the Jacobi functions suggests the possibility that $\mathscr{P}(u, \omega_1, \omega_3)$ may be real for real u, when ω_1 is real and $\omega_3 = \tau\omega_1$, where $\tau = \frac{1}{2}(1 + iv)(v > 0)$. This will now be proved to be so.

We first note that, in this case, new primitive half-periods ω_1', ω_3', can be constructed thus:

$$\omega_1' = \omega_1 - \omega_3 = \frac{1}{2}(1 - iv)\omega_1, \qquad \omega_3' = \omega_3 = \frac{1}{2}(1 + iv)\omega_1, \qquad (6.15.1)$$

where ω_1', ω_3' are seen to be complex conjugates. Nevertheless, we prefer to retain ω_1, ω_3 as the half-periods in the calculations which follow.

The nome q is defined in the usual way and we note that

$$q = e^{i\pi\tau} = ie^{-(1/2)\pi v}, \qquad q^* = -q, \qquad q^{1/4} = \exp\{\tfrac{1}{8}\pi(i - v)\},$$

$$(q^{1/4})^* = \exp\{-\tfrac{1}{8}\pi(i + v)\} = q^{1/4}\exp(-\tfrac{1}{4}i\pi) = \frac{1}{\sqrt{2}}(1 - i)q^{1/4}. \qquad (6.15.2)$$

Equation (1.2.11) gives

$$\theta_1(z|\tau) = 2q^{1/4} \sum_{n=0}^{\infty} (-1)^n q^{n(n+1)} \sin(2n + 1)z. \qquad (6.15.3)$$

Hence, provided z is real, taking conjugates

$$\theta_1^*(z|\tau) = 2(q^{1/4})^* \sum_{n=0}^{\infty} (-1)^n (q^*)^{n(n+1)} \sin(2n+1)z$$

$$= 2(1-i)q^{1/4} \sum_{n=0}^{\infty} (-1)^n q^{n(n+1)} \sin(2n+1)z$$

$$= \frac{1}{\sqrt{2}}(1-i)\theta_1(z|\tau), \tag{6.15.4}$$

since $n(n+1)$ is necessarily even.

Similarly, using equation (1.2.12), we can show that

$$\theta_2^*(z|\tau) = \frac{1}{\sqrt{2}}(1-i)\theta_2(z|\tau). \tag{6.15.5}$$

Furthermore, by equations (1.2.13) and (1.2.14),

$$\theta_3^*(z|\tau) = 1 + 2\sum_{n=1}^{\infty} (q^*)^{n^2} \cos 2nz = 1 + 2\sum_{n=1}^{\infty} (-1)^n q^{n^2} \cos 2nz$$

$$= \theta_4(z|\tau), \tag{6.15.6}$$

after noting that n and n^2 are even or odd together.

Since we are assuming z is real, $(d\theta_1/dz)^* = d\theta_1^*/dz$ and it follows from equation (6.15.4) that

$$(D^r\theta_1)^* = \frac{1}{\sqrt{2}}(1-i)D^r\theta_1, \tag{6.15.7}$$

where $D = d/dz$. The identities (6.15.5) and (6.15.6) can be differentiated similarly.

We now refer to equation (6.7.28) defining $\mathscr{P}(u)$. Since ω_1 is real, z is real with u and, taking conjugates, we find (using (6.15.7)) that

$$\mathscr{P}^*(u) = \mathscr{P}(u), \tag{6.15.8}$$

factors $(1-i)/\sqrt{2}$ canceling everywhere. This proves that $\mathscr{P}(u)$ is real for real values of u.

The values of $\mathscr{P}(u)$ at the zeros of $\mathscr{P}'(u)$ can be calculated from equations (6.7.29)–(6.7.31). Taking conjugates of both members of equation (6.7.29), we show that

$$e_1^* = e_1, \tag{6.15.9}$$

after again canceling factors $(1-i)/\sqrt{2}$. Thus, e_1 is real. However, taking conjugates in equation (6.7.30), we conclude that

$$e_2^* = e_3, \tag{6.15.10}$$

proving that e_2, e_3 are complex conjugates (they are not real and equal, since $\theta_3''(0)/\theta_3(0) \neq \theta_4''(0)/\theta_4(0)$). Hence, in the case we are studying, the cubic equation (6.7.27) has one real root e_1 and a pair of conjugate complex roots e_2, e_3. Since $e_1 + e_2 + e_1 = 0$, the real parts of e_2, e_3 must both be $-\frac{1}{2}e_1$.

As u increases from 0 to $\omega_1, \mathscr{P}(u)$ decreases from $+\infty$ to e_1. Since $\mathscr{P}(2\omega_1 - u) = \mathscr{P}(u)$, as u increases further from ω_1 to $2\omega_1, \mathscr{P}(u)$ increases back toward $+\infty$. A table of values of $\omega_1^2 \mathscr{P}(u, \omega_1, \omega_3)$ for $u/\omega_1 = 0.02(0.02)1$, $v = 0.2(0.2)4$ is given in Table J (p. 311). In Table K (p. 315), values of $\omega_1^2 e_1, \omega_1^2 e_2, \omega_1^4 g_2, \omega_1^6 g_3$, are also provided for these values of v. As $v \to +\infty$, $\omega_1^2 \mathscr{P}(u) \to \frac{1}{4}\pi^2 \{\operatorname{cosec}^2(\pi u/2\omega_1) - \frac{1}{3}\}, \omega_1^2 e_1 \to \pi^2/6$, etc., as explained in sections 6.6 and 6.7.

$\mathscr{P}(u)$ also assumes real values when u is purely imaginary. This follows from the identity (6.10.11), if we work with the conjugate primitive periods, thus:

$$\mathscr{P}(iu, \tfrac{1}{2}\omega_1(1 - iv), \tfrac{1}{2}\omega_1(1 + iv)) = -\mathscr{P}(u, \tfrac{1}{2}\omega_1(v - i), \tfrac{1}{2}\omega_1(v + i))$$
$$= -\mathscr{P}(u, \tfrac{1}{2}\omega_1 v(1 - i/v), \tfrac{1}{2}\omega_1 v(1 + i/v)).$$
$$(6.15.11)$$

The last function is of the type we have been studying, except that ω_1 is replaced by $\omega_1 v$ and v is replaced by $1/v$. It takes real values when u is real, decreasing from $+\infty$ to a minimum at $u = \omega_1 v$ and, thereafter, increasing to $+\infty$ as u increases further to $2\omega_1 v$. We conclude that $\mathscr{P}(iu)$ takes real values, increasing from $-\infty$ at $u = 0$ to a maximum at $u = \omega_1 v$ and thereafter decreasing to $-\infty$ again at $u = 2\omega_1 v$. Its maximum value is

$$\mathscr{P}(i\omega_1 v, \omega_1, \omega_3) = \mathscr{P}(2\omega_3 - \omega_1, \omega_1, \omega_3) = \mathscr{P}(\omega_1, \omega_1, \omega_3) = e_1. \quad (6.15.12)$$

We note that, for real u, the full range $(-\infty, +\infty)$ is covered by the values of $\mathscr{P}(u)$ and $\mathscr{P}(iu)$.

6.16. Extraction of $\mathscr{P}(u|g_2,g_3)$ from Tables

Clearly, the invariants g_2 and g_3 are both functions of the half-periods ω_1, ω_3 and, as noted earlier at (6.7.36),

$$\omega_1^4 g_2(\omega_1, \omega_3) = g_2(1, \tau), \qquad \omega_1^6 g_3(\omega_1, \omega_3) = g_3(1, \tau). \quad (6.16.1)$$

Thus, defining another invariant G by the equation

$$G = g_2^3/27g_3^2, \quad (6.16.2)$$

we note that

$$G = g_2^3(1, \tau)/\{27g_3^2(1, \tau)\}, \quad (6.16.3)$$

i.e., $G = G(\tau)$ is a function of the ratio $\tau = \omega_3/\omega_1$ alone.

G is closely related to the discriminant Δ of the cubic equation

$$4s^3 - g_2 s - g_3 = 0. \quad (6.16.4)$$

Assuming g_2 and g_3 are real, we have $\Delta = g_2^3 - 27g_3^2$ and it is known that the cubic has (i) three distinct real roots if $\Delta > 0$, (ii) three real roots which are not all distinct if $\Delta = 0$, (iii) a real root and a pair of complex conjugate roots if $\Delta < 0$. Since $\Delta = 27g_3^2(G - 1)$, cases (i)–(iii) correspond to $G > 1$, $G = 1$, and $G < 1$ respectively.

If ω_1 is real and ω_3 is purely imaginary, we know that the roots e_α of (6.16.4) are all real; hence, $G > 1$ in this case. If ω_1 is real and $\omega_3 = \frac{1}{2}\omega_1(1 + iv)$ ($v > 0$), only one of these roots is real and, then, $G < 1$. Both these inequalities can be verified by reference to Table K.

If ω_1 is real and $\omega_3 = i\kappa\omega_1$ then, as proved above, G will depend on κ alone, i.e., $G = G(\kappa)$. But we have already remarked that $\omega_1^2 \mathscr{P}(u, \omega_1, i\kappa\omega_1)$ can be tabulated against u/ω_1 and κ and it now follows that $\omega_1^2 \mathscr{P}$ can also be tabulated against u/ω_1 and G. However, the inverse functional relationship $\kappa = \kappa(G)$ is double-valued, since a change in sign of g_3 does not affect G, but leads to a new value of κ (refer to Table K). Accordingly, two cases need considering, viz. $g_3 > 0$ and $g_3 < 0$. Table L (p. 316) provides values of $\omega_1^2 \mathscr{P}$ for $u/\omega_1 = 0.02(0.02)1$ and $G = -6(1) - 2(0.2)2(1)6$ in these two cases.

Clearly, by choosing real values for κ, only values $G > 1$ can be generated. For $G < 1$ in Table L, we need to take $\omega_3 = \frac{1}{2}\omega_1(1 + iv)$. Again, the relationship $v = v(G)$ is double-valued and the cases $g_3 > 0$, $g_3 < 0$ have to be allocated separate tables.

The two values of κ corresponding to a value of $G(>1)$ are reciprocals of one another. For consider the relationship between the functions $\mathscr{P}(u, \omega_1, \tau\omega_1)$ and $\mathscr{P}(u, \omega_1, \tau'\omega_1)$, when $\tau\tau' = -1$. Using the identities (6.6.17) and (6.10.1), we have that

$$\mathscr{P}(u, \omega_1, \tau'\omega_1) = \tau^2 \mathscr{P}(\tau u, \tau\omega_1, -\omega_1) = \tau^2 \mathscr{P}(\tau u, \omega_1, \tau\omega_1). \quad (6.16.5)$$

If e_α, e_α' are the stationary values for the two functions, then

$$\left.\begin{array}{l}
e_1' = \mathscr{P}(\omega_1, \omega_1, \tau'\omega_1) = \tau^2 \mathscr{P}(\tau\omega_1, \omega_1, \tau\omega_1) = \tau^2 e_3, \\
e_2' = \mathscr{P}(\omega_1 + \tau'\omega_1, \omega_1, \tau'\omega_1) = \tau^2 \mathscr{P}(\tau\omega_1 - \omega_1, \omega_1, \tau\omega_1) = \tau^2 e_2, \\
e_3' = \mathscr{P}(\tau'\omega_1, \omega_1, \tau'\omega_1) = \tau^2 \mathscr{P}(\omega_1, \omega_1, \tau\omega_1) = \tau^2 e_1.
\end{array}\right\} \quad (6.16.6)$$

Hence,

$$g_2' = \tau^4 g_2, \qquad g_3' = \tau^6 g_3, \qquad G' = G. \quad (6.16.7)$$

In particular, if ω_1 is real and $\tau = i\kappa$, $\tau' = i\kappa'$ are purely imaginary, with $\kappa\kappa' = 1$, then $g_2' = \kappa^4 g_2$, $g_3' = -\kappa^6 g_3$, $G' = G$; i.e., $G(1/\kappa) = G(\kappa)$ and the signs of $g_3(1/\kappa)$ and $g_3(\kappa)$ are opposite.

For the special case $\kappa = \kappa' = 1$, we have $g_3(1) = -g_3(1)$, showing that $g_3(1) = 0$. Since g_2 is positive, $G(1) = +\infty$, as shown in Table K.

As $\kappa \to +\infty$, the limiting values of g_2 and $g_3(>0)$ are given at (6.7.39). The corresponding limiting value of G proves to be 1, i.e., $G(+\infty) = 1$ when $g_3 > 0$. Since $\kappa' \to 0$, we also deduce that $G(0) = 1$, when $g_3 < 0$.

It is also true that the pair of values of v corresponding to a given value of $G(<1)$ are reciprocals of one another. This follows when we refer $\mathscr{P}$ to its primitive half-periods $\frac{1}{2}\omega_1(1-iv)$, $\frac{1}{2}\omega_1(1+iv)$ by the following argument: If $vv'=1$, then, using the identities (6.10.11), (6.6.17),

$$\mathscr{P}(u, \tfrac{1}{2}\omega_1(1-iv'), \tfrac{1}{2}\omega_1(1+iv')) = -\mathscr{P}(iu, \tfrac{1}{2}\omega_1(v'-i), \tfrac{1}{2}\omega_1(v'+i))$$
$$= -v^2\mathscr{P}(ivu, \tfrac{1}{2}\omega_1(1-iv), \tfrac{1}{2}\omega_1(1+iv)),$$

(6.16.8)

from which it follows that

$$\left.\begin{aligned}
e_1' &= \mathscr{P}(\tfrac{1}{2}\omega_1(1-iv'), \tfrac{1}{2}\omega_1(1-iv'), \tfrac{1}{2}\omega_1(1+iv')) = -v^2 e_3,\\
e_2' &= \mathscr{P}(\omega_1, \tfrac{1}{2}\omega_1(1-iv'), \tfrac{1}{2}\omega_1(1+iv')) = -v^2 e_2,\\
e_3' &= \mathscr{P}(\tfrac{1}{2}\omega_1(1+iv'), \tfrac{1}{2}\omega_1(1-iv'), \tfrac{1}{2}\omega_1(1+iv')) = -v^2 e_1,
\end{aligned}\right\}$$

(6.16.9)

and, hence, that

$$g_2' = v^4 g_2, \qquad g_3' = -v^6 g_3, \qquad G' = G. \tag{6.16.10}$$

The special case $v = v' = 1$ requires that $g_3 = 0$. Since g_2 is negative in this case, $G = -\infty$ as shown in Table K.

As $v \to +\infty$, $G \to 1$. Also, $v' \to 0$ and hence $G(+\infty) = G(0) = 1$.

For convenience, the quantities $\omega_1^2 e_a$, $\omega_1^4 g_2$, and $\omega_1^6 g_3$ have been tabulated against G in Table M (p. 328), for the two cases $g_3 > 0$ and $g_3 < 0$.

To extract values of $\mathscr{P}(u|g_2, g_3)$ from Table L, we first calculate G and then, after noting the sign of g_3, refer to Table M for the corresponding values of $\omega_1^4 g_2$ and $\omega_1^6 g_3$. Knowing g_2 and g_3, the value of ω_1 can then be computed from either of these. u/ω_1 is next found and the value of $\omega_1^2 \mathscr{P}$ then determined from Table L. Interpolation difficulties will be encountered, but the principle is valid and a sufficiently motivated reader may compute more extensive tables to overcome this restriction.

6.17. Elliptic Integrals (Negative Discriminant)

If $t = \mathscr{P}(u, \omega_1, \tfrac{1}{2}\omega_1(1+iv))$, where u, ω_1 are real and $v > 0$, then (equation (6.7.26))

$$\frac{dt}{du} = -\sqrt{(4t^3 - g_2 t - g_3)}, \tag{6.17.1}$$

the negative root being appropriate if we take u in the interval $(0, \omega_1)$. Integration from $u = 0$, $t = +\infty$, yields the elliptic integral

$$\int_t^\infty (4s^3 - g_2 s - g_3)^{-1/2}\, ds = u = \mathscr{P}^{-1}(t, \omega_1, \tfrac{1}{2}\omega_1(1+iv)). \tag{6.17.2}$$

As already stated, the inverse $\mathscr{P}$-function is to be given a value in the interval $(0, \omega_1)$. Also, we have seen that, as u increases from 0 to ω_1, t decreases from $+\infty$ to e_1 and it follows that $t \geq e_1$ in equation (6.17.2). As proved in the last section, $\Delta < 0$ in this case.

The result just obtained enables us to calculate an elliptic integral of the first kind when the cubic under the root sign possesses a negative discriminant and the coefficient of the cube term is positive. As an example, provided $t \geqslant 1$,

$$\int_t^\infty (s^3 - 1)^{-1/2}\, ds = 2\mathscr{P}^{-1}(t|0, 4). \tag{6.17.3}$$

When $u = \omega_1$, then $t = e_1$ and, hence,

$$\omega_1 = \int_{e_1}^\infty (4s^3 - g_2 s - g_3)^{-1/2}\, ds. \tag{6.17.4}$$

It now follows from the results (6.17.2) and (6.17.4) that

$$\int_{e_1}^t (4s^3 - g_2 s - g_3)^{-1/2}\, ds = \omega_1 - \mathscr{P}^{-1}(t|g_2, g_3). \tag{6.17.5}$$

Using the identity (6.8.11), we note that an alternative form of this result is

$$\int_{e_1}^t (4s^3 - g_2 s - g_3)^{-1/2}\, ds = \mathscr{P}^{-1}\left[e_1 + \frac{(e_1 - e_2)(e_1 - e_3)}{t - e_1} \right], \tag{6.17.6}$$

where e_2, e_3 are the conjugate complex zeros of the cubic.

Suppose we apply these results to the function $t = \mathscr{P}(u|g_2, -g_3)$. Equation (6.17.2) transforms to

$$\int_t^\infty (4s^3 - g_2 s + g_3)^{-1/2}\, ds = \mathscr{P}^{-1}(t|g_2, -g_3), \tag{6.17.7}$$

where $t \geqslant -e_1$ (the zeros of the cubic are now $-e_1, -e_2, -e_3$). Changing the variable of integration by $s = -r$ and replacing t by $-t$, this takes the form

$$\int_{-\infty}^t (g_3 + g_2 r - 4r^3)^{-1/2}\, dr = \mathscr{P}^{-1}(-t|g_2, -g_3), \tag{6.17.8}$$

where $t \leqslant e_1$. This result covers the case when the coefficient of the cube term is negative.

For example,

$$\int_{-\infty}^t (1 - r^3)^{-1/2}\, dr = 2\mathscr{P}^{-1}(-t|0, -4). \tag{6.17.9}$$

Since, by equation (6.10.14), $\mathscr{P}(u|g_2, -g_3) = -\mathscr{P}(iu|g_2, g_3)$ and, as proved in section 6.15, $\mathscr{P}(iu|g_2, g_3)$ increases from $-\infty$ to a maximum e_1 as u increases from 0 to $\omega_1 v$, we have $\mathscr{P}^{-1}(-e_1|g_2, -g_3) = \omega_1 v$. Hence, putting $t = e_1$ in equation (6.17.8), we obtain the result

$$\int_{-\infty}^{e_1} (g_3 + g_2 r - 4r^3)^{-1/2}\, dr = \omega_1 v. \tag{6.17.10}$$

It now follows that

$$\int_t^{e_1} (g_3 + g_2 r - 4r^3)^{-1/2}\, dr = \omega_1 v - \mathscr{P}^{-1}(-t|g_2, -g_3). \tag{6.17.11}$$

EXERCISES

1. Prove the identities:

$$\sigma(u+\omega_1) = \frac{2\omega_1}{\pi\theta_3\theta_4}\exp\{\eta_1(u+\tfrac{1}{2}\omega_1)\}\sigma_1(u),$$

$$\sigma(u+\omega_2) = -\frac{2\omega_1}{\pi\theta_2\theta_4}\exp\{\eta_2(u+\tfrac{1}{2}\omega_2)+\tfrac{1}{4}i\pi\}\sigma_2(u)$$

$$\sigma(u+\omega_3) = \frac{2i\omega_1}{\pi\theta_2\theta_3}\exp\{\eta_3(u+\tfrac{1}{2}\omega_3)\}\sigma_3(u)$$

$$\sigma_1(u+\omega_1) = -\frac{\pi\theta_3\theta_4}{2\omega_1}\exp\{\eta_1(u+\tfrac{1}{2}\omega_1)\}\sigma(u)$$

$$\sigma_1(u+\omega_2) = \frac{\theta_4}{\theta_2}\exp\{\eta_2(u+\tfrac{1}{2}\omega_2)-\tfrac{1}{4}i\pi\}\sigma_3(u)$$

$$\sigma_1(u+\omega_3) = \frac{\theta_3}{\theta_2}\exp\{\eta_3(u+\tfrac{1}{2}\omega_3)\}\sigma_2(u)$$

$$\sigma_2(u+\omega_1) = \frac{\theta_4}{\theta_3}\exp\{\eta_1(u+\tfrac{1}{2}\omega_1)\}\sigma_3(u)$$

$$\sigma_2(u+\omega_2) = \frac{\pi\theta_2\theta_4}{2\omega_1}\exp\{\eta_2(u+\tfrac{1}{2}\omega_2)-\tfrac{1}{4}i\pi\}\sigma(u)$$

$$\sigma_2(u+\omega_3) = \frac{\theta_2}{\theta_3}\exp\{\eta_3(u+\tfrac{1}{2}\omega_3)\}\sigma_1(u)$$

$$\sigma_3(u+\omega_1) = \frac{\theta_3}{\theta_4}\exp\{\eta_1(u+\tfrac{1}{2}\omega_1)\}\sigma_2(u)$$

$$\sigma_3(u+\omega_2) = \frac{\theta_2}{\theta_4}\exp\{\eta_2(u+\tfrac{1}{2}\omega_2)+\tfrac{1}{4}i\pi\}\sigma_1(u)$$

$$\sigma_3(u+\omega_3) = \frac{i\pi\theta_2\theta_3}{2\omega_1}\exp\{\eta_3(u+\tfrac{1}{2}\omega_3)\}\sigma(u).$$

2. Show that, in a neighborhood of $u = 0$,

(i) $\sigma(u) = u - \frac{1}{240}g_2 u^5 - \frac{1}{840}g_3 u^7 + O(u^9)$,

(ii) $\zeta(u) = \frac{1}{u} - \frac{1}{60}g_2 u^3 - \frac{1}{140}g_3 u^5 + O(u^7)$.

3. Prove that

$$\mathscr{P}(u+\omega_3) = \left(\frac{\pi}{2\omega_1}\right)^2\left[\frac{1}{3}\cdot\frac{\theta_1'''(0)}{\theta_1'(0)} - \frac{d}{dz}\left\{\frac{\theta_4'(z)}{\theta_4(z)}\right\}\right],$$

where $z = \pi u/2\omega_1$.

4. With the usual notation, prove that

$$e_1 = \frac{\pi^2}{12\omega_1^2}(\theta_3^4 + \theta_4^4), \qquad e_2 = \frac{\pi^2}{12\omega_1^2}(\theta_2^4 - \theta_4^4), \qquad e_3 = -\frac{\pi^2}{12\omega_1^2}(\theta_2^4 + \theta_3^4),$$

$$g_2 = \frac{\pi^4}{24\omega_1^4}(\theta_2^8 + \theta_3^8 + \theta_4^8), \qquad g_3 = \frac{\pi^6}{432\omega_1^6}(\theta_2^4 + \theta_3^4)(\theta_3^4 + \theta_4^4)(\theta_4^4 - \theta_2^4).$$

5. Prove that

(i) $\quad \mathscr{P}(u+v) + \mathscr{P}(u-v) = \dfrac{[\mathscr{P}(u) + \mathscr{P}(v)][2\mathscr{P}(u)\mathscr{P}(v) - \frac{1}{2}g_2] - g_3}{[\mathscr{P}(u) - \mathscr{P}(v)]^2},$

(ii) $\quad \mathscr{P}(u-v) - \mathscr{P}(u+v) = \dfrac{\mathscr{P}'(u)\mathscr{P}'(v)}{[\mathscr{P}(u) - \mathscr{P}(v)]^2}.$

6. Obtain the duplication formulae

$$\mathscr{P}(2u) = \frac{1}{4}\left[\frac{\mathscr{P}''(u)}{\mathscr{P}'(u)}\right]^2 - 2\mathscr{P}(u) = \frac{\mathscr{P}^4(u) + \frac{1}{2}g_2\mathscr{P}^2(u) + 2g_3\mathscr{P}(u) + \frac{1}{16}g_2^2}{4\mathscr{P}^3(u) - g_2\mathscr{P}(u) - g_3}.$$

7. Show that $\mathscr{P}(\frac{2}{3}\omega_i) = \mathscr{P}(\frac{4}{3}\omega_i)$ $(i = 1, 2, 3, 4)$, where $\omega_4 = \omega_1 - \omega_3$, and deduce that $\mathscr{P}(\frac{2}{3}\omega_i)$ are the four roots of the quartic

$$x^4 - \tfrac{1}{2}g_2 x^2 - g_3 x - \tfrac{1}{48}g_2^2 = 0.$$

8. Show that

$$\mathscr{P}(\tfrac{1}{2}\omega_1) = e_1 + \{(e_1 - e_2)(e_1 - e_3)\}^{1/2},$$

$$\mathscr{P}(\tfrac{1}{2}\omega_1 + \omega_3) = e_1 - \{(e_1 - e_2)(e_1 - e_3)\}^{1/2}.$$

(Hint: Use equation (6.7.10) and properties of the sigma functions.)

9. Show that

$$\frac{\mathscr{P}'(u + \omega_1)}{\mathscr{P}'(u)} = -\left[\frac{\mathscr{P}(\tfrac{1}{2}\omega_1) - \mathscr{P}(\omega_1)}{\mathscr{P}(u) - \mathscr{P}(\omega_1)}\right]^2.$$

(Hint: Use a result proved in the previous exercise.)

10. Show that

$$\mathscr{P}(2u) - \mathscr{P}(\omega_1) = \frac{\{\mathscr{P}(u) - \mathscr{P}(\tfrac{1}{2}\omega_1)\}^2\{\mathscr{P}(u) - \mathscr{P}(\tfrac{1}{2}\omega_1 + \omega_3)\}^2}{\{\mathscr{P}'(u)\}^2}.$$

(Hint: Use Exercise 6.)

11. Show that

$$\mathscr{P}(u+v)\mathscr{P}(u-v) = \frac{[\mathscr{P}(u)\mathscr{P}(v) + \frac{1}{4}g_2]^2 + g_3[\mathscr{P}(u) + \mathscr{P}(v)]}{[\mathscr{P}(u) - \mathscr{P}(v)]^2}.$$

12. Show that

$$2\zeta(2u) - 4\zeta(u) = \frac{\mathscr{P}''(u)}{\mathscr{P}'(u)}.$$

(Hint: Use the identity (6.8.4).)

13. Show that

$$\left[\frac{\mathscr{P}''(u)}{\mathscr{P}'(u)}\right]^2 = 6\mathscr{P}(u) + \sum_{i,j,k} \frac{\{\mathscr{P}(u) - e_i\}\{\mathscr{P}(u) - e_j\}}{\mathscr{P}(u) - e_k},$$

the summation being extended over the three cyclic permutations of $(1,2,3)$. Deduce that

(i) $\left[\dfrac{\mathscr{P}''(u)}{\mathscr{P}'(u)}\right]^2 = 9\mathscr{P}(u) + \mathscr{P}(u + \omega_1) + \mathscr{P}(u + \omega_2) + \mathscr{P}(u + \omega_3),$

(ii) $\mathscr{P}(2u) = \frac14\{\mathscr{P}(u) + \mathscr{P}(u + \omega_1) + \mathscr{P}(u + \omega_2) + \mathscr{P}(u + \omega_3)\}.$

14. Obtain the symmetrical identity

$$\mathscr{P}'(u + v) = \frac{3\{\mathscr{P}'(u) - \mathscr{P}'(v)\}\{\mathscr{P}(u) + \mathscr{P}(v)\}}{2\{\mathscr{P}(u) - \mathscr{P}(v)\}} - \frac14\left[\frac{\mathscr{P}'(u) - \mathscr{P}'(v)}{\mathscr{P}(u) - \mathscr{P}(v)}\right]^3 - \frac12\{\mathscr{P}'(u) + \mathscr{P}'(v)\}.$$

15. Verify that

$$\frac{\partial}{\partial u} \mathscr{P}(u)\{\mathscr{P}(u + v) - \mathscr{P}(v)\}$$

is symmetric in u and v. (Hint: Use the result of the previous exercise.) Deduce that

$$\frac{\mathscr{P}'(u) - \mathscr{P}'(v)}{\mathscr{P}(u) - \mathscr{P}(v)} = \frac{\mathscr{P}'(v) + \mathscr{P}'(u + v)}{\mathscr{P}(v) - \mathscr{P}(u + v)}.$$

16. Prove that

$$\mathscr{P}'(u)\mathscr{P}'(u + \omega_1)\mathscr{P}'(u + \omega_2)\mathscr{P}'(u + \omega_3) = 16(e_1 - e_2)^2(e_2 - e_3)^2(e_3 - e_1)^2 = g_2^3 - 27g_3^2$$

and deduce that

$$\frac{\mathscr{P}''(u)}{\mathscr{P}'(u)} + \frac{\mathscr{P}''(u + \omega_1)}{\mathscr{P}'(u + \omega_1)} + \frac{\mathscr{P}''(u + \omega_2)}{\mathscr{P}'(u + \omega_2)} + \frac{\mathscr{P}''(u + \omega_3)}{\mathscr{P}'(u + \omega_3)} = 0.$$

17. Prove that

$$\{\mathscr{P}(u) + \mathscr{P}(u + \omega_2)\}\{\mathscr{P}(u + \omega_1) + \mathscr{P}(u + \omega_3)\} = -4e_2\mathscr{P}(2u) - 4e_1 e_3.$$

18. Express the identity (6.5.2) in the alternative form

$$\sigma(u) = \frac{2\omega_1}{\pi} \exp\left(\frac{\eta_1 u^2}{2\omega_1}\right) \sin\frac{\pi u}{2\omega_1} \prod_{n=1}^{\infty}\left[1 - \frac{\sin^2(\pi u/2\omega_1)}{\sin^2(n\pi\omega_3/\omega_1)}\right].$$

19. Obtain the expansion

$$\zeta(u) = \frac{\eta_1 u}{\omega_1} + \frac{\pi}{2\omega_1} \sum_{n=-\infty}^{\infty} \cot(z - n\pi\tau), \qquad z = \pi u/2\omega_1.$$

(Hint: Refer to the previous exercise and equation (6.6.1).)

20. By differentiating the expansion found in the last exercise, prove that

$$\mathscr{P}(u) = -\frac{\eta_1}{\omega_1} + \left(\frac{\pi}{2\omega_1}\right)^2 \sum_{n=-\infty}^{\infty} \operatorname{cosec}^2(z - n\pi\tau).$$

Show that

$$\mathscr{P}(u) - \left(\frac{\pi}{2\omega_1}\right)^2 \operatorname{cosec}^2 z \to -\frac{1}{3}\left(\frac{\pi}{2\omega_1}\right)^2$$

as $u \to 0$, and deduce that

$$\eta_1 = \frac{\pi^2}{2\omega_1}\left[\frac{1}{6} + \sum_{n=1}^{\infty} \operatorname{cosec}^2 n\pi\tau\right].$$

21. Prove that

(i) $\displaystyle\int \mathscr{P}^2(u)\,du = \frac{1}{6}\mathscr{P}'(u) + \frac{1}{12}g_2 u,$

(ii) $\displaystyle\int \mathscr{P}^3(u)\,du = \frac{1}{120}\mathscr{P}'''(u) - \frac{3}{20}g_2\zeta(u) + \frac{1}{10}g_3 u.$

22. Prove that

$$\mathscr{P}(u, \omega_1, \tfrac{1}{2}\omega_1(1 + iv)) = \left(\frac{\pi}{2\omega_1}\right)^2\left[\frac{1}{3}\cdot\frac{\psi'''(0)}{\psi'(0)} + \left(\frac{\psi'(z)}{\psi(z)}\right)^2 - \frac{\psi''(z)}{\psi(z)}\right],$$

where $z = \pi u/2\omega_1$ and

$$\psi(z) = \sum_{n=0}^{\infty} sp^{n(n+1)}\sin(2n + 1)z,$$

where $p = \exp(-\frac{1}{2}\pi v)$ and $s = i^{n(n-1)}$ is a sign factor (viz. $+, +, -, -$, etc.).

23. If $w = \mathscr{P}(u|0, g)$, show that

$$(w'')^3 = \frac{27}{2}(w'^2 + g)^2.$$

Deduce that the general solution of the equation

$$\left(\frac{dx}{du}\right)^3 = (ax^2 + bx + c)^2$$

is

$$x = \frac{27}{2a^2}\mathscr{P}'(u + A|0, g) - \frac{b}{2a},$$

where A is an arbitrary constant and $g = a^2(4ac - b^2)/729$.

Substituting $x = -1/y$, deduce that the general solution of the equation

$$\left(\frac{dy}{du}\right)^3 = y^2(ay^2 + by + c)^2$$

is

$$y = \lambda/\{\mu - \mathscr{P}'(u + A|0, g)\},$$

where $\lambda = 2c^2/27$, $\mu = -bc/27$, $g = c^2(4ac - b^2)/729$.

24. If $w = \mathscr{P}^2(u|g, 0)$, show that

$$w'^4 = 16w^3(4w - g)^2.$$

Deduce that the general solution of the equation

$$\left(\frac{dx}{du}\right)^4 = (x+a)^2(x+b)^3$$

is

$$x = 256\mathscr{P}^2(u + A|g, 0) - b,$$

where $g = (b - a)/64$.

25. If $\omega_1 = \omega$, $\omega_3 = \omega^2$, where $1, \omega, \omega^2$, are the cube roots of unity, show that $e_1 = \omega e_2, e_3 = \omega^2 e_2$, where e_2 is real. Deduce that $g_2 = 0$, $g_3 = 4e_2^3$ and show that

$$e_2 = \frac{4\pi^2}{\sqrt{3}}\lambda(1 - \lambda^2 - \lambda^6 + \lambda^{12} + \lambda^{20} - \cdots)^4,$$

where $\lambda = \exp(-\sqrt{3}\pi/2)$. (Hint: Refer to Exercise 4 above.)

CHAPTER 7

Applications of the Weierstrass Functions

7.1. Orthogonal Families of Cartesian Ovals

Consider the transformation

$$z = \mathscr{P}(w), \tag{7.1.1}$$

mapping a period parallelogram of $\mathscr{P}$ in the w-plane onto the z-plane. We shall suppose ω_1 to be real and ω_3 to be imaginary, so that the points $e_\alpha = \mathscr{P}(\omega_\alpha)$ all lie on the real axis in the z-plane (in the order e_3, e_2, e_1 from the left). Also, since $\mathscr{P}(w)$ is then real for real values of w, we shall have $\mathscr{P}(w^*) = [\mathscr{P}(w)]^*$ (for the Taylor expansion of $\mathscr{P}$ about any point on the real axis must have real coefficients).

Let r_α be the distance of the point z from the point e_α. Then

$$r_\alpha = |z - e_\alpha| = \sqrt{[\mathscr{P}(w) - e_\alpha][\mathscr{P}(w^*) - e_\alpha]} = [\mathscr{P}(w) - e_\alpha]^{1/2}[\mathscr{P}(w^*) - e_\alpha]^{1/2}$$
$$= \frac{\sigma_\alpha(w)\sigma_\alpha(w^*)}{\sigma(w)\sigma(w^*)}, \tag{7.1.2}$$

by equations (6.9.1) (the root sign taken must be correct, since the stated result makes r_α positive).

Equations (6.9.3) indicate that the identity (6.4.9) can be expressed in the form

$$\frac{\sigma_1(u)\sigma_1(v)}{\sigma(u)\sigma(v)}\sigma_2(u+v) - \frac{\sigma_2(u)\sigma_2(v)}{\sigma(u)\sigma(v)}\sigma_1(u+v) = (e_1 - e_2)\sigma_3(u+v). \tag{7.1.3}$$

Thus, replacing u by w and v by w^*, we deduce that

$$r_1\sigma_2(2u) - r_2\sigma_1(2u) = (e_1 - e_2)\sigma_3(2u), \tag{7.1.4}$$

where $w = u + iv$. This shows that the lines $u = $ constant in the w-plane transform into a family of Cartesian ovals (see section 4.6.) in the z-plane, their bipolar equation being (7.1.4).

Alternatively, we can make use of the identities (6.4.10) and (6.4.11) to give the equations

$$r_2\sigma_3(2u) - r_3\sigma_2(2u) = (e_2 - e_3)\sigma_1(2u),\qquad(7.1.5)$$

$$r_1\sigma_3(2u) - r_3\sigma_1(2u) = (e_1 - e_3)\sigma_2(2u),\qquad(7.1.6)$$

which must be the equations for the same family of Cartesian ovals, but with respect to the poles (e_2, e_3) and (e_1, e_3) respectively. This feature that a Cartesian oval possesses three collinear poles, any two of which may be used to yield its bipolar equation, has already been remarked in section 4.6.

If, in the identity (7.1.3), we replace u by w and v by $-w^*$, we are led to the equation

$$-r_1\sigma_2(2iv) + r_2\sigma_1(2iv) = (e_1 - e_2)\sigma_3(2iv).\qquad(7.1.7)$$

This proves that the lines $v = $ constant in the w-plane also transform into a family of Cartesian ovals. Since the lines $u = $ constant and $v = $ constant intersect orthogonally in the w-plane, the family of ovals (7.1.7) must also be orthogonal to the family (7.1.4).

By using the identities (6.4.10) and (6.4.11), alternative equations for this second family of ovals are found in the form

$$-r_2\sigma_3(2iv) + r_3\sigma_2(2iv) = (e_2 - e_3)\sigma_1(2iv).\qquad(7.1.8)$$

$$-r_1\sigma_3(2iv) + r_3\sigma_1(2iv) = (e_1 - e_3)\sigma_2(2iv).\qquad(7.1.9)$$

The similarity between these results and those of section 4.6 suggests that the transformations (4.6.1) and (7.1.1) must be closely related. That this is so follows immediately from equation (6.9.11). Thus, if $z = \mathscr{P}(w)$ and ω_1, ω_3 are identified with the quarter-periods of the Jacobi functions (equations (6.9.5)), then the first of equations (6.9.11) shows that

$$z = -\tfrac{1}{3}(1 + k^2) + \mathrm{ns}^2 w = k^2\,\mathrm{sn}^2(w + iK') - \tfrac{1}{3}(1 + k^2),\qquad(7.1.10)$$

which is simply related to the transformation (4.6.1).

7.2. Solution of Euler's Equations for Body Rotation

Euler's equations for the angular velocity $\boldsymbol{\omega} = (\omega_1, \omega_2, \omega_3)^*$ of a body rotating about a fixed point O, under forces having zero moment about this point, have been given at (5.7.3). The square of the magnitude of the angular velocity

---

*The periods $2\omega_a$ of $\mathscr{P}$ will not be referred to in this section, leaving the symbols ω_a free to represent components of angular velocity throughout.

is given by the equation

$$\omega_1^2 + \omega_2^2 + \omega_3^2 = \omega^2 \tag{7.2.1}$$

and we have two first integrals, viz.

$$A\omega_1^2 + B\omega_2^2 + C\omega_3^2 = T, \tag{7.2.2}$$

$$A^2\omega_1^2 + B^2\omega_2^2 + C^2\omega_3^2 = DT, \tag{7.2.3}$$

where $H^2 = DT$ is the square of the magnitude of the angular momentum. These three equations can be solved for $\omega_1^2, \omega_2^2, \omega_3^2$, to yield

$$\left.\begin{aligned}
\omega_1^2 &= \frac{BC}{(A-B)(A-C)}(\omega^2 - \alpha), \\[2mm]
\omega_2^2 &= \frac{CA}{(B-C)(B-A)}(\omega^2 - \beta), \\[2mm]
\omega_3^2 &= \frac{AB}{(C-A)(C-B)}(\omega^2 - \gamma),
\end{aligned}\right\} \tag{7.2.4}$$

where

$$\alpha = \frac{T}{BC}(B+C-D), \qquad \beta = \frac{T}{CA}(C+A-D), \qquad \gamma = \frac{T}{AB}(A+B-D). \tag{7.2.5}$$

It remains to derive an equation determining ω^2.

Using Euler's equations, we calculate that

$$\frac{d}{dt}(\omega^2) = 2\omega_1\dot{\omega}_1 + 2\omega_2\dot{\omega}_2 + 2\omega_3\dot{\omega}_3$$

$$= -\frac{2}{ABC}(B-C)(C-A)(A-B)\omega_1\omega_2\omega_3. \tag{7.2.6}$$

Substituting for $\omega_1, \omega_2, \omega_3$ from equations (7.2.4), we now find that

$$\frac{d}{dt}(\omega^2) = -2[-(\omega^2 - \alpha)(\omega^2 - \beta)(\omega^2 - \gamma)]^{1/2} \tag{7.2.7}$$

determines ω^2.

As in section 5.7, we shall assume that the principal moments of inertia (A, B, C) are in descending order of magnitude. Then, eliminating ω_1^2 between equations (7.2.2), (7.2.3), we obtain the equation

$$B(A-B)\omega_2^2 + C(A-C)\omega_3^2 = (A-D)T. \tag{7.2.8}$$

This shows that $A > D$. Again, eliminating ω_3^2 between these equations, we get

$$A(A-C)\omega_1^2 + B(B-C)\omega_2^2 = (D-C)T, \tag{7.2.9}$$

proving that $D > C$. There are, therefore, only two cases to consider:

$$\text{(a)} \quad A > D > B > C, \tag{7.2.10}$$

$$\text{(b)} \quad A > B > D > C. \tag{7.2.11}$$

In case (a), we find from equations (7.2.5) that

$$\text{(a)} \quad \beta > \gamma > \alpha. \tag{7.2.12}$$

In case (b), we find from the same equations that

$$\text{(b)} \quad \beta > \alpha > \gamma. \tag{7.2.13}$$

Since $A > D$, in both cases we have

$$\gamma > 0. \tag{7.2.14}$$

Equations (7.2.4) require that $\omega^2 \leqslant \beta$, $\omega^2 \geqslant \alpha$, $\omega^2 \geqslant \gamma$. Thus, in the cases (a) and (b) we shall have

$$\text{(a)} \quad \beta \geqslant \omega^2 \geqslant \gamma, \tag{7.2.15}$$

$$\text{(b)} \quad \beta \geqslant \omega^2 \geqslant \alpha, \tag{7.2.16}$$

respectively. In both cases, the expression under the square root sign in equation (7.2.7) is positive and the equation requires that ω^2 should oscillate between the extreme values (β, γ) in case (a) and (β, α) in case (b).

To bring the equation (7.2.7) to the canonical form (6.7.16), we need to change the dependent variable to χ by the transformation

$$\omega^2 = \chi + \tfrac{1}{3}(\alpha + \beta + \gamma). \tag{7.2.17}$$

Then

$$d\chi/dt = -2[(e_1 - \chi)(\chi - e_2)(\chi - e_3)]^{1/2}, \tag{7.2.18}$$

where, in conformity with the established notation, $e_1 > e_2 > e_3$ and

$$\text{(a)} \quad \begin{cases} e_1 = \beta - \tfrac{1}{3}(\alpha + \beta + \gamma) \\ e_2 = \gamma - \tfrac{1}{3}(\alpha + \beta + \gamma) \\ e_3 = \alpha - \tfrac{1}{3}(\alpha + \beta + \gamma) \end{cases}, \quad \text{(b)} \quad \begin{cases} e_1 = \beta - \tfrac{1}{3}(\alpha + \beta + \gamma) \\ e_2 = \alpha - \tfrac{1}{3}(\alpha + \beta + \gamma) \\ e_3 = \gamma - \tfrac{1}{3}(\alpha + \beta + \gamma) \end{cases}, \tag{7.2.19}$$

in the two cases (a) and (b) respectively.

We now have $e_1 + e_2 + e_3 = 0$ and the results given at (6.12.25) become applicable. Thus, integrating equation (7.2.18) for χ over the range (χ, e_1) and t over the range (t, t_0), where t_0 is the time when $\chi = e_1$, we find that

$$t_0 - t = -\frac{1}{2} \int_\chi^{e_1} [(e_1 - \chi)(\chi - e_2)(\chi - e_3)]^{-1/2} \, d\chi, \tag{7.2.20}$$

and, using the third of the standard integrals (6.12.25), this leads to the equation

$$-e_1 - \frac{(e_1 - e_2)(e_1 - e_3)}{\chi - e_1} = \mathscr{P}(t - t_0 | g_2, -g_3). \tag{7.2.21}$$

Solving for χ and choosing $t_0 = 0$, this yields the result

$$\omega^2 = \beta - \frac{(\beta - \gamma)(\beta - \alpha)}{\mathscr{P}(t|g_2, -g_3) + e_1}, \qquad (7.2.22)$$

valid for both cases (a) and (b). If we now substitute for ω^2 into equations (7.2.4), the results (5.7.7) and (5.7.18) are derived once again thus:

First, we note from equation (6.7.37) that

$$\mathscr{P}(t|g_2, -g_3) = p^2 \mathscr{P}(pt|p^{-4}g_2, -p^{-6}g_3). \qquad (7.2.23)$$

We now choose $p = (e_1 - e_3)^{1/2}$, i.e.,

$$(a) \quad p^2 = \beta - \alpha, \qquad (b) \quad p^2 = \beta - \gamma, \qquad (7.2.24)$$

in the two cases. The stationary values of $\mathscr{P}(t|g_2, -g_3)$ are $-e_3, -e_2, -e_1$ (equations (6.10.12)) and the stationary values of $\mathscr{P}(pt|p^{-4}g_2, -p^{-6}g_3)$ are accordingly $e_1' = -p^{-2}e_3, e_2' = -p^{-2}e_2, e_3' = -p^{-2}e_1$. It follows that

$$e_1' - e_3' = p^{-2}(e_1 - e_3) = 1 \qquad (7.2.25)$$

and equations (6.9.9) are therefore valid in respect of $\mathscr{P}(pt)$. We can now calculate the modulus k of the associated Jacobi functions thus,

$$k^2 = e_2' - e_3' = p^{-2}(e_1 - e_2), \qquad (7.2.26)$$

or

$$(a) \quad k^2 = \frac{\beta - \gamma}{\beta - \alpha}, \qquad (b) \quad k^2 = \frac{\beta - \alpha}{\beta - \gamma}. \qquad (7.2.27)$$

Also, the first of equations (6.9.11) yields

$$\mathscr{P}(t) = p^2 \mathscr{P}(pt) = p^2 \{ -\tfrac{1}{3}(1 + k^2) + \mathrm{ns}^2 pt \}. \qquad (7.2.28)$$

Substituting for $\mathscr{P}(t)$ in equation (7.2.22), we now calculate that

$$(a) \quad \omega^2 = \beta - (\beta - \gamma) \mathrm{sn}^2(pt, k), \qquad (7.2.29)$$

$$(b) \quad \omega^2 = \beta - (\beta - \alpha) \mathrm{sn}^2(pt, k). \qquad (7.2.30)$$

Equations (7.2.29), (7.2.30) confirm that ω^2 oscillates between the extreme values (β, γ) in case (a) and between the extreme values (β, α) in case (b).

Substituting for ω^2 from equation (7.2.29) into the first of equations (7.2.4), we now show that

$$\omega_1^2 = \frac{BC}{(A - B)(A - C)}(\beta - \alpha)\{1 - k^2 \mathrm{sn}^2(pt, k)\} = \frac{T(D - C)}{A(A - C)} \mathrm{dn}^2(pt, k), \qquad (7.2.31)$$

where

$$p^2 = \beta - \alpha = \frac{T}{ABC}(A - B)(D - C), \qquad (7.2.32)$$

$$k^2 = \frac{\beta - \gamma}{\beta - \alpha} = \frac{(A-D)(B-C)}{(A-B)(D-C)}. \tag{7.2.33}$$

These results are equivalent to the first of equations (5.7.18). It is left for the reader to verify the remaining equations (5.7.7) (case (b)) and (5.7.18) (case (a)).

7.3. Spherical Pendulum

In this section, we shall calculate the motion of a particle P, having mass m, suspended from a fixed point O by a light wire of length l, and acted upon by a constant weight mg and the thrust or tension in the wire. Thus, the trajectory of the bob lies on a sphere of radius l and center O.

Let AOA' be the vertical diameter of the sphere on which the mass moves (Fig. 7.1) and let PN be the perpendicular from P to this diameter. Then, cylindrical polar coordinates for P are (r, θ, z), where $r = PN$, θ is the angle between the meridian APA' and a datum meridian AMA', and $z = ON$ (positive for N below O). We can now write down the energy equation for the bob in the form

$$\tfrac{1}{2}m(\dot{r}^2 + r^2\dot{\theta}^2 + \dot{z}^2) - mgz = \tfrac{1}{2}mv_0^2 - mgz_0, \tag{7.3.1}$$

where v_0 is the velocity and z_0 the depth of the particle below O at the instant of projection.

Since both the forces acting upon P have zero moment about the z-axis, the particle's angular momentum about this axis is conserved. Thus,

$$mr^2\dot{\theta} = mv_0 r_0 \sin \alpha, \tag{7.3.2}$$

where r_0 is the initial value of r and α is the angle made by the velocity of projection with the meridian. The case when the bob swings in a vertical plane has already been treated in section 5.1 and we shall here assume, therefore, that α does not equal zero.

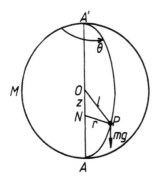

Figure 7.1. Spherical pendulum.

Writing

$$c = v_0^2 - 2gz_0, \qquad h = v_0 r_0 \sin\alpha, \tag{7.3.3}$$

we can eliminate $\dot\theta$ between equations (7.3.1) and (7.3.2) to give

$$\dot r^2 + \dot z^2 = c + 2gz - h^2/r^2. \tag{7.3.4}$$

But, $r^2 + z^2 = l^2$ and, hence, $r\dot r + z\dot z = 0$. Thus, eliminating r and $\dot r$, we arrive at the result

$$l^2\dot z^2 = (c + 2gz)(l^2 - z^2) - h^2 = \phi(z), \tag{7.3.5}$$

determining the variation of z with t.

Now note the following particular values of the cubic $\phi(z)$:

$$\phi(-\infty) = +\infty, \qquad \phi(-l) = -h^2, \qquad \phi(z_0) > 0, \qquad \phi(l) = -h^2. \tag{7.3.6}$$

Clearly, $\phi(z)$ must have a zero in each of the intervals $(-\infty, -l), (-l, z_0)$, and (z_0, l); we shall denote these by z_1, z_2, z_3 respectively. Hence,

$$z_1 < -l < z_2 < z_0 < z_3 < l \tag{7.3.7}$$

and equation (7.3.5) can be written in the form

$$l^2\dot z^2 = 2g(z - z_1)(z - z_2)(z_3 - z). \tag{7.3.8}$$

Since z must lie in the interval $(-l, l)$, for $\dot z$ to be real we require that

$$z_2 \leqslant z \leqslant z_3, \tag{7.3.9}$$

i.e., the bob moves between the horizontal planes $z = z_2$ (upper) and $z = z_3$ (lower), touching each plane in turn when $\dot z = 0$.

Since $\dot z^2$ equals a cubic in z, z must be an elliptic function of t. To bring the equation (7.3.5) to the canonical form (6.7.26), we transform the dependent variable z linearly by

$$z = -2ls - c/6g \tag{7.3.10}$$

to a new dimensionless dependent variable s and, at the same time, transform the independent variable t to a dimensionless time τ ($\tau \neq \omega_3/\omega_1$ in this section) by the equation

$$\tau = \lambda t, \quad \text{where } \lambda^2 = g/l. \tag{7.3.11}$$

Equation (7.3.5) then takes the form

$$(ds/d\tau)^2 = 4s^3 - g_2 s - g_3, \tag{7.3.12}$$

where

$$g_2 = 1 + \frac{c^2}{12g^2 l^2}, \qquad g_3 = \frac{h^2}{4gl^3} + \frac{c^3}{216g^3 l^3} - \frac{c}{6gl} \tag{7.3.13}$$

and are dimensionless invariants. It now follows that

$$s = \mathscr{P}(\tau + A | g_2, g_3) \tag{7.3.14}$$

where A is the integration constant. Since the zeros of the cubic in s are all real, ω_1 is real and ω_3 is purely imaginary.

The roots of the cubic equation $\phi(z) = 0$ have been taken to be $z_1 < z_2 < z_3$. The roots of the equation $4s^3 - g_2 s - g_3 = 0$ will be denoted, as usual, by e_α ($\alpha = 1, 2, 3$), where $e_3 < e_2 < e_1$. Thus, $s = e_1$ when $z = z_1$, $s = e_2$ when $z = z_2$, and $s = e_3$ when $z = z_3$. If, therefore, we choose $t = 0$ to be an instant when $z = z_3$, i.e., the bob is at its lowest level, then $s = e_3$ at this time. Substituting $\tau = 0, s = e_3$ in equation (7.3.14), we find $\mathscr{P}(A) = e_3$ and it follows that $A = \omega_3$ (i.e., A is imaginary, but the values of $\mathscr{P}(\tau + A)$ are real, as shown in section 6.11). Our solution for z is accordingly

$$z = -2l\mathscr{P}(\tau + \omega_3) - \frac{c}{6g}$$

$$= -2l\left[e_3 + \frac{(e_3 - e_1)(e_3 - e_2)}{\mathscr{P}(\tau) - e_3}\right] - \frac{c}{6g}$$

$$= z_3 - \frac{1}{2l} \cdot \frac{(z_3 - z_1)(z_3 - z_2)}{\mathscr{P}(\lambda t) - e_3}, \tag{7.3.15}$$

having used equation (6.8.13).

When $\tau = 0$, $\mathscr{P}(\tau) = +\infty$ and, thus, $z = z_3$. When $\tau = \omega_1$, $\mathscr{P}(\tau) = e_1$ and so $z = z_2$. $\lambda\omega_1$, therefore, is the time taken by the bob to move from its lowest level to its highest (or *vice versa*). $2\lambda\omega_1$ is the period of the complete z-oscillation.

To calculate the angle θ, we refer to equation (7.3.2). This yields

$$\dot{\theta} = h/r^2 = h/(l^2 - z^2) = \frac{h}{2l}\left[\frac{1}{l - z} + \frac{1}{l + z}\right]. \tag{7.3.16}$$

Substituting for z from equation (7.3.10) and defining a, b such that

$$\mathscr{P}(a) = -\frac{1}{2} - \frac{c}{12gl}, \qquad \mathscr{P}(b) = \frac{1}{2} - \frac{c}{12gl}, \tag{7.3.17}$$

we can express

$$d\theta/d\tau = \frac{h}{4\sqrt{(gl^3)}}\left[\frac{1}{s - \mathscr{P}(a)} - \frac{1}{s - \mathscr{P}(b)}\right]. \tag{7.3.18}$$

$\mathscr{P}(a), \mathscr{P}(b)$ are the values of s corresponding to $z = l, -l$ respectively. Thus, as z decreases through the values $l, z_3, z_2, -l, z_1$ (see (7.3.7)), s increases through the corresponding values $\mathscr{P}(a), e_3, e_2, \mathscr{P}(b), e_1$. Hence

$$\mathscr{P}(a) < e_3 < e_2 < \mathscr{P}(b) < e_1. \tag{7.3.19}$$

In section 6.11, we showed that $\mathscr{P}(u)$ increases from $-\infty$ to e_3 as u varies from 0 to ω_3 along the imaginary axis. Hence, we can take $a = \pm \alpha\omega_3$, where $0 < \alpha < 1$. We also proved that $\mathscr{P}(\omega_1 + u)$ decreases from e_1 to e_2 for the same variation of u. Thus, we can take $b = \pm(\omega_1 + \beta\omega_3)$, where $0 < \beta < 1$. The signs of a and b are open to choice and will now be determined.

We note that

$$[\mathscr{P}'(\tau)]^2 = 4\mathscr{P}^3(\tau) - g_2\mathscr{P}(\tau) - g_3 = \frac{1}{4gl^3}\phi[-2l\mathscr{P}(\tau) - c/6g], \quad (7.3.20)$$

is valid for all complex values of τ (except for the poles). Hence

$$\left.\begin{array}{l} [\mathscr{P}'(a)]^2 = \dfrac{1}{4gl^3}\phi(l) = -\dfrac{h^2}{4gl^3}, \\[4mm] [\mathscr{P}'(b)]^2 = \dfrac{1}{4gl^3}\phi(-l) = -\dfrac{h^2}{4gl^3}. \end{array}\right\} \qquad (7.3.21)$$

We now assign the signs of a and b so that $\mathscr{P}'(a) = \mathscr{P}'(b) = +ih/\sqrt{(4gl^3)}$. Equation (7.3.18) can now be written in the form

$$\begin{aligned} i(d\theta/d\tau) &= \frac{1}{2}\left[\frac{\mathscr{P}'(a)}{\mathscr{P}(\tau + \omega_3) - \mathscr{P}(a)} - \frac{\mathscr{P}'(b)}{\mathscr{P}(\tau + \omega_3) - \mathscr{P}(b)}\right], \\ &= \tfrac{1}{2}[-\zeta(\tau + \omega_3 + a) + \zeta(\tau + \omega_3 - a) + \zeta(\tau + \omega_3 + b) \\ &\quad - \zeta(\tau + \omega_3 - b)] + \zeta(a) - \zeta(b), \end{aligned} \qquad (7.3.22)$$

the last step being validated by the identity (6.8.2). Using equation (6.6.1), equation (7.3.22) integrates immediately to give

$$i\theta = \frac{1}{2}\ln\left[\frac{\sigma(\tau + \omega_3 - a)\sigma(\tau + \omega_3 + b)}{\sigma(\tau + \omega_3 + a)\sigma(\tau + \omega_3 - b)}\right] + [\zeta(a) - \zeta(b)]\tau + \text{constant.} \quad (7.3.23)$$

The integration constant can be fixed by the condition $\theta = 0$ at $\tau = 0$.

If we increment τ by $2\omega_1$ in the right-hand member of equation (7.3.23), it follows from equation (6.2.10) that θ increments by ε, where

$$i\varepsilon = 2\eta_1(b - a) - 2\omega_1[\zeta(b) - \zeta(a)]. \qquad (7.3.24)$$

ε is the angle through which the vertical plane containing the pendulum rotates as z makes one complete oscillation. It must, of course, be real, as indeed follows from the fact that the expression on the right-hand side of (7.3.24) has zero real part: Thus, $b - a = \omega_1 + (\beta - \alpha)\omega_3$ and the real part of $2\eta_1(b - a)$ is therefore $2\eta_1\omega_1$. $\zeta(a) = \zeta(\alpha\omega_3)$ is purely imaginary. Using the identity (6.8.4), we find that

$$\begin{aligned} \zeta(b) = \zeta(\omega_1 + \beta\omega_3) &= \zeta(\omega_1) + \zeta(\beta\omega_3) + \frac{1}{2}\cdot\frac{\mathscr{P}'(\omega_1) - \mathscr{P}'(\beta\omega_3)}{\mathscr{P}(\omega_1) - \mathscr{P}(\beta\omega_3)} \\ &= \eta_1 + \zeta(\beta\omega_3) - \frac{1}{2}\cdot\frac{\mathscr{P}'(\beta\omega_3)}{e_1 - \mathscr{P}(\beta\omega_3)}. \end{aligned} \qquad (7.3.25)$$

Since $\zeta(\beta\omega_3), \mathscr{P}'(\beta\omega_3)$ are imaginary and $\mathscr{P}(\beta\omega_3)$ is real, the real part of $\zeta(b)$ is seen to be η_1. It now follows that the real parts of the two terms on the right-hand side of (7.3.24) cancel and that ε is real.

To calculate the projection onto a horizontal plane of the motion of the bob, we define Cartesian coordinates (x, y) in such a plane by equations $x = r\cos\theta$, $y = r\sin\theta$. Then

$$X = x + iy = re^{i\theta}, \qquad Y = x - iy = re^{-i\theta}, \qquad (7.3.26)$$

$$
\begin{aligned}
XY = r^2 &= l^2 - z^2 = (l - z)(l + z) \\
&= 4l^2[\mathscr{P}(\tau + \omega_3) - \mathscr{P}(a)][\mathscr{P}(b) - \mathscr{P}(\tau + \omega_3)] \\
&= -4l^2\frac{\sigma(\tau + \omega_3 + a)\sigma(\tau + \omega_3 - a)\sigma(\tau + \omega_3 + b)\sigma(\tau + \omega_3 - b)}{\sigma^2(a)\sigma^2(b)\sigma^4(\tau + \omega_3)},
\end{aligned} \qquad (7.3.27)
$$

having appealed to the identity (6.7.5). Also, using the result (7.3.23),

$$X/Y = e^{2i\theta} = C\frac{\sigma(\tau + \omega_3 - a)\sigma(\tau + \omega_3 + b)}{\sigma(\tau + \omega_3 + a)\sigma(\tau + \omega_3 - b)}e^{2\{\zeta(a) - \zeta(b)\}}. \qquad (7.3.28)$$

From the last two equations, we now deduce immediately that

$$
\left.
\begin{aligned}
X &= 2ilC^{1/2}\frac{\sigma(\tau + \omega_3 - a)\sigma(\tau + \omega_3 + b)}{\sigma(a)\sigma(b)\sigma^2(\tau + \omega_3)}e^{\{\zeta(a) - \zeta(b)\}\tau}, \\
Y &= 2ilC^{-1/2}\frac{\sigma(\tau + \omega_3 + a)\sigma(\tau + \omega_3 - b)}{\sigma(a)\sigma(b)\sigma^2(\tau + \omega_3)}e^{\{\zeta(b) - \zeta(a)\}\tau}.
\end{aligned}
\right\} \qquad (7.3.29)
$$

By plotting the complex number X in the Argand plane, for a succession of values of τ, the projected trajectory can be traced out. There are two cases: (i) when $z_3 > z_2 > 0$ and (ii) when $z_3 > 0 > z_2$. In the first case, the bob never rises above the level of O; in the second case, part of the motion is above O and part is below. The form of the projected trajectory in each case is indicated in Fig. 7.2.

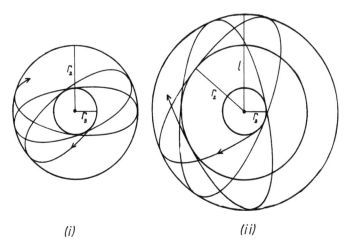

(i)   (ii)

Figure 7.2. Trajectories of a pendulum bob.

To calculate the tension T in the suspending wire, we resolve "force = mass × acceleration" in this direction to give the equation

$$T - mg \cos \psi = mv^2/l, \tag{7.3.30}$$

where ψ is the angle made by the wire with the downward vertical. Since $\cos \psi = z/l$ and $v^2 = 2gz + c$ is the energy equation, then

$$T = m(3gz + c)/l = -6mg \mathscr{P}(\tau + \omega_3) + mc/2l. \tag{7.3.31}$$

7.4. Motion of a Spinning Top or Gyroscope

We shall suppose that a point O on the top's axis is held fixed, but that it is otherwise free to rotate about this point without frictional resistance being generated. In Fig. 7.3, $Oxyz$ is a nonrotating (inertial) rectangular frame, with Oz directed vertically upward. $O123$ is a rotating rectangular frame, such that $O3$ is the top's axis of symmetry, $O13$ is the vertical plane through this axis, and $O2$ lies in the horizontal plane Oxy. θ is the angle made by the top's axis with the upward vertical Oz and ϕ is the angle made by the vertical plane through this axis with the plane Oxz.

The frame $O123$ has simultaneous angular velocities $\dot\theta$ about $O2$ and $\dot\phi$ about Oz. Thus, the components of its vector angular velocity $\boldsymbol{\Omega}$ taken in the directions of the axes of the frame $O123$ are as follows:

$$\boldsymbol{\Omega} = (-\dot\phi \sin \theta, \dot\theta, \dot\phi \cos \theta). \tag{7.4.1}$$

Relative to the frame $O123$, the top has an angular velocity $\dot\psi$ about $O3$. Hence, its net vector angular velocity $\boldsymbol{\omega}$ relative to a fixed frame has components

$$\boldsymbol{\omega} = (-\dot\phi \sin \theta, \dot\theta, n) \tag{7.4.2}$$

along the axes $O1, O2, O3$, where $n = \dot\phi \cos \theta + \dot\psi$.

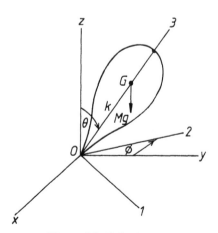

Figure 7.3. Spinning top.

Because of the symmetry of the top about $O3$, the axes of the frame $O123$ are principal axes of inertia for the top with respect to O. We denote the principal moments of inertia by (A, A, C). Then, the vector angular momentum of the top's motion about O is given by

$$\mathbf{h} = (-A\dot{\phi}\sin\theta, A\dot{\theta}, Cn). \tag{7.4.3}$$

Let G be the center of mass of the top and let $OG = k$. Then the vector moment $\mathbf{L}$ of the weight Mg of the top about O has components

$$\mathbf{L} = (0, Mgk\sin\theta, 0) \tag{7.4.4}$$

in the frame $O123$.

As in section 5.7, $\partial\mathbf{h}/\partial t$ will denote the rate of change of $\mathbf{h}$ relative to the moving frame $O123$ and $d\mathbf{h}/dt$ will denote the rate of change of $\mathbf{h}$ relative to a fixed frame instantaneously coincident with $O123$. Then, the equation of angular momentum can be written

$$d\mathbf{h}/dt = \partial\mathbf{h}/\partial t + \boldsymbol{\Omega} \times \mathbf{h} = \mathbf{L}. \tag{7.4.5}$$

We shall only make use of the $O3$-component of this equation, viz.

$$C\dot{n} = 0. \tag{7.4.6}$$

This demonstrates that n is constant (assumed positive).

If (h_x, h_y, h_z) are the components of $\mathbf{h}$ along the axes of the fixed frame $Oxyz$, the equation of angular momentum has components

$$\dot{h}_x = L_x, \qquad \dot{h}_y = L_y, \qquad \dot{h}_z = L_z. \tag{7.4.7}$$

But $L_z = 0$ (i.e., the weight has no moment about Oz) and it follows that h_z is constant. Thus, further resolving the components (7.4.3), we find

$$h_z = A\dot{\phi}\sin^2\theta + Cn\cos\theta = H, \tag{7.4.8}$$

where H is constant.

Finally, we need the energy equation

$$\tfrac{1}{2}(A\dot{\phi}^2\sin^2\theta + A\dot{\theta}^2 + Cn^2) + Mgk\cos\theta = E, \tag{7.4.9}$$

where E is constant.

Eliminating $\dot{\phi}$ between the last two equations and putting $\lambda = \cos\theta$, we derive the equation

$$A^2\dot{\lambda}^2 = A(1-\lambda^2)(K - 2Mgk\lambda) - (H - Cn\lambda)^2 = f(\lambda), \tag{7.4.10}$$

where

$$K = 2E - Cn^2. \tag{7.4.11}$$

K must be positive, for otherwise $f(\lambda) < 0$ when $0 < |\lambda| < 1$. $f(\lambda)$ is a cubic in λ and it follows that, in general, λ is an elliptic function of t.

Suppose that, initially, $\lambda = \lambda_0$; then $f(\lambda_0) \geqslant 0$. Also, $f(+\infty) = +\infty$, $f(1) < 0$ (the special case $H = Cn$ will not be treated), $f(\lambda_0) \geqslant 0, f(-1) < 0$, showing that $f(\lambda)$ has two zeros between -1 and 1 (the special case of coincident zeros will be ignored) and a third real zero greater than 1. Denoting these zeros by $\lambda_1 > 1 > \lambda_2 > \lambda_3 > -1$, equation (7.4.10) can be written

$$A\dot{\lambda}^2 = 2Mgk(\lambda_1 - \lambda)(\lambda_2 - \lambda)(\lambda - \lambda_3). \qquad (7.4.12)$$

Since $|\lambda| \leqslant 1$ during the motion, this equation shows that λ must oscillate between the values λ_2 and λ_3.

The analysis now proceeds as in the last section. We first transform the dependent variable to s by the linear transformation

$$\lambda = Ps + Q, \quad \text{where } P = 2A/Mgk, Q = (AK + C^2n^2)/6AMgk. \quad (7.4.13)$$

This brings (7.4.10) to the canonical form

$$\dot{s}^2 = 4s^3 - g_2 s - g_3, \qquad (7.4.14)$$

where

$$\left. \begin{array}{l} A^4 g_2 = \frac{1}{12}(AK + C^2n^2)^2 + AMgk(AMgk - HCn), \\ A^6 g_3 = \frac{1}{216}(AK + C^2n^2)^3 + \frac{1}{12}AMgk(AMgk - HCn)(AK + C^2n^2) \\ \qquad + \frac{1}{4}(AMgk)^2(H^2 - AK). \end{array} \right\} \quad (7.4.15)$$

The zeros e_1, e_2, e_3 of the cubic in s correspond to the zeros $\lambda_1, \lambda_2, \lambda_3$ respectively. Taking $t = 0$ when $\lambda = \lambda_3$ (i.e., θ a maximum), then $s = e_3$ at $t = 0$ and the appropriate solution of equation (7.4.14) is $s = \mathscr{P}(t + \omega_3 | g_2, g_3)$. We conclude that

$$\lambda = P\mathscr{P}(t + \omega_3) + Q, \qquad (7.4.16)$$

where ω_1 is real and ω_3 is purely imaginary.

To obtain $\phi(t)$, we use equation (7.4.8). Thus,

$$\dot{\phi} = (H - Cn\cos\theta)/A\sin^2\theta = \frac{H - Cn\lambda}{A(1 - \lambda^2)} = \frac{L}{\lambda + 1} - \frac{M}{\lambda - 1}, \quad (7.4.17)$$

where

$$L = (H + Cn)/2A, \qquad M = (H - Cn)/2A. \qquad (7.4.18)$$

Substituting for λ from equation (7.4.16), we can express the equation (7.4.17) in the form

$$\dot{\phi} = \frac{L/P}{\mathscr{P}(t + \omega_3) - \mathscr{P}(a)} - \frac{M/P}{\mathscr{P}(t + \omega_3) - \mathscr{P}(b)}, \qquad (7.4.19)$$

where

$$\mathscr{P}(a) = -(Q + 1)/P, \qquad \mathscr{P}(b) = -(Q - 1)/P. \qquad (7.4.20)$$

$\mathscr{P}(a)$ is the value of s when $\lambda = -1$ and $\mathscr{P}(b)$ is the value of s when $\lambda = 1$. It follows that

$$\mathscr{P}(a) < e_3 < e_2 < \mathscr{P}(b) < e_1 \qquad (7.4.21)$$

and, hence, that we can take $a = i\alpha, b = \omega_1 + i\beta$ (α, β real). Since

$$[\mathscr{P}'(t)]^2 = 4\mathscr{P}^3(t) - g_2\mathscr{P}(t) - g_3 = \frac{1}{A^2P^2} f(P\mathscr{P}(t) + Q) \qquad (7.4.22)$$

for values of t real or complex, we calculate that

$$\left.\begin{array}{l} [\mathscr{P}'(a)]^2 = \dfrac{1}{A^2 P^2} f(-1) = -(H + Cn)^2/A^2 P^2 = -4L^2/P^2, \\[3mm] [\mathscr{P}'(b)]^2 = \dfrac{1}{A^2 P^2} f(+1) = -(H - Cn)^2/A^2 P^2 = -4M^2/P^2. \end{array}\right\} \qquad (7.4.23)$$

We now fix the signs of a, b by choosing

$$\mathscr{P}'(a) = 2\mathrm{i}L/P, \qquad \mathscr{P}'(b) = 2\mathrm{i}M/P. \qquad (7.4.24)$$

Thus, equation (7.4.19) now reduces to the form

$$\mathrm{i}\dot{\phi} = \frac{1}{2}\left[\frac{\mathscr{P}'(a)}{\mathscr{P}(t + \omega_3) - \mathscr{P}(a)} - \frac{\mathscr{P}'(b)}{\mathscr{P}(t + \omega_3) - \mathscr{P}(b)}\right] \qquad (7.4.25)$$

and can be integrated easily, as in the previous section, to yield the result

$$\mathrm{i}\phi = \frac{1}{2}\ln\left[\frac{\sigma(t + \omega_3 - a)\sigma(t + \omega_3 + b)}{\sigma(t + \omega_3 + a)\sigma(t + \omega_3 - b)}\right] + [\zeta(a) - \zeta(b)]t + \text{const.} \quad (7.4.26)$$

7.5. Projectile Subject to Resistance Proportional to Cube of Speed

If m is the mass of the projectile and w is its speed, the resistive force opposing the motion will be assumed to be mkw^3, where k is constant. The weight mg will be assumed constant. Ultimately, supposing the projectile does not strike the ground, it will fall vertically with the constant terminal velocity W given by

$$W^3 = g/k, \qquad (7.5.1)$$

at which speed the weight and resistance balance one another. Clearly, any point on a possible trajectory can be regarded as the point of initial projection and, therefore, no matter what the initial conditions, it is possible to imagine the trajectory produced backward from the point until the speed becomes infinite and to adopt the point at which this occurs as the starting point for our integration of the equations of motion. Thus, it will be convenient to take $w = +\infty$ at $t = 0$. Then, the only initial condition which remains variable is the direction of projection and a single infinity of trajectories will embrace all possibilities.

For further convenience, we shall employ oblique axes, the x-axis being in the direction of projection and the y-axis vertically downward. Then, if $\mathbf{i}, \mathbf{j}$ are unit vectors along these axes respectively, the position vector $\mathbf{r}$ of the projectile, when its coordinates are (x, y), is given by

$$\mathbf{r} = x\mathbf{i} + y\mathbf{j}. \qquad (7.5.2)$$

Thus, the velocity **w** of the projectile is given by

$$\mathbf{w} = \dot{\mathbf{r}} = \dot{x}\mathbf{i} + \dot{y}\mathbf{j} = u\mathbf{i} + v\mathbf{j}, \tag{7.5.3}$$

where $u = \dot{x}, v = \dot{y}$ are the components of velocity along the axes.

The resistive force acting on the projectile when its velocity is **w** can be conveniently expressed in the vector form $-mkw^2\mathbf{w}$. Thus, the equation of motion is

$$m\dot{\mathbf{w}} = -mkw^2\mathbf{w} + mg\mathbf{j}. \tag{7.5.4}$$

Substituting for **w** from equation (7.5.3) and equating coefficients of **i** and **j** in the two members, we obtain the following two components of the equation of motion:

$$\dot{u} = -kw^2u, \qquad \dot{v} = -kw^2v + g. \tag{7.5.5}$$

Suppose that the line of projection is at an angle α above the horizontal. Then $\mathbf{i} \cdot \mathbf{j} = -\sin\alpha$. Thus, squaring equation (7.5.3), we calculate that

$$w^2 = u^2 + v^2 - 2uv\sin\alpha. \tag{7.5.6}$$

Next, division of the equations (7.5.5) leads to the equation

$$\frac{dv}{du} = \frac{v}{u} - \frac{g}{kuw^2}. \tag{7.5.7}$$

Transforming to a new dependent variable p by putting $v = pu$, so that

$$p = v/u = \dot{y}/\dot{x} = dy/dx, \tag{7.5.8}$$

equation (7.5.7) can be expressed in the form

$$u\frac{dp}{du} = -\frac{g}{ku^3(p^2 - 2p\sin\alpha + 1)}, \tag{7.5.9}$$

after using equation (7.5.6). The variables p and u are now separable and the last equation can be integrated to yield

$$\tfrac{1}{3}p^3 - p^2\sin\alpha + p = W^3/3u^3 + \text{constant}. \tag{7.5.10}$$

But, initially $v = \dot{y} = 0$, i.e., $p = 0$, and $u = \infty$. Hence, the constant vanishes and we can solve for u in the form

$$u = W/P^{1/3}, \quad \text{where } P = p^3 - 3p^2\sin\alpha + 3p. \tag{7.5.11}$$

It then follows that

$$v = pu = Wp/P^{1/3}. \tag{7.5.12}$$

Differentiating equation (7.5.8) with respect to t, we calculate that

$$dp/dt = (u\dot{v} - \dot{u}v)/u^2 = g/u, \tag{7.5.13}$$

by equations (7.5.5). Hence,

$$t = \frac{W}{g}\int_0^p P^{-1/3}\, dp \tag{7.5.14}$$

expresses the time t in terms of the parameter p.

We now have

$$
\left.
\begin{aligned}
\frac{dx}{dp} &= \frac{dx}{dt}\cdot\frac{dt}{dp} = \frac{u^2}{g} = \frac{W^2}{g}P^{-2/3}, \\
\frac{dy}{dp} &= \frac{dy}{dt}\cdot\frac{dt}{dp} = \frac{uv}{g} = \frac{W^2}{g}pP^{-2/3}.
\end{aligned}
\right\}
\tag{7.5.15}
$$

These equations integrate to yield parametric equations for the trajectory, viz.

$$
x = \frac{W^2}{g}\int_0^p P^{-2/3}\, dp, \qquad y = \frac{W^2}{g}\int_0^p pP^{-2/3}\, dp.
\tag{7.5.16}
$$

It remains to evaluate the integrals in (7.5.14) and (7.5.16).

Firstly, to integrate $P^{-2/3}$, we change the variable by $q = P^{1/3}/p$. Then

$$
\begin{aligned}
dq/dp &= -p^{-2}P^{1/3} + p^{-1}P^{-2/3}(p^2 - 2p\sin\alpha + 1) \\
&= P^{-2/3}(\sin\alpha - 2/p),
\end{aligned}
\tag{7.5.17}
$$

and

$$
\begin{aligned}
q^3 = P/p^3 &= 1 - \frac{3}{p}\sin\alpha + \frac{3}{p^2} \\
&= \tfrac{1}{4}[4 - 3\sin^2\alpha + 3(\sin\alpha - 2/p)^2].
\end{aligned}
\tag{7.5.18}
$$

It follows from these two equations that

$$
-\sqrt{3}(4q^3 - 4 + 3\sin^2\alpha)^{-1/2}\, dq = P^{-2/3}\, dp
\tag{7.5.19}
$$

(N.B. q decreases from $+\infty$ as p increases from 0) and, hence, that

$$
\begin{aligned}
x &= \frac{\sqrt{3}W^2}{g}\int_q^\infty (4q^3 - 4 + 3\sin^2\alpha)^{-1/2}\, dq \\
&= \frac{W^2}{g}\int_r^\infty (4r^3 - g_3)^{-1/2}\, dr,
\end{aligned}
\tag{7.5.20}
$$

where

$$
r = q/3 = P^{1/3}/3p, \qquad g_3 = (4 - 3\sin^2\alpha)/27.
\tag{7.5.21}
$$

It now follows immediately that

$$
x = \frac{W^2}{g}\mathscr{P}^{-1}(P^{1/3}/3p\,|\,0, g_3).
\tag{7.5.22}
$$

Inverting the last result, we obtain

$$
\tfrac{1}{3}q = P^{1/3}/3p = \mathscr{P}(gx/W^2).
\tag{7.5.23}
$$

Differentiating with respect to p, this gives

$$
\frac{1}{3}\frac{dq}{dp} = \frac{q}{W^2}\mathscr{P}'(gx/W^2)\left(\frac{dx}{dp}\right).
\tag{7.5.24}
$$

But

$$dq/dp = P^{-2/3}(\sin\alpha - 2/p) \quad \text{and} \quad dx/dp = \frac{W^2}{g}P^{-2/3}. \qquad (7.5.25)$$

We conclude that

$$\mathscr{P}'(gx/W^2) = (p\sin\alpha - 2)/3p. \qquad (7.5.26)$$

As $p \to \infty$, the tangent to the trajectory approaches the vertical and the projectile ultimately falls along a vertical asymptote, whose equation we shall take to be $x = a$. Thus, substituting these values for p and x in the last equation, we find that

$$\mathscr{P}'(ga/W^2) = \tfrac{1}{3}\sin\alpha. \qquad (7.5.27)$$

This now permits us to write equation (7.5.26) in the form

$$p = \frac{2}{3\{\mathscr{P}'(ga/W^2) - \mathscr{P}'(gx/W^2)\}}. \qquad (7.5.28)$$

Observe that equation (7.5.18) requires that $q \to 1$ as $p \to \infty$. Hence, by equation (7.5.23),

$$\mathscr{P}(ga/W^2) = 1/3. \qquad (7.5.29)$$

Since $G = 0$, $g_3 > 0$, reference to Table M shows that $\omega_1^6 g_3 = 12.825$. Thus, $\omega_1 = 1.530 g_3^{-1/6}$ and, hence, $\omega_1^2\mathscr{P}(ga/W^2) = 0.780 g_3^{-1/3}$. In particular, if $\alpha = 0$ (i.e., horizontal projection), then $g_3 = 4/27$ and, thus, $\omega_1^2\mathscr{P}(ga/W^2) = 1.475$. Referring to Table L, we find that $ga/(W^2\omega_1) = 1$, or $a = 2.103 W^2/g$. For $\alpha = 45°$, $g_3 = 5/54$, $\omega_1^2\mathscr{P}(ga/W^2) = 1.724$, $ga/(W^2\omega_1) = 0.787$, $a = 1.790 W^2/g$.

Since $p = dy/dx$, we can now calculate the equation of the trajectory, thus:

$$y = \int_0^x p\,dx = \frac{2W^2}{3g}\int_0^{gx/W^2} \frac{d\xi}{\mathscr{P}'(\eta) - \mathscr{P}'(\xi)}, \qquad (7.5.30)$$

where $\eta = ga/W^2$. But, we know that

$$[\mathscr{P}'(\xi)]^2 = 4\mathscr{P}^3(\xi) - g_3, \qquad (7.5.31)$$

from which it follows that

$$y = \frac{W^2}{6g}\int_0^{gx/W^2} \frac{\mathscr{P}'(\eta) + \mathscr{P}'(\xi)}{\mathscr{P}^3(\eta) - \mathscr{P}^3(\xi)}\,d\xi. \qquad (7.5.32)$$

We now factorize the denominator of the integrand, thus:

$$\mathscr{P}^3(\eta) - \mathscr{P}^3(\xi) = (\mathscr{P}(\eta) - \mathscr{P}(\xi))(\omega\mathscr{P}(\eta) - \mathscr{P}(\xi))(\omega^2\mathscr{P}(\eta) - \mathscr{P}(\xi)), \qquad (7.5.33)$$

where $1, \omega, \omega^2$ are the three cube roots of unity. The integrand then separates out into partial fractions to give

$$y = \frac{W^2}{2g} \int_0^{gx/W^2} \{\mathscr{P}'(\eta) + \mathscr{P}'(\xi)\}$$

$$\times \left[\frac{1}{\mathscr{P}(\eta) - \mathscr{P}(\xi)} + \frac{\omega}{\omega\mathscr{P}(\eta) - \mathscr{P}(\xi)} + \frac{\omega^2}{\omega^2\mathscr{P}(\eta) - \mathscr{P}(\xi)} \right] d\xi. \quad (7.5.34)$$

(Note: $\mathscr{P}(\eta) = 1/3$ by equation (7.5.29).)

Putting $g_2 = 0, \lambda = \omega$, in the identity (6.7.37), we deduce that

$$\mathscr{P}(\omega u | 0, g_3) = \omega\mathscr{P}(u | 0, g_3). \quad (7.5.35)$$

Similarly, with $\lambda = \omega^2$, we calculate that

$$\mathscr{P}(\omega^2 u | 0, g_3) = \omega^2 \mathscr{P}(u | 0, g_3). \quad (7.5.36)$$

Differentiation of these two identities provides the further results

$$\mathscr{P}'(\omega u | 0, g_3) = \mathscr{P}'(u | 0, g_3), \qquad \mathscr{P}'(\omega^2 u | 0, g_3) = \mathscr{P}'(u | 0, g_3). \quad (7.5.37)$$

It follows that the result (7.5.34) can now be expressed in the form

$$y = \frac{W^2}{2g} \int_0^{gx/W^2} \left[\frac{\mathscr{P}'(\eta) + \mathscr{P}'(\xi)}{\mathscr{P}(\eta) - \mathscr{P}(\xi)} + \omega \frac{\mathscr{P}'(\omega\eta) + \mathscr{P}'(\xi)}{\mathscr{P}(\omega\eta) - \mathscr{P}(\xi)} + \omega^2 \frac{\mathscr{P}'(\omega^2\eta) + \mathscr{P}'(\xi)}{\mathscr{P}(\omega^2\eta) - \mathscr{P}(\xi)} \right] d\xi$$

$$= \frac{W^2}{g} \int_0^{gx/W^2} [\zeta(\eta - \xi) - \zeta(\eta) + \zeta(\xi) + \omega\zeta(\omega\eta - \xi) - \omega\zeta(\omega\eta) + \omega\zeta(\xi)$$

$$+ \omega^2\zeta(\omega^2\eta - \xi) - \omega^2\zeta(\omega^2\eta) + \omega^2\zeta(\xi)] d\xi, \quad (7.5.38)$$

having used the identity (6.8.4) (with v replaced by $-v$). Since $\omega^2 + \omega + 1 = 0$, the terms in $\zeta(\xi)$ now cancel. Also, on account of the results (6.7.35), the identity (6.6.16) can be reexpressed in the form $\zeta(\lambda u | g_2, g_3) = \lambda^{-1}\zeta(u | \lambda^4 g_2, \lambda^6 g_3)$. Taking $g_2 = 0, \lambda = \omega, \omega^2$, we deduce that

$$\zeta(\omega u | 0, g_3) = \omega^2\zeta(u | 0, g_3), \qquad \zeta(\omega^2 u | 0, g_3) = \omega\zeta(u | 0, g_3). \quad (7.5.39)$$

Thus

$$\zeta(\eta) + \omega\zeta(\omega\eta) + \omega^2\zeta(\omega^2\eta) = 3\zeta(\eta), \quad (7.5.40)$$

and equation (7.5.38) reduces to

$$y = \frac{W^2}{g} \int_0^{gx/W^2} [\zeta(\eta - \xi) + \omega\zeta(\omega\eta - \xi) + \omega^2\zeta(\omega^2\eta - \xi) - 3\zeta(\eta)] d\xi. \quad (7.5.41)$$

This integral can now be evaluated in terms of the sigma function by use of the identity (6.6.1). We obtain

$$gy/W^2 = \ln\frac{\sigma(\eta)}{\sigma(\eta - \xi)} + \omega\ln\frac{\sigma(\omega\eta)}{\sigma(\omega\eta - \xi)} + \omega^2\ln\frac{\sigma(\omega^2\eta)}{\sigma(\omega^2\eta - \xi)} - 3\zeta(\eta)\xi, \quad (7.5.42)$$

where $\xi = gx/W^2, \eta = ga/W^2$. This is the equation of the trajectory.

To calculate t, we use equation (7.5.14) in the form

$$t = \frac{W}{g} \int_0^p \frac{P^{1/3}}{P} \, pP^{-2/3} \, dp = \frac{1}{W} \int_0^x qp \, dx, \tag{7.5.43}$$

having used the first of equations (7.5.15). Substituting for p and q from equations (7.5.28) and (7.5.23) respectively, we show that

$$t = \frac{2W}{g} \int_0^{gx/W^2} \frac{\mathscr{P}(\xi)}{\mathscr{P}'(\eta) - \mathscr{P}'(\xi)} \, d\xi. \tag{7.5.44}$$

This integral can be treated in a similar manner to the integral (7.5.30) to give

$$t = \frac{W}{2g} \int_0^{gx/W^2} \left[\frac{\mathscr{P}'(\eta) + \mathscr{P}'(\xi)}{\mathscr{P}(\eta) - \mathscr{P}(\xi)} + \omega^2 \frac{\mathscr{P}'(\omega\eta) + \mathscr{P}'(\xi)}{\mathscr{P}(\omega\eta) - \mathscr{P}(\xi)} + \omega \frac{\mathscr{P}'(\omega^2\eta) + \mathscr{P}'(\xi)}{\mathscr{P}(\omega^2\eta) - \mathscr{P}(\xi)} \right] d\xi$$

$$= \frac{W}{g} \int_0^{gx/W^2} [\zeta(\eta - \xi) + \omega^2 \zeta(\omega\eta - \xi) + \omega \zeta(\omega^2\eta - \xi)] \, d\xi$$

$$= \frac{W}{g} \left[\ln \frac{\sigma(\eta)}{\sigma(\eta - \xi)} + \omega^2 \ln \frac{\sigma(\omega\eta)}{\sigma(\omega\eta - \xi)} + \omega \ln \frac{\sigma(\omega^2\eta)}{\sigma(\omega^2\eta - \xi)} \right]. \tag{7.5.45}$$

The formulae (7.5.42), (7.5.45) suffer from the disadvantage that complex quantities ω, ω^2 are involved. These can be eliminated in the following manner:

First, note that equation (6.6.1) implies that

$$\int_u^v \zeta(s) \, ds = \ln \frac{\sigma(v)}{\sigma(u)}. \tag{7.5.46}$$

Then, integrating the Taylor expansion

$$\zeta(s) = \zeta(u) + (s - u)\zeta'(u) + \frac{1}{2!}(s - u)^2 \zeta''(u) + \cdots, \tag{7.5.47}$$

we prove that

$$\ln \frac{\sigma(v)}{\sigma(u)} = (v - u)\zeta(u) - \frac{1}{2!}(v - u)^2 \mathscr{P}(u) - \frac{1}{3!}(v - u)^3 \mathscr{P}'(u) + \cdots, \tag{7.5.48}$$

after using the identity (6.6.6). The following expansions are now obtainable:

$$\ln \frac{\sigma(\eta)}{\sigma(\eta - \xi)} = \xi\zeta(\eta) + \frac{1}{2!}\xi^2 \mathscr{P}(\eta) - \frac{1}{3!}\xi^3 \mathscr{P}'(\eta) + \frac{1}{4!}\xi^4 \mathscr{P}''(\eta) - \cdots,$$

$$\ln \frac{\sigma(\omega\eta)}{\sigma(\omega\eta - \xi)} = \omega^2 \xi\zeta(\eta) + \frac{1}{2!}\omega\xi^2 \mathscr{P}(\eta) - \frac{1}{3!}\xi^3 \mathscr{P}'(\eta) + \frac{1}{4!}\omega^2 \xi^4 \mathscr{P}''(\eta) - \cdots,$$

$$\ln \frac{\sigma(\omega^2\eta)}{\sigma(\omega^2\eta - \xi)} = \omega\xi\zeta(\eta) + \frac{1}{2!}\omega^2 \xi^2 \mathscr{P}(\eta) - \frac{1}{3!}\xi^3 \mathscr{P}'(\eta) + \frac{1}{4!}\omega\xi^4 \mathscr{P}''(\eta) - \cdots,$$

$$\tag{7.5.49}$$

after use of the results (7.5.35), (7.5.36), and (7.5.39). Substitution in equations (7.5.42), (7.5.45) and use of the result $1 + \omega + \omega^2 = 0$ now yield the formulae

$$gy/3W^2 = \frac{1}{4!}\xi^4 \mathscr{P}''(\eta) - \frac{1}{7!}\xi^7 \mathscr{P}^{(5)}(\eta) + \frac{1}{10!}\xi^{10}\mathscr{P}^{(8)}(\eta) - \cdots, \qquad (7.5.50)$$

$$gt/3W^2 = \frac{1}{2!}\xi^2 \mathscr{P}(\eta) - \frac{1}{5!}\xi^5 \mathscr{P}^{(3)}(\eta) + \frac{1}{8!}\xi^8 \mathscr{P}^{(6)}(\eta) - \cdots. \qquad (7.5.51)$$

The values of the numerical coefficients $\mathscr{P}^{(n)}(\eta)$ can be found by repeated differentiation of the identity (7.5.31) and use of the results (7.5.29), (7.5.27), i.e., $\mathscr{P}(\eta) = 1/3$, $\mathscr{P}'(\eta) = (\sin \alpha)/3$. It will be found that

$$\left.\begin{array}{l} \mathscr{P}''(\eta) = \tfrac{2}{3}, \qquad \mathscr{P}^{(3)}(\eta) = \tfrac{4}{3}\sin \alpha, \qquad \mathscr{P}^{(4)}(\eta) = \tfrac{4}{3}(2 + \sin^2\alpha), \\[2mm] \mathscr{P}^{(5)}(\eta) = \tfrac{40}{3}\sin \alpha, \qquad \mathscr{P}^{(6)}(\eta) = \tfrac{80}{3}(1 + \sin^2\alpha), \\[2mm] \mathscr{P}^{(7)}(\eta) = \tfrac{80}{3}\sin \alpha(8 + \sin^2\alpha), \qquad \mathscr{P}^{(8)}(\eta) = \tfrac{1280}{3} + 800 \sin^2\alpha. \end{array}\right\} \qquad (7.5.52)$$

EXERCISES

1. A particle moves under gravity in a smooth circular tube of radius a, which is constrained to rotate about a vertical axis inclined at an acute angle α to its plane and through its center with constant angular velocity ω. If θ is the angular distance of the particle from the lowest point of the circle, obtain the tangential component of the equation of motion in the form

$$a(\ddot{\theta} - \omega^2 \cos^2\alpha \sin \theta \cos \theta) = -g \cos \alpha \sin \theta.$$

If the particle is initially at rest relative to the tube with $\theta = \tfrac{1}{2}\pi$, show that θ oscillates between the values $\tfrac{1}{2}\pi$ and γ, where $\cos \gamma = 2g/(a\omega^2 \cos \alpha)$ (assume $\omega^2 > 2g \sec \alpha/a$). At time t after $\theta = \tfrac{1}{2}\pi$, prove that

$$\sec \theta = \tfrac{1}{3}\sec \gamma + \mathscr{P}(\lambda t),$$

where $\lambda^2 = g \cos \alpha/2a$ and the stationary values of $\mathscr{P}$ are given by

$$e_1 = \tfrac{2}{3}\sec \gamma, \qquad e_2 = 1 - \tfrac{1}{3}\sec \gamma, \qquad e_3 = -1 - \tfrac{1}{3}\sec \gamma.$$

(r, ϕ, z) are cylindrical polar coordinates, the z-axis pointing vertically upward. A particle moves on the smooth inner surface of the paraboloid of revolution $r^2 = 4az$. Initially, it is projected horizontally with speed $\sqrt{(6ag)}$ at a height a above the vertex. Show that its subsequent motion under gravity is confined between the planes $z = a$, $3a$. If t denotes the time measured from an instant when the particle touches the plane $z = 3a$ and u is such that $du/dt = \sqrt{(\tfrac{1}{2}ag)}(a + z)^{-1}$, $u = 0$ at $z = 3a$, obtain the equations

$$z = a\{1 - \mathscr{P}(u + \omega_3)\}, \qquad t = \sqrt{(2a/g)}\{2u + \zeta(u + \omega_3) - \eta_3\},$$

where the invariants of the Weierstrass function are given by $g_2 = 16$, $g_3 = 0$.

3. The motion of the bob of a spherical pendulum of length a is wholly between levels $3a/5$ and $4a/5$ below the support. At a time t after passing its greatest depth, show that the depth of the bob is given by

$$z = a\left(\frac{4}{5} - \frac{13}{35\mathscr{P}(\omega t) + 24}\right),$$

where $\omega = \sqrt{(g/2a)}$ and the invariants of the Weierstrass function are $g_2 = 5092/1225$, $g_3 = 66912/42875$.

4. A particle is free to move in a plane under an attractive force of μ/r^4 per unit mass toward a point O, r being its distance from the pole. It is projected in a direction perpendicular to the radius from O with speed $\sqrt{(7\mu/9a^3)}$ when $r = a$. Show that it falls into O along a trajectory whose polar equation is

$$r = \frac{18a}{7 + 18\mathscr{P}(\gamma\theta)},$$

where $\gamma^2 = 3/14$ and the invariants of the Weierstrass function are $g_2 = 147/81$, $g_3 = -143/729$.

5. Show that equation (7.3.23) can be expressed in the form

$$i\theta = \frac{1}{2}\ln\left[\frac{\theta_3(x + z)\theta_4(x - y)}{\theta_3(x - z)\theta_4(x + y)}\right]$$

$$+ [\zeta(a) - \zeta(b) + \eta_1 + \eta_1\omega_3(\beta - \alpha)/\omega_1]\tau + \text{const.},$$

where $x = \pi\tau/2\omega_1$, $y = \pi a\omega_3/2\omega_1$, $z = \pi\beta\omega_3/2\omega_1$. (Take $a = \alpha\omega_3$, $b = \omega_1 + \beta\omega_3$, with $|\alpha| < 1$, $|\beta| < 1$, as in section 7.3.)

6. If, for the top treated in section 7.4, we have $H = Cn$, show that when (i) $K < 2Mgk$, λ oscillates between values λ_2, λ_3 ($-1 < \lambda_3 < \lambda_2 < 1$), and when (ii) $K > 2Mgk$, λ oscillates between values λ_2 and 1 ($-1 < \lambda_2 < 1$). If (iii) $K = 2Mgk$, prove that

$$\lambda = 1 - \alpha^2 \operatorname{sech}^2\mu(t - t_0), \qquad \phi = \Omega(t - t_0) + \tan^{-1}\left[\frac{\mu}{\Omega}\tanh\mu(t - t_0)\right]$$

where $\alpha^2 = 2 - C^2n^2/(AK)$, $\mu = (\alpha/2)\sqrt{K/A}$, $\Omega = Cn/2A$. (t_0 is a constant and ϕ is taken to be zero at $t = t_0$.)

CHAPTER 8

Complex Variable Analysis

8.1. Existence of Primitive Periods

As remarked in section 2.2, the general definition of an elliptic function is that it is a doubly periodic function, all of whose singularities (except at infinity) are poles. In section 2.3, we commented on the existence of primitive periods characterized by the property that any period is expressible as the sum of multiples of these primitive periods; we also distinguished between primitive periods and fundamental periods, a fundamental period being defined to be such that no submultiple is a period. We shall commence this chapter by proving the existence of primitive periods.

We first observe that no analytic function, except a constant, can possess arbitrarily small periods. For, suppose $f(z)$ has periods $2\omega_r (r = 1, 2, \dots)$, where $\omega_r \to 0$ as $r \to \infty$. Then, if $f(z)$ is regular at $z = a$, $f(z) - f(a)$ is regular at this point and has zeros at $z = a + 2\omega_r$. Thus, $f(z) - f(a)$ has an infinity of zeros in any neighborhood of the point and cannot be regular at $z = a$, unless it is constant (zero) (see, e.g., E.T. Copson, *Introduction to the Theory of Functions of a Complex Variable*, Oxford University Press, 1935, p. 74). It follows, immediately, that if an analytic function has a period 2ω, then either this period or some submultiple is fundamental.

Thus, given a pair of periods for a certain elliptic function (assumed nonconstant for the remainder of this section), we can always replace these by fundamental periods $2\omega_1$ and $2\omega_3$.

We next show that the ratio ω_3/ω_1 must be complex, for suppose instead $\omega_3 = \lambda\omega_1$, where λ is real. λ cannot be an integer for, otherwise, the period $2\omega_3$ would not be fundamental. Let $\lambda = m + \alpha$, where m is an integer and $0 < \alpha < 1$. Then $2\omega_3 = 2m\omega_1 + 2\alpha\omega_1$ and, hence, $2\alpha\omega_1$ is a period; i.e., there is certainly a period whose value is a fraction of $2\omega_1$. There cannot be an infinity of periods of the type $2\alpha\omega_1$, for if there were, their set would possess a limit point (Bolzano-Weierstrass theorem) and the difference of two such periods could be made arbitrarily small; i.e., the elliptic function would have arbitrarily small periods, contrary to what we proved earlier. Let $\alpha = \alpha_1$ be the smallest value of α for which $2\alpha\omega_1$ is a period.

Since $2\omega_1$ is not a multiple of $2\alpha_1\omega_1$, by repeating the above argument (with $2\omega_3$ replaced by $2\omega_1$ and $2\omega_1$ replaced by $2\alpha_1\omega_1$) we can demonstrate the existence of a period $2\alpha_2\omega_1$, with $0 < \alpha_2 < \alpha_1$ and so contradict our postulate that α_1 is minimal. Our assumption that ω_3/ω_1 is real must, accordingly, be rejected.

Given that $2\omega_1$, $2\omega_3$ are fundamental and that ω_3/ω_1 is complex, we next prove the existence of primitive periods with the property that any period is a sum of multiples of these.

Since the complex numbers ω_1, ω_3 are represented by nonparallel vectors in the complex plane, any complex number can be expressed uniquely in the form $x\omega_1 + y\omega_3$, where x and y are real. In particular, any period can be written in the form $2\lambda\omega_1 + 2\mu\omega_3$. If λ, μ are invariably integers or zero, then $2\omega_1$, $2\omega_3$ are primitive.

However, suppose there is a period for which λ, μ are not both integers or zero. Then

$$\lambda = m + \alpha, \qquad \mu = n + \beta, \tag{8.1.1}$$

where m, n, are integers or zero, $0 \leqslant \alpha < 1$, $0 \leqslant \beta < 1$, and both α, β cannot vanish. It follows that $2\alpha\omega_1 + 2\beta\omega_3$ is a period.

If $\alpha = 0$, then $2\omega_3$, $2\beta\omega_3$ are periods and, as proved earlier, $2\omega_3$ cannot be fundamental. Similarly, if $\alpha \neq 0$, $\beta = 0$, $2\omega_1$ is not fundamental. Hence, $0 < \alpha < 1, 0 < \beta < 1$.

Consider the period parallelogram with vertices 0, $2\omega_1$, $2\omega_1 + 2\omega_3$, $2\omega_3$. The period $2\alpha\omega_1 + 2\beta\omega_3$ is represented by a point inside this parallelogram. There cannot be an infinity of periods represented by such points, for if so, the set would have a limit point and the difference between two such periods could be made as small as we please, contrary to the prohibition against arbitrarily small periods. Choose the point (α_1, β_1) for which β has least value and write $2\omega_3' = 2\alpha_1\omega_1 + 2\beta_1\omega_3$. Then $2\omega_3'$ is a fundamental period, since otherwise a smaller value of β than β_1 would exist.

Every period can now be expressed as a sum of multiples of $2\omega_1$, $2\omega_3'$. For if not, by repetition of the above argument, we can prove the existence of a period $2\alpha'\omega_1 + 2\beta'\omega_3'$, where $0 < \alpha' < 1$, $0 < \beta' < 1$. Then, since

$$2\alpha'\omega_1 + 2\beta'\omega_3' = 2(\alpha' + \alpha_1\beta')\omega_1 + 2\beta'\beta_1\omega_3, \tag{8.1.2}$$

and $\beta'\beta_1 < \beta_1$, this contradicts our hypothesis that β_1 is a minimum (if $\alpha' + \alpha_1\beta' > 1$, remove a period $2\omega_1$ to bring the period inside the period parallelogram). Thus, $2\omega_1$, $2\omega_3'$ must be primitive periods.

As previously, we shall always choose our notation so that, if $2\omega_1$ and $2\omega_3$ are primitive periods, then the imaginary part of ω_3/ω_1 is positive. We recall from section 2.3 that $2\Omega_1$ and $2\Omega_3$ will also be primitive periods such that $\mathscr{I}(\Omega_3/\Omega_1) > 0$, if

$$\Omega_1 = a\omega_1 + b\omega_3, \qquad \Omega_3 = c\omega_1 + d\omega_3, \qquad (8.1.3)$$

where a, b, c, d are integers (or zeros) such that $ad - bc = 1$.

Note that, if $2\omega_r$ $(r = 1, 2, 3)$ are periods of an elliptic function and $2\Omega_1, 2\Omega_3$ are primitive periods, then

$$2\omega_r = 2\lambda_r\Omega_1 + 2\mu_r\Omega_3, \qquad (8.1.4)$$

where λ_r, μ_r are integers or zeros. Eliminating Ω_1, Ω_3, we find

$$2l\omega_1 + 2m\omega_2 + 2n\omega_3 = 0, \qquad (8.1.5)$$

where l, m, n are integers (or zeros). Thus, no elliptic function can possess three independent periods.

8.2. General Properties of Elliptic Functions

We first observe that no elliptic function (except a constant) can be bounded over the interior and perimeter of a period parallelogram, for if it were, it would be bounded over the whole complex plane and hence, by Liouville's theorem, would be constant. It must, accordingly, have some poles inside or on the parallelogram and the number of these must be finite for, otherwise, there would be a limit point of such singularities and this would be an essential singularity. It is always possible, therefore, to translate the parallelogram in the complex plane so as to create a cell for which there are no singularities on its perimeter. The number of poles within the cell (each multiple pole being counted according to its order) is termed the *order* of the elliptic function. Referring to section 2.8, we note that the Jacobian functions each have a pair of simple poles within a cell and so are of second order. $\mathscr{P}(u)$ has a double pole at each of the congruent points $u = 2m\omega_1 + 2n\omega_3$ and hence the Weierstrass function is also of order two. Its derivative $\mathscr{P}'(u)$ has a triple pole at $u = 0$ and is of third order.

Next, if $f(z)$ is an elliptic function, consider the contour integral

$$\oint f(z)\,dz \qquad (8.2.1)$$

taken around the perimeter of a cell $ABCD$. Since the values of $f(z)$ along AB are repeated along DC and these sides are traversed in opposite senses, their contributions to the integral cancel. So, also, for the same reason, do the contributions of BC and AD. Thus, the integral vanishes and, by Cauchy's theorem, the sum of the residues at the poles within the cell must be zero. As a corollary, it follows that an elliptic function cannot be of unit order, since such a function would have zero residue at its pole and would therefore be regular everywhere. The result we have just proved is readily verified for the case of the Jacobian functions by reference to section 2.8. The Weierstrass function has only one pole within each cell but, by equation (6.6.13), its residue at this pole is zero, in accordance with the theorem.

Now consider the integral

$$\oint \frac{f'(z)}{f(z) - \alpha} \, dz \tag{8.2.2}$$

where α is an arbitrary complex number and the contour is the perimeter of a cell (it is assumed the cell is chosen so that $f(z)$ does not take the value α on its perimeter—this must be possible, for if not there is a limit point of points at which $f(z) = \alpha$ and this limit point is an essential singularity).

Suppose $f(z) = \alpha$ at a point $z = a$ within the cell. Then by Taylor's theorem

$$f(z) = \alpha + (z - a)^m \phi(z) \tag{8.2.3}$$

in some neighborhood of $z = a$, where m is the multiplicity of the zero of $f(z) - \alpha$, $\phi(z)$ is regular, and $\phi(a) \neq 0$. Hence

$$f'(z) = m(z - a)^{m-1} \phi(z) + (z - a)^m \phi'(z) \tag{8.2.4}$$

and it then follows that

$$\frac{f'(z)}{f(z) - \alpha} = \frac{m}{z - a} + \frac{\phi'(z)}{\phi(z)}. \tag{8.2.5}$$

Since $\phi'(z)/\phi(z)$ is regular in the neighborhood of $z = a$, we see that the integrand has a simple pole at this point with residue m.

If $z = b$ is a pole of $f(z)$ within the cell, of order n, then by Laurent's theorem

$$f(z) = B_1(z - b)^{-1} + B_2(z - b)^{-2} + \cdots + B_n(z - b)^{-n} + \psi(z), \tag{8.2.6}$$

where $\psi(z)$ is regular in some neighborhood of $z = b$. Hence,

$$\frac{f'(z)}{f(z) - \alpha} = \frac{-n + O(z - b)}{(z - b) + O(z - b)^2} \tag{8.2.7}$$

and the integrand is seen to have a simple pole at $z = b$ with residue $-n$.

We have now identified all possible singularities of the integrand within a cell and conclude that, by Cauchy's residue theorem,

$$\oint \frac{f'(z)}{f(z) - \alpha} \, dz = 2\pi i (\sum m - \sum n). \tag{8.2.8}$$

But, once again, the integrand assumes the same values at corresponding points on opposite sides of the cell and the integral accordingly vanishes. Thus $\sum m = \sum n = $ order N of $f(z)$ and we have proved that $f(z)$ takes any arbitrary value α N times within a cell, it being understood that multiple zeros of $f(z) - \alpha$ are to be counted according to their order.

In particular, the number of zeros of an elliptic function inside a cell is equal to the order of the function. Examples: $\operatorname{sn} u$ has a pair of simple zeros at $u = 0$,

$2K$, within the cell whose vertices are $-K-iK', 3K-iK', 3K+iK'$, $-K+iK'$. $\mathscr{P}'(u)$ has three simple zeros at $u = \omega_1, \omega_3, \omega_1 + \omega_3$, which all lie within the cell whose vertices are $-\frac{1}{2}(\omega_1 + \omega_3), \frac{1}{2}(3\omega_1 - \omega_3), \frac{3}{2}(\omega_1 + \omega_3),$ $\frac{1}{2}(-\omega_1 + 3\omega_3)$; $\mathscr{P}(u)$ takes the value e_1 once only within the same cell, viz. at $u = \omega_1$—but $\mathscr{P}(u) - e_1$ has a double zero at this point, since its derivative $\mathscr{P}'(u)$ also vanishes there.

Another important general result is reached by integrating $zf'(z)/f(z)$ around the perimeter of a cell. This integrand has singularities at the zeros and poles of $f(z)$. Thus, if $z = a$ is a zero of $f(z)$, then as shown at equation (8.2.5) (with $\alpha = 0$), $f'(z)/f(z)$ has a simple pole at $z = a$ with residue m (the multiplicity of the zero). It follows that $zf'(z)/f(z)$ has a simple pole at $z = a$ with residue ma. If $z = b$ is a pole of $f(z)$, we have shown at equation (8.2.7) that $f'(z)/f(z)$ has a simple pole at this point with residue $-n$ (n the multiplicity of the pole). Hence, $zf'(z)/f(z)$ has a simple pole at $z = b$ with residue $-nb$. By Cauchy's residue theorem, therefore,

$$\oint \frac{zf'(z)}{f(z)} \, dz = 2\pi i \sum (ma - nb), \tag{8.2.9}$$

the summation being over all zeros and poles of $f(z)$ inside the cell.

But, if $\zeta, \zeta + 2\omega_1, \zeta + 2\omega_1 + 2\omega_3, \zeta + 2\omega_3$ are the vertices of the cell, the integral separates into four components, thus:

$$\omega_1 \int_0^2 \frac{(\zeta + t\omega_1)f'(\zeta + t\omega_1)}{f(\zeta + t\omega_1)} \, dt + \omega_3 \int_0^2 \frac{(\zeta + 2\omega_1 + t\omega_3)f'(\zeta + 2\omega_1 + t\omega_3)}{f(\zeta + 2\omega_1 + t\omega_3)} \, dt$$

$$- \omega_1 \int_0^2 \frac{(\zeta + 2\omega_3 + t\omega_1)f'(\zeta + 2\omega_3 + t\omega_1)}{f(\zeta + 2\omega_3 + t\omega_1)} \, dt$$

$$- \omega_3 \int_0^2 \frac{(\zeta + t\omega_3)f'(\zeta + t\omega_3)}{f(\zeta + t\omega_3)} \, dt. \tag{8.2.10}$$

Combining the first and third components, and the second and fourth, and making use of the periodicity property of $f(z)$, this expression reduces to the form

$$- 2\omega_1\omega_3 \int_0^2 \frac{f'(\zeta + t\omega_1)}{f(\zeta + t\omega_1)} \, dt + 2\omega_1\omega_3 \int_0^2 \frac{f'(\zeta + t\omega_3)}{f(\zeta + t\omega_3)} \, dt$$

$$= |2\omega_1 \ln\{f(\zeta + t\omega_3)\} - 2\omega_3 \ln\{f(\zeta + t\omega_1)\}|_0^2$$

$$= 2\pi i(2r\omega_1 + 2s\omega_3) \qquad (r, s = \text{integers}), \tag{8.2.11}$$

since $f(\zeta + 2\omega_1) = f(\zeta + 2\omega_3) = f(\zeta)$ and the logarithms of these complex quantities can only differ by integral multiples of $2\pi i$, therefore.

It now follows that $\sum(ma)$ and $\sum(nb)$ can only differ by a period, $2r\omega_1 + 2s\omega_3$.

Suppose the elliptic function $f(z)$ is odd and that $z = \omega_1$ is not a pole. Then $f(-\omega_1) = -f(\omega_1)$, since the function is odd, and $f(-\omega_1) = f(\omega_1)$, since $2\omega_1$

is a period. Thus, $f(\omega_1) = 0$ and ω_1 must be a zero of $f(z)$. A similar argument is applicable to the points $z = \omega_2, \omega_3$ and also to the points $2\omega_1, 2\omega_2, 2\omega_3$. Thus, the lattice and midlattice points of an odd elliptic function are either zeros or poles.

Further, suppose α is a lattice or midlattice point which is a zero of order m of the odd elliptic function $f(z)$. Then α is neither a zero nor a pole of the mth derivative $f^{(m)}(z)$, which accordingly cannot be an odd elliptic function. Hence, m must be odd.

If, however, α is a pole of order m of $f(z)$, then it is a zero of the same order of the odd elliptic function $1/f(z)$ and again m must be odd.

We conclude that an odd elliptic function has a zero or a pole of odd order at all its lattice and midlattice points.

If $f(z)$ is even, its value is not restricted at the lattice and midlattice points. However, if α is such a point and it happens that it is a zero of $f(z)$ of order m, then this point is a zero of order $(m - 1)$ of the odd elliptic function $f'(z)$ and m must therefore be even. Similarly, if α is a pole of order m of $f(z)$, it is a pole of order $(m + 1)$ of $f'(z)$ and, again, m must be even. Thus, if an even elliptic function has a zero or a pole at a lattice or midlattice point, the order of the zero or pole is even.

If $f(z)$ is even, choose a cell for which $f(z)$ has no zeros or poles on its perimeter. Then, if α is a zero inside the cell, $-\alpha$ is also a zero and the point congruent to $-\alpha$ inside the cell will be a zero. Thus, the zeros inside the cell can be arranged in pairs, which will be distinct unless $\alpha = 2r\omega_1 + 2s\omega_3 - \alpha$ for some integers r and s. In the latter case, $\alpha = r\omega_1 + s\omega_3$ and α is either a lattice point or a midlattice point. Since the order of the zero at such a point is even, in all cases the total number of zeros within the cell (counting each according to its multiplicity) must be even. We conclude that $f(z)$ is an elliptic function of even order. If $f(z)$ is odd, note that it is possible for there to be a pair of such zeros inside the cell, in which case the total number of zeros is even; e.g., sn u is an odd elliptic function, but its order is 2.

8.3. Partial Fraction Expansion of $\mathscr{P}(u)$

Let C_n $(n = 0, 1, 2, \ldots)$ be a sequence of period parallelograms of $\mathscr{P}(z)$ with vertices $-(2n + 1)(\omega_1 + \omega_3)$, $(2n + 1)(\omega_1 - \omega_3)$, $(2n + 1)(\omega_1 + \omega_3)$, $(2n + 1)(-\omega_1 + \omega_3)$. The positions of these in relation to the lattice of primitive period parallelograms are indicated in Fig. 8.1; note that they avoid the poles of $\mathscr{P}(z)$. We shall prove that the limit

$$\lim_{n \to \infty} \oint_{C_n} \frac{\mathscr{P}'(z)}{u - z} \, dz, \tag{8.3.1}$$

where u takes any complex value which is not one of the poles $2r\omega_1 + 2s\omega_3$ of $\mathscr{P}(z)$, vanishes.

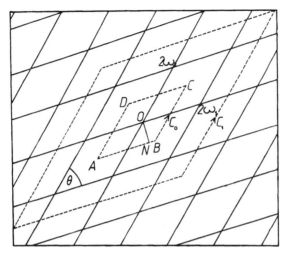

Figure 8.1. Integration contours $C_0, C_1, \ldots$.

Consider the contribution to the integral of the side of C_n parallel to AB (see figure). On this side, $z = -(2n+1)\omega_3 + t\omega_1 (-(2n+1) \leqslant t \leqslant (2n+1))$ and the contribution is accordingly

$$
\omega_1 \int_{-(2n+1)}^{(2n+1)} \frac{\mathscr{P}'\{-(2n+1)\omega_3 + t\omega_1\}}{u + (2n+1)\omega_3 - t\omega_1} \, dt
$$

$$
= \omega_1 \int_{-(2n+1)}^{(2n+1)} \frac{\mathscr{P}'(t\omega_1 + \omega_3)}{u + (2n+1)\omega_3 - t\omega_1} \, dt. \tag{8.3.2}
$$

Similarly, on the side parallel to CD, $z = (2n+1)\omega_3 - t\omega_1$ with the same limits for t and its contribution is therefore

$$
-\omega_1 \int_{-(2n+1)}^{(2n+1)} \frac{\mathscr{P}'\{(2n+1)\omega_3 - t\omega_1\}}{u - (2n+1)\omega_3 + t\omega_1} \, dt
$$

$$
= \omega_1 \int_{-(2n+1)}^{(2n+1)} \frac{\mathscr{P}'(t\omega_1 + \omega_3)}{u - (2n+1)\omega_3 + t\omega_1} \, dt, \tag{8.3.3}
$$

since $\mathscr{P}'\{(2n+1)\omega_3 - t\omega_1\} = \mathscr{P}'(-t\omega_1 - \omega_3) = -\mathscr{P}'(t\omega_1 + \omega_3)$, $\mathscr{P}'(z)$ being odd. The sum of these contributions is now found to be

$$
P_n = 2u\omega_1 \int_{-(2n+1)}^{(2n+1)} \frac{\mathscr{P}'(t\omega_1 + \omega_3)}{u^2 - \{(2n+1)\omega_3 - t\omega_1\}^2} \, dt. \tag{8.3.4}
$$

As we traverse the contour C_n, $\mathscr{P}'(z)$ continually repeats the values it takes on the primitive parallelogram $ABCD$, on which it is bounded. Thus, we can assume $|\mathscr{P}'(t\omega_1 + \omega_3)| \leqslant M$ on C_n, M being independent of n. Let $ON = p$ be the perpendicular from the origin O on to the side AB of C_0. Then $(2n+1)p$ is the minimum distance from O of points on the sides of C_n parallel to AB and

CD. Hence

$$|(2n+1)\omega_3 - t\omega_1| \geqslant (2n+1)p. \tag{8.3.5}$$

We now calculate that (for sufficiently large n)

$$|P_n| \leqslant 2|u\omega_1| \left| \int_{-(2n+1)}^{(2n+1)} \frac{\mathscr{P}'(t\omega_1 + \omega_3)}{u^2 - \{(2n+1)\omega_3 - t\omega_1\}^2} \, dt \right|$$

$$\leqslant 2|u\omega_1| \frac{M(4n+2)}{(2n+1)^2 p^2 - |u|^2}. \tag{8.3.6}$$

Letting $n \to \infty$, we see that $P_n \to 0$.

Similarly, the sum of the contributions of the sides of C_n parallel to BC and AD tends to zero as $n \to \infty$ and we have established what we set out to prove.

Now consider the singularities of the integrand $\mathscr{P}'(z)/(u-z)$. These occur at the poles of $\mathscr{P}'(z)$ and at $z = u$. The residue at $z = u$ is clearly $-\mathscr{P}'(u)$. In the neighborhood of $z = 2r\omega_1 + 2s\omega_3 = \Omega_{rs}$, let $z = \Omega_{rs} + \varepsilon$. Then, using the expansion (6.7.21),

$$\frac{\mathscr{P}'(z)}{u-z} = \frac{\mathscr{P}'(\varepsilon)}{u - \Omega_{rs} - \varepsilon}$$

$$= \frac{1}{u - \Omega_{rs}} \{-2\varepsilon^{-3} + O(\varepsilon)\} \left[1 + \frac{\varepsilon}{u - \Omega_{rs}} + \frac{\varepsilon^2}{(u - \Omega_{rs})^2} + O(\varepsilon^3)\right]$$

$$= -\frac{2}{u - \Omega_{rs}} \frac{1}{\varepsilon^3} - \frac{2}{(u - \Omega_{rs})^2} \frac{1}{\varepsilon^2} - \frac{2}{(u - \Omega_{rs})^3} \frac{1}{\varepsilon} + O(1). \tag{8.3.7}$$

Thus, the residue at Ω_{rs} is $-2(u - \Omega_{rs})^{-3}$. Cauchy's residue theorem now shows that

$$\oint_{C_n} \frac{\mathscr{P}'(z)}{u-z} \, dz = 2\pi i \left\{-\mathscr{P}'(u) - 2 \sum_{r,s} (u - \Omega_{rs})^{-3}\right\}, \tag{8.3.8}$$

the summation being performed for all poles Ω_{rs} lying within C_n. Letting $n \to \infty$, it follows that in the limit

$$\mathscr{P}'(u) = -2 \sum_{r,s} (u - \Omega_{rs})^{-3}, \tag{8.3.9}$$

the summation now being over all positive and negative integral values of r and s, including zero.

This result provides an expansion for $\mathscr{P}'(u)$ in terms of partial fractions of the form $(u - \Omega_{rs})^{-3}$. However, the method of proof only establishes its validity for a particular order of summation of the terms—it is assumed that we commence with the term $-2/u^3$, then add the terms associated with points Ω_{rs} between C_0 and C_1, then add terms associated with points between C_1 and C_2, and so on. To prove that the order of summation is immaterial, we need to establish that the series is absolutely convergent for at least one order of summation.

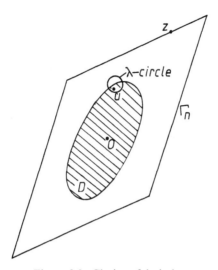

Figure 8.2. Choice of λ-circle.

The points Ω_{rs} lying between C_n and C_{n-1} all lie on a parallelogram which we shall denote by Γ_n. Let us suppose that u is confined to some finite closed domain D. Then, for some $N, \Gamma_N, \Gamma_{N+1}, \ldots$ will all enclose D. Consider the locus of points z for which $|u - z|/|z| = \lambda$ (a positive constant). This is a circle and, by varying the parameter λ, we generate a system of coaxial circles with limit points at O and u. As $\lambda \to 0$, the λ-circle collapses upon the limit point at u. It follows, therefore, that we can choose $\lambda > 0$ so that the corresponding circle is enclosed by Γ_n for $n \geqslant N$ and *any value of u* in D (Fig. 8.2). With this value of λ, any point z on Γ_n $(n \geqslant N)$ will be such that

$$|u - z|/|z| > \lambda. \tag{8.3.10}$$

In particular,

$$|u - \Omega_{rs}| > \lambda|\Omega_{rs}|, \tag{8.3.11}$$

for points Ω_{rs} on Γ_N, Γ_{N+1}, etc.

If p is the length of the perpendicular from O to AB or AD, whichever is the smaller, then the distance of any lattice point on Γ_n from O cannot be less than $2np$. There are $8n$ such lattice points. Hence, summing over such points alone,

$$\sum |\Omega_{rs}|^{-3} < 8n(2np)^{-3} = 1/(n^2 p^3). \tag{8.3.12}$$

It now follows from this result and the inequality (8.3.11) that, if we sum over the lattice points on $\Gamma_N, \Gamma_{N+1}, \ldots$ ad infinitum, then

$$\sum |u - \Omega_{rs}|^{-3} < \lambda^{-3} \sum |\Omega_{rs}|^{-3} < \frac{1}{(\lambda p)^3} \sum_{n=N}^{\infty} \frac{1}{n^2}. \tag{8.3.13}$$

But $\sum 1/n^2$ is known to converge and $\sum |u - \Omega_{rs}|^{-3}$ is a series of positive terms which either converges or diverges to $+\infty$. Clearly $\sum |u - \Omega_{rs}|^{-3}$ must converge, therefore, and $\sum(u - \Omega_{rs})^{-3}$ is absolutely convergent (the omission of a finite number of terms never affects the validity of this type of argument).

Having proved absolute convergence for a special order of summation, it now follows by a well-known theorem that the sum of the series cannot be affected by any reordering of the sequence of summation. But, if u is replaced by $u + 2\omega_1$ or by $u + 2\omega_3$ in the series (8.3.9), the only effect is a rearrangement of the order of the terms and this does not change its sum. Thus $\mathscr{P}'(u + 2\omega_1) = \mathscr{P}'(u + 2\omega_3) = \mathscr{P}'(u)$ and we have recovered the periodicity property of this function. This series is, accordingly, used by some authors to define a doubly periodic function upon which to base the theory of elliptic functions (e.g., see E.T. Copson, *Theory of Functions of a Complex Variable*, Oxford University Press, 1935).

Further, since $|u - \Omega_{rs}|^{-3} < \lambda^{-3}|\Omega_{rs}|^{-3}$, by Weierstrass's M-test, the convergence is always uniform with respect to a variable u confined to a finite domain of the complex plane (it is not necessary to exclude lattice points from this domain, since these are involved in only a finite number of terms of the series).

The uniform convergence permits us to integrate the series (8.3.9) term by term along any contour which does not pass through a lattice point. Thus, taking the end points of the contour to be $u = a$ and u, we deduce that

$$\mathscr{P}(u) - \mathscr{P}(a) = \sum_{r,s}\left[\frac{1}{(u - \Omega_{rs})^2} - \frac{1}{(a - \Omega_{rs})^2}\right], \tag{8.3.14}$$

which can be written in the form

$$\mathscr{P}(u) = \mathscr{P}(a) - \frac{1}{a^2} + \frac{1}{u^2} + \sum_{r,s}'\left[\frac{1}{(u - \Omega_{rs})^2} - \frac{1}{(a - \Omega_{rs})^2}\right], \tag{8.3.15}$$

where $\sum_{r,s}'$ indicates a summation which excludes the term $r = s = 0$. Letting $a \to 0$, by equation (6.6.13) $\mathscr{P}(a) - 1/a^2 \to 0$ and we have derived the following partial fraction expansion for $\mathscr{P}(u)$:

$$\mathscr{P}(u) = \frac{1}{u^2} + \sum_{r,s}'\left[\frac{1}{(u - \Omega_{rs})^2} - \frac{1}{\Omega_{rs}^2}\right]. \tag{8.3.16}$$

An alternative method of establishing this expansion is to rely upon Liouville's theorem. Thus, the expansion is first proved to define a doubly periodic analytic function $F(u)$ with double poles at $u = \Omega_{rs}$, by establishing absolute and uniform convergence in any closed domain excluding the lattice points. The function $\mathscr{P}(u) - F(u)$ is then easily shown to be bounded throughout a period parallelogram (the infinities cancel) and hence over the whole complex plane. It then follows by Liouville's theorem that $\mathscr{P}(u) - F(u) = $ constant. That the constant is zero is proved by letting $u \to 0$. However, this method depends upon the form of the expansion being known at the outset and we have preferred a more direct proof.

8.4. Expansions for $\zeta(u)$ and $\sigma(u)$

That the expansion just found for $\mathscr{P}(u)$ converges absolutely and uniformly can be demonstrated by a method very similar to that employed to establish these properties of the series for $\mathscr{P}'(u)$. Hence, using equation (6.6.6), we are permitted to integrate the series (8.3.16) term by term to yield an expansion for Weierstrass's zeta function, viz.

$$\zeta(u) = \frac{1}{u} + \sum_{r,s}' \left[\frac{1}{u - \Omega_{rs}} + \frac{1}{\Omega_{rs}} + \frac{u}{\Omega_{rs}^2} \right]. \tag{8.4.1}$$

(Take the lower limit of integration to be a, and let $a \to 0$ noting that $\zeta(a) - 1/a = O(a^3)$—see Exercise 2 in Chapter 6.)

This series is also absolutely and uniformly convergent and thus, using equation (6.6.1), another term-by-term integration leads to the sigma function expansion

$$\ln \sigma(u) = \ln u + \sum_{r,s}' \left[\ln\left(1 - \frac{u}{\Omega_{rs}}\right) + \frac{u}{\Omega_{rs}} + \frac{u^2}{2\Omega_{rs}^2} \right]. \tag{8.4.2}$$

(Note that $\ln \sigma(a) - \ln a = O(a^4)$—see Exercise 2 in Chapter 6.) This is equivalent to the infinite product

$$\sigma(u) = u \prod_{r,s}' \left(1 - \frac{u}{\Omega_{rs}}\right) \exp\left(\frac{u}{\Omega_{rs}} + \frac{u^2}{2\Omega_{rs}^2}\right). \tag{8.4.3}$$

By making use of the relationship (6.2.9) between sigma and theta functions, we can now deduce the following infinite product formula for $\theta_1(z)$:

$$\theta_1(z) = \theta_1'(0)z \prod_{r,s}' \left(1 - \frac{z}{\alpha_{rs}}\right) \exp\left[\frac{z}{\alpha_{rs}} + z^2\left(\frac{1}{2\alpha_{rs}^2} + \frac{\theta_1'''(0)}{6\theta_1'(0)}\right)\right], \tag{8.4.4}$$

where

$$\alpha_{rs} = \pi(r + s\tau). \tag{8.4.5}$$

8.5. Invariants Expressed in Terms of the Periods

The expansion of $\mathscr{P}(u)$ derived in section 8.3 is of little use for numerical calculation, since its rate of convergence is very slow. However, it enables us to obtain expressions for the invariants g_2 and g_3 in terms of the periods $2\omega_1$, $2\omega_3$, thus:

The function $\mathscr{P}(u) - 1/u^2$ is regular and $O(u^2)$ in a neighborhood of the origin and so has a Taylor expansion

$$\mathscr{P}(u) - \frac{1}{u^2} = au^2 + bu^4 + \cdots. \tag{8.5.1}$$

The coefficients a and b are given by

$$a = \tfrac{1}{2} \lim_{u \to 0} \{\mathscr{P}''(u) - 6/u^4\}, \qquad b = \tfrac{1}{24} \lim_{u \to 0} \{\mathscr{P}''''(u) - 120/u^6\}. \qquad (8.5.2)$$

Differentiating the expansion (8.3.9) term-by-term (justified by its uniform convergence), we calculate that

$$\mathscr{P}''(u) = 6 \sum (u - \Omega_{rs})^{-4}, \qquad \mathscr{P}''''(u) = 120 \sum (u - \Omega_{rs})^{-6}. \qquad (8.5.3)$$

It follows that

$$a = 3 \sum{}' \Omega_{rs}^{-4}, \qquad b = 5 \sum{}' \Omega_{rs}^{-6}. \qquad (8.5.4)$$

Comparing this expansion with that given at equation (6.7.25), we now deduce that

$$g_2 = 60 \sum{}' \Omega_{rs}^{-4}, \qquad g_3 = 140 \sum{}' \Omega_{rs}^{-6}. \qquad (8.5.5)$$

8.6. Fourier Expansions for Certain Ratios of Theta Functions

The theta functions have been defined by their Fourier expansions in section 1.2. In this section, we shall derive Fourier expansions for the ratios $\theta_i'(z)/\theta_i(z)$ ($i = 1, 2, 3, 4$). The case $i = 4$ is the simplest and we shall obtain this first.

$\theta_4(z)$ has simple zeros where $z = x + iy = m\pi + (n + \tfrac{1}{2})\pi\tau$. It follows that $\theta_4'(z)/\theta_4(z)$ has simple poles along the parallel lines $y = \pm \tfrac{1}{2}\pi \mathscr{I}\tau$ and is regular throughout the domain D between these lines. Since $\theta_4'(z)/\theta_4(z)$ is odd and of period π, the results found in Appendix B are applicable, showing that a Fourier sine expansion exists in the form

$$\theta_4'(z)/\theta_4(z) = \sum_{n=1}^{\infty} b_n \sin 2nz, \qquad (8.6.1)$$

where

$$b_n = \frac{2i}{\pi} \int_{-\pi/2}^{\pi/2} \frac{\theta_4'(x)}{\theta_4(x)} e^{-2inx} \, dx, \qquad (8.6.2)$$

the integration being carried out along the real axis. The expansion is absolutely convergent provided z is confined to a closed region within D.

To calculate b_n, we shall integrate $e^{-2inz}\theta_4'(z)/\theta_4(z)$ around the parallelogram $ABCD$ with vertices $-\tfrac{1}{2}\pi, \tfrac{1}{2}\pi, \tfrac{1}{2}\pi + \pi\tau, -\tfrac{1}{2}\pi + \pi\tau$, respectively. Since the integrand is of period π, it takes the same values over BC as it does over AD and the contributions to the integral of these two sides accordingly cancel. On CD, we take $z = x + \pi\tau$, where x is real and decreases from $\tfrac{1}{2}\pi$ to $-\tfrac{1}{2}\pi$ in passing from C to D. Thus, the contribution to the integral of this side is

$$\int_{\pi/2}^{-\pi/2} \frac{\theta_4'(x + \pi\tau)}{\theta_4(x + \pi\tau)} e^{-2in(x + \pi\tau)} \, dx. \qquad (8.6.3)$$

But, referring to equations (1.3.5), we note that

$$\theta_4(x + \pi\tau) = -q^{-1}e^{-2ix}\theta_4(x). \tag{8.6.4}$$

Differentiating with respect to x, we find that

$$\theta_4'(x + \pi\tau) = -q^{-1}e^{-2ix}\{\theta_4'(x) - 2i\theta_4(x)\}. \tag{8.6.5}$$

Hence, the contribution of CD is

$$-q^{-2n}\int_{-\pi/2}^{\pi/2}\left[\frac{\theta_4'(x)}{\theta_4(x)} - 2i\right]e^{-2inx}\,dx = -q^{-2n}I, \tag{8.6.6}$$

where I is the integral appearing in (8.6.2). The integral taken around the whole parallelogram therefore has value

$$(1 - q^{-2n})I. \tag{8.6.7}$$

The only zero of $\theta_4(z)$ within the parallelogram is at $z = \frac{1}{2}\pi\tau$ and the residue of $e^{-2inz}\theta_4'(z)/\theta_4(z)$ at this point is $\exp\{-2in(\frac{1}{2}\pi\tau)\} = q^{-n}$. We conclude that

$$(1 - q^{-2n})I = 2\pi i q^{-n} \tag{8.6.8}$$

and it then follows from equation (8.6.2) that

$$b_n = 4q^n/(1 - q^{2n}). \tag{8.6.9}$$

Thus,

$$\frac{\theta_4'(z)}{\theta_4(z)} = 4\sum_{n=1}^{\infty}\frac{q^n}{1 - q^{2n}}\sin 2nz. \tag{8.6.10}$$

Replacing z by $z - \frac{1}{2}\pi\tau$, we can derive a Fourier expansion for $\theta_1'(z)/\theta_1(z)$ which will be valid in the strip obtained from D by displacement $\frac{1}{2}\pi\mathscr{I}\tau$ parallel to the y-axis, i.e., the strip between the lines $y = 0, \pi\mathscr{I}\tau$.

We first write the result (8.6.10) in the equivalent form

$$\frac{\theta_4'(z)}{\theta_4(z)} = \frac{2}{i}\sum_{n=-\infty}^{\infty}{}'\frac{q^n}{1 - q^{2n}}e^{2inz}, \tag{8.6.11}$$

the term $n = 0$ being omitted from the summation. Then, using an identity (1.3.9), we have

$$\theta_4(z - \frac{1}{2}\pi\tau) = -iq^{-1/4}e^{iz}\theta_1(z) \tag{8.6.12}$$

which, on differentiation, yields

$$\theta_4'(z - \frac{1}{2}\pi\tau) = -iq^{-1/4}e^{iz}\{\theta_1'(z) + i\theta_1(z)\}. \tag{8.6.13}$$

Thus, replacement of z by $z - \frac{1}{2}\pi\tau$ in the identity (8.6.11) leads to the result

$$\frac{\theta_1'(z)}{\theta_1(z)} + i = \frac{2}{i}\sum_{n=-\infty}^{\infty}{}'\frac{1}{1 - q^{2n}}e^{2inz}$$

$$= \frac{2}{i}\sum_{n=1}^{\infty}\frac{1}{1 - q^{2n}}e^{2inz} + \frac{2}{i}\sum_{n=1}^{\infty}\frac{q^{2n}}{q^{2n} - 1}e^{-2inz}$$

$$= \frac{2}{i} \sum_{n=1}^{\infty} \left[1 + \frac{q^{2n}}{1 - q^{2n}} \right] e^{2inz} - \frac{2}{i} \sum_{n=1}^{\infty} \frac{q^{2n}}{1 - q^{2n}} e^{-2inz}$$

$$= \frac{2}{i} \sum_{n=1}^{\infty} e^{2inz} + 4 \sum_{n=1}^{\infty} \frac{q^{2n}}{1 - q^{2n}} \sin 2nz$$

$$= \frac{2}{i} \cdot \frac{e^{2iz}}{1 - e^{2iz}} + 4 \sum_{n=1}^{\infty} \frac{q^{2n}}{1 - q^{2n}} \sin 2nz, \qquad (8.6.14)$$

the geometric progression being summable since $|e^{2iz}| = e^{-2y} < 1$ if $y > 0$. The last identity is now easily simplified to the form

$$\frac{\theta_1'(z)}{\theta_1(z)} = \cot z + 4 \sum_{n=1}^{\infty} \frac{q^{2n}}{1 - q^{2n}} \sin 2nz, \qquad (8.6.15)$$

valid in the strip $0 < y < \pi \mathscr{I} \tau$.

Replacing z by $z + \frac{1}{2}\pi$ in the identities (8.6.10) and (8.6.15) and referring to the identities (1.3.7), (1.3.8), we can now derive the further results

$$\frac{\theta_3'(z)}{\theta_3(z)} = 4 \sum_{n=1}^{\infty} \frac{(-q)^n}{1 - q^{2n}} \sin 2nz, \qquad (8.6.16)$$

$$\frac{\theta_2'(z)}{\theta_2(z)} = -\tan z + 4 \sum_{n=1}^{\infty} \frac{(-1)^n q^{2n}}{1 - q^{2n}} \sin 2nz, \qquad (8.6.17)$$

valid in the strips $-\frac{1}{2}\pi \mathscr{I} \tau < y < \frac{1}{2}\pi \mathscr{I} \tau$ and $0 < y < \pi \mathscr{I} \tau$ respectively.

The reader may verify that the expansion (8.6.15) can also be derived by replacing z by $z + \frac{1}{2}\pi\tau$ in (8.6.11), thus proving that the former expansion is also valid in the strip $-\pi \mathscr{I} \tau < y < 0$. Letting z tend to any real value not a multiple of π, we can also prove that the expansion is valid on the real axis except at the zeros $z = m\pi$ of $\theta_1(z)$. (Note: The expansion can be proved uniformly convergent in a neighborhood of such a real value and hence continuous.) We conclude finally, therefore, that the expansion (8.6.15) is valid in the double strip $-\pi \mathscr{I} \tau < y < \pi \mathscr{I} \tau$, except at the points $z = m\pi$. A similar extension then applies to the expansion (8.6.17), the exceptional points being the real zeros of $\theta_2(z)$, viz. $z = (m + \frac{1}{2})\pi$.

The uniform convergence property of the expansions permits us to integrate term by term to generate further expansions as follows:

$$\left. \begin{array}{l} \ln \theta_1(z) = \ln(\sin z) - 2 \sum_{n=1}^{\infty} \frac{q^{2n}}{n(1 - q^{2n})} \cos 2nz + \text{constant}, \\[3mm] \ln \theta_2(z) = \ln(\cos z) - 2 \sum_{n=1}^{\infty} \frac{(-1)^n q^{2n}}{n(1 - q^{2n})} \cos 2nz + \text{constant}, \\[3mm] \ln \theta_3(z) = -2 \sum_{n=1}^{\infty} \frac{(-q)^n}{n(1 - q^{2n})} \cos 2nz + \text{constant}, \\[3mm] \ln \theta_4(z) = -2 \sum_{n=1}^{\infty} \frac{q^n}{n(1 - q^{2n})} \cos 2nz + \text{constant}. \end{array} \right\} \qquad (8.6.18)$$

The values of the constants depend on the branches of the logarithmic functions chosen.

Using equation (6.6.8), we can now derive a Fourier expansion for Weierstrass's zeta function. Thus, substituting from the first of the results (8.6.18), we obtain

$$\zeta(u) = \frac{2\eta_1}{\pi}z + \frac{\pi}{2\omega_1}\left[\cot z + 4\sum_{n=1}^{\infty}\frac{q^{2n}}{1-q^{2n}}\sin 2nz\right], \qquad (8.6.19)$$

where $z = \pi u/2\omega_1$. Differentiating this last expansion with respect to u, we are led to the following Fourier expansion of $\mathscr{P}(u)$:

$$\mathscr{P}(u) = -\zeta'(u) = -\frac{\eta_1}{\omega_1} + \left(\frac{\pi}{2\omega_1}\right)^2\left[\operatorname{cosec}^2 z - 8\sum_{n=1}^{\infty}\frac{nq^{2n}}{1-q^{2n}}\cos 2nz\right]. \qquad (8.6.20)$$

(Compare these expansions with those of Exercises 19 and 20 in Chapter 6.)

8.7. Fourier Expansions of the Jacobian Functions

The function sn u has two parallel lines of poles $2nK \pm iK'$ ($n = 0, \pm 1, \pm 2$, etc.) and is regular throughout the strip of the u-plane lying between these lines. Since $iK' = \tau K$, these poles are at the points $K(2n \pm \tau)$. Transforming to a new complex variable $z(= x + iy)$ by $u = 2Kz/\pi$, in the z-plane the poles lie at the points $\frac{1}{2}\pi(2n \pm \tau)$, i.e., on the lines $y = \pm\frac{1}{2}\pi\mathscr{I}\tau$ and, regarded as a function of z, sn u is regular in the strip $|y| < \frac{1}{2}\pi\mathscr{I}\tau$. sn u is also an odd function of z and is periodic with period 2π. Hence, as shown in Appendix A, it can be expanded in a Fourier sine series thus;

$$\operatorname{sn} u = \sum_{n=1}^{\infty} b_n \sin nz, \qquad (8.7.1)$$

where

$$b_n = \frac{i}{\pi}\int_{-\pi}^{\pi} \operatorname{sn} u\, e^{-nix}\, dx, \qquad (8.7.2)$$

the integration being along the real axis. The expansion is valid throughout the strip $|y| < \frac{1}{2}\pi\mathscr{I}\tau$.

To calculate the integral, we integrate $e^{-niz}\operatorname{sn}(2Kz/\pi)$ around the parallelogram $ABCD$ with vertices $-\pi, \pi, \pi\tau, \pi\tau - 2\pi$, respectively. Since $\operatorname{sn}(2Kz/\pi)$ has period 2π, the integrand takes the same values over the opposite sides AD and BC, and their contributions to the integral accordingly cancel. On CD, $z = x + \pi(\tau - 1)$, where x decreases from π to $-\pi$ as we move from C to D. Hence, the contribution of this side to the integral is

$$\int_{\pi}^{-\pi} \exp\{-ni(x + \pi\tau - \pi)\}\operatorname{sn}\left(\frac{2K}{\pi}x + 2iK' - 2K\right)dx$$

$$= (-1)^n q^{-n}\int_{-\pi}^{\pi} \operatorname{sn} u\, e^{-nix}\, dx. \qquad (8.7.3)$$

The integral taken around the whole parallelogram accordingly has value

$$\{1+(-1)^n q^{-n}\}I, \tag{8.7.4}$$

where I is the integral appearing in (8.7.2).

Inside the parallelogram, there are two poles of $\operatorname{sn}(2Kz/\pi)$ and these are at the points $z = \frac{1}{2}\pi\tau$ and $z = \frac{1}{2}\pi\tau - \pi$. The corresponding values of u are iK' and $iK' - 2K$ respectively and, as shown in section 2.8, the residues of $\operatorname{sn} u$ at these points are $1/k$ and $-1/k$ respectively. The corresponding residues of $\operatorname{sn}(2Kz/\pi)$ are $\pi/(2kK)$ and $-\pi/(2kK)$, and the residues of the complete integrand $e^{-niz}\operatorname{sn}(2Kz/\pi)$ are therefore $q^{-n/2}\pi/(2kK)$, $-(-1)^n q^{-n/2}\pi/(2kK)$, respectively. By Cauchy's residue theorem, we conclude that the integral around $ABCD$ must be zero if n is even and must be $2\pi^2 iq^{-n/2}/(kK)$ if n is odd. Taking this result in conjunction with that given at (8.7.4), we have proved that

$$\left.\begin{aligned} I &= \frac{2\pi^2 iq^{-n/2}}{kK(1-q^{-n})}, \quad n \text{ odd,}\\[2mm] &= 0, \qquad\qquad\quad\ n \text{ even.} \end{aligned}\right\} \tag{8.7.5}$$

b_n can now be calculated from equation (8.7.2) and the Fourier expansion (8.7.1) written down in the form

$$\operatorname{sn} u = \frac{2\pi}{kK}\left[\frac{q^{1/2}}{1-q}\sin z + \frac{q^{3/2}}{1-q^3}\sin 3z + \frac{q^{5/2}}{1-q^5}\sin 5z + \cdots\right]. \tag{8.7.6}$$

The same method leads to the following Fourier expansions for $\operatorname{cn} u$ and $\operatorname{dn} u$:

$$\operatorname{cn} u = \frac{2\pi}{kK}\left[\frac{q^{1/2}}{1+q}\cos z + \frac{q^{3/2}}{1+q^3}\cos 3z + \frac{q^{5/2}}{1+q^5}\cos 5z + \cdots\right], \tag{8.7.7}$$

$$\operatorname{dn} u = \frac{\pi}{2K} + \frac{2\pi}{K}\left[\frac{q}{1+q^2}\cos 2z + \frac{q^2}{1+q^4}\cos 4z + \frac{q^3}{1+q^6}\cos 6z + \cdots\right]. \tag{8.7.8}$$

The verifications are left as exercises for the reader. (Note: $\operatorname{dn}(2Kz/\pi)$ is of period π. Integrate around a parallelogram with vertices $-\frac{1}{2}\pi, \frac{1}{2}\pi, \frac{1}{2}\pi + \pi\tau, -\frac{1}{2}\pi + \pi\tau$.) All these expansions are valid in the strip $|y| < \frac{1}{2}\pi\mathscr{I}\tau$.

Replacing u by $u + K$ (or z by $z + \frac{1}{2}\pi$) in these three expansions, we show that

$$\operatorname{cd} u = \frac{2\pi}{kK}\sum_{n=0}^{\infty}\frac{(-1)^n q^{n+1/2}}{1-q^{2n+1}}\cos(2n+1)z, \tag{8.7.9}$$

$$\operatorname{sd} u = \frac{2\pi}{kk'K}\sum_{n=0}^{\infty}\frac{(-1)^n q^{n+1/2}}{1+q^{2n+1}}\sin(2n+1)z, \tag{8.7.10}$$

$$\operatorname{nd} u = \frac{\pi}{2k'K} + \frac{2\pi}{k'K}\sum_{n=1}^{\infty}\frac{(-1)^n q^n}{1+q^{2n}}\cos 2nz, \tag{8.7.11}$$

all valid in the strip $|y| < \frac{1}{2}\pi\mathscr{I}\tau$.

Replacing u by $u - iK'$ (or z by $z - \frac{1}{2}\pi\tau$) in the six Fourier expansions now available, we are led to expansions for the remaining Jacobian functions. The manipulations are similar to those performed in the last section and are left for the reader to supply. The results are as follows:

$$\operatorname{ns} u = \frac{\pi}{2K}\operatorname{cosec} z + \frac{2\pi}{K}\sum_{n=0}^{\infty}\frac{q^{2n+1}}{1-q^{2n+1}}\sin(2n+1)z, \qquad (8.7.12)$$

$$\operatorname{ds} u = \frac{\pi}{2K}\operatorname{cosec} z - \frac{2\pi}{K}\sum_{n=0}^{\infty}\frac{q^{2n+1}}{1+q^{2n+1}}\sin(2n+1)z, \qquad (8.7.13)$$

$$\operatorname{cs} u = \frac{\pi}{2K}\cot z - \frac{2\pi}{K}\sum_{n=1}^{\infty}\frac{q^{2n}}{1+q^{2n}}\sin 2nz, \qquad (8.7.14)$$

$$\operatorname{dc} u = \frac{\pi}{2K}\sec z + \frac{2\pi}{K}\sum_{n=0}^{\infty}\frac{(-1)^n q^{2n+1}}{1-q^{2n+1}}\cos(2n+1)z, \qquad (8.7.15)$$

$$\operatorname{nc} u = \frac{\pi}{2k'K}\sec z - \frac{2\pi}{k'K}\sum_{n=0}^{\infty}\frac{(-1)^n q^{2n+1}}{1+q^{2n+1}}\cos(2n+1)z, \qquad (8.7.16)$$

$$\operatorname{sc} u = \frac{\pi}{2k'K}\tan z + \frac{2\pi}{k'K}\sum_{n=1}^{\infty}\frac{(-1)^n q^{2n}}{1+q^{2n}}\sin 2nz. \qquad (8.7.17)$$

More easily, the last three of these expansions can be obtained from the first three by replacement of u by $u + K$. All expansions are valid in the "double" strip $|y| < \pi\mathscr{I}\tau$, except at the poles of cosec z, etc., on the real axis (see previous section for the method of justification).

8.8. Other Trigonometric Expansions of Jacobi's Functions

We consider $f(z) = \operatorname{sn}(2Kz/\pi)\operatorname{cosec}(z - \zeta)$ as a function of z, ζ taking an arbitrary complex value. We integrate it around a parallelogram $ABCD$ whose sides AB, CD are parallel to the real axis, and whose sides BC, AD are parallel to a vector representing the complex number τ and pass through the points $\frac{1}{2}\pi$, $-\frac{1}{2}\pi$ respectively on the real axis (Fig. 8.3). C will be taken to have affix $\frac{1}{2}\pi + M\pi\tau$, where M is a large positive integer, and thus D has affix $-\frac{1}{2}\pi + M\pi\tau$. Similarly, A, B will be taken to have affixes $-\frac{1}{2}\pi - N\pi\tau$, $\frac{1}{2}\pi - N\pi\tau$ respectively, with N large and positive but independent of M.

It is easily verified that $f(z)$ is of period π and so takes the same values along BC as along AD. The contributions of these sides to the complex integral accordingly cancel. On CD, $z = M\pi\tau + x$, where x decreases from $\frac{1}{2}\pi$ to $-\frac{1}{2}\pi$ as we move from C to D; thus, on this side, putting $\zeta = \xi + i\eta$, $\tau = \sigma + iv$ ($v > 0$),

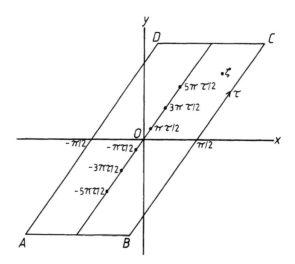

Figure 8.3. Integration contour for $\operatorname{sn}(2Kz/\pi)\operatorname{cosec}(z-\zeta)$.

we have

$$|f(z)| = |\operatorname{sn}(2Kx/\pi)\operatorname{cosec}(M\pi\tau + x - \zeta)|$$
$$\leqslant |\operatorname{cosec}\{x - \xi + M\pi\sigma + i(M\pi v - \eta)\}|$$

$$= \frac{2}{|\exp\{-M\pi v + \eta + i(x - \xi + M\pi\sigma)\} - \exp\{M\pi v - \eta - i(x - \xi + M\pi\sigma)\}|}$$

$$\leqslant \frac{2}{\exp(M\pi v - \eta) - \exp(-M\pi v + \eta)}, \tag{8.8.1}$$

showing that $f(z) \to 0$, uniformly in x, as $M \to \infty$. It follows that the integral of $f(z)$ along CD tends to zero as $M \to \infty$. Similarly, the contribution of AB to the integral can be shown to vanish in the limit as $N \to \infty$. Hence, in the double limit, the integral of $f(z)$ around the parallelogram is zero.

We next determine the positions of the singularities of $f(z)$. As we have seen earlier, $\operatorname{sn}(2Kz/\pi)$ has simple poles at $z = \pi\{m + (n + \frac{1}{2})\tau\}$. Of these points, only the poles at $z = (n + \frac{1}{2})\pi\tau$ lie inside $ABCD$ (see Fig. 8.3). The residues at these points are all the same, viz. $\pi/(2kK)$. Thus, assuming the point ζ does not coincide with a pole, the residue of $f(z)$ at the singularity $z = (n + \frac{1}{2})\pi\tau$ is $(\pi/2kK)\operatorname{cosec}\{(n + \frac{1}{2})\pi\tau - \zeta\}$. Assuming ζ lies inside $ABCD$, the only singularity of $\operatorname{cosec}(z - \zeta)$ in this region is at $z = \zeta$ and here it has a simple pole with residue 1. Thus, $f(z)$ has a pole at $z = \zeta$ with residue $\operatorname{sn}(2K\zeta/\pi)$.

But, since the integral around $ABCD$ vanishes (in the limit), the sum of the residues must also be zero, i.e.,

$$\operatorname{sn}(2K\zeta/\pi) + \sum_{n=-\infty}^{\infty} \frac{\pi}{2kK}\operatorname{cosec}\{(n + \frac{1}{2})\pi\tau - \zeta\} = 0. \tag{8.8.2}$$

Putting $2K\zeta/\pi = u, \tau = iK'/K$, we have arrived at the expansion

$$\operatorname{sn} u = \frac{\pi}{2kK} \sum_{n=-\infty}^{\infty} \operatorname{cosec} \frac{\pi}{2K} \{u - (2n+1)iK'\}, \qquad (8.8.3)$$

where, since M, N were independent, the limits $\pm\infty$ can be approached independently. The periodicity of the two members of this equation requires that the expansion remain valid when the point ζ lies outside $ABCD$.

Similarly, by integrating $\operatorname{cn}(2Kz/\pi)\operatorname{cosec}(z-\zeta)$ around the same contour, we are led to the expansion

$$\operatorname{cn} u = \frac{i\pi}{2kK} \sum_{n=-\infty}^{\infty} (-1)^{n-1} \operatorname{cosec} \frac{\pi}{2K} \{u - (2n+1)iK'\}. \qquad (8.8.4)$$

Since $\operatorname{dn}(2Kz/\pi)\operatorname{cosec}(z-\zeta)$ is of period 2π (not π), the method has to be amended to provide a similar expansion for $\operatorname{dn} u$. The function to be integrated is taken to be $\operatorname{dn}(2Kz/\pi)\cot(z-\zeta)$. This is of period π so that the contributions of the sides BC, DA of the parallelogram again cancel. However, to ensure convergence, it is necessary to assume $z = x - N\pi\tau$ on AB and $z = x + (N+1)\pi\tau$ on CD, i.e., AB and CD cannot be permitted to move to infinity independently; the reason for this is that their individual contributions to the integral no longer approach zero, but the sum of their contributions can be made to have this limit if AB and CD are related as just described. Then, this sum is found to be

$$(-1)^N \int_{-\pi/2}^{\pi/2} \operatorname{dn}(2Kx/\pi)\{\cot(x - \zeta - N\pi\tau) + \cot(x - \zeta + \overline{N+1}\pi\tau)\}\, dx,$$

$$= 2(-1)^N \int_{-\pi/2}^{\pi/2} \operatorname{dn}(2Kx/\pi) \frac{\sin(2x - 2\zeta + \pi\tau)}{\cos(2N+1)\pi\tau - \cos(2x - 2\zeta + \pi\tau)}\, dx \qquad (8.8.5)$$

which is easily shown to approach zero as $N \to \infty$, since $|\cos(2N+1)\pi\tau| \to \infty$. Thus, in the limit, the integral taken around $ABCD$ is zero.

Reference to section 2.8 reveals that $\operatorname{dn}(2Kz/\pi)$ has simple poles inside $ABCD$ at the points $z = (n+\tfrac{1}{2})\pi\tau$ with residues $(-1)^{n-1}i\pi/2K, n = 0, \pm 1, \pm 2$, etc. The corresponding residues of the complete integrand are $(-1)^{n-1}(i\pi/2K)\cot\{(n+\tfrac{1}{2})\pi\tau - \zeta\}$. The pole at $z = \zeta$ has residue $\operatorname{dn}(2K\zeta/\pi)$. For an arbitrary value of N, the sum of the residues inside $ABCD$ is therefore

$$\operatorname{dn}(2K\zeta/\pi) + \frac{i\pi}{2K} \sum_{n=-N}^{N} (-1)^{n-1} \cot\{(n+\tfrac{1}{2})\pi\tau - \zeta\}. \qquad (8.8.6)$$

As $N \to \infty$, this must tend to zero. Hence, with $2K\zeta/\pi = u$,

$$\operatorname{dn} u = \frac{i\pi}{2K} \sum_{n=-\infty}^{\infty} (-1)^{n-1} \cot \frac{\pi}{2K} \{u - (2n+1)iK'\}, \qquad (8.8.7)$$

it being understood that the limits $\pm\infty$ must be approached together.

8.9. Representation of a General Elliptic Function with Theta and Sigma Functions

If a pair of primitive periods $2\omega_1, 2\omega_3$ of an elliptic function $F(u)$ are known and the positions and multiplicities of all its zeros and poles inside some cell are also known, then the function is determined to within a constant factor. We shall prove this theorem by showing how to construct $F(u)$ from the theta function $\theta_1(z|\tau)$, where $\tau = \omega_3/\omega_1$.

Suppose $F(u)$ is of order N and let u_i $(i = 1, 2, \ldots, N)$ be its zeros inside some cell and v_i its poles, each zero and pole being repeated in these sequences a number of times equal to its multiplicity. Then, as proved in section 8.2, $\sum u_i$ and $\sum v_i$ can differ only by a period. It is possible (in many ways), therefore, to replace one or more of the zeros and poles by congruent points in the lattice, in such a way that $\sum u_i = \sum v_i$. We shall assume this has been done. Then, introducing a new variable by $z = \pi u/2\omega_1$, in terms of this variable F will have periods π and $\pi\tau$, and we shall denote the zeros by z_i and poles by p_i.

We now construct the function $\Theta(z)$ by the formula

$$\Theta(z) = \prod_{i=1}^{N} \frac{\theta_1(z - z_i)}{\theta_1(z - p_i)}, \tag{8.9.1}$$

the parameter for all the theta functions being τ. Referring to equations (1.3.2), it follows that

$$\Theta(z + \pi) = \Theta(z), \tag{8.9.2}$$

$$\Theta(z + \pi\tau) = \exp\{2i(\sum z_i - \sum p_i)\}\Theta(z) = \Theta(z),$$

since we have chosen the zeros and poles so that $\sum z_i = \sum p_i$. Hence, $\Theta(z)$ is an elliptic function with the same periods as $F(u)$.

Further, $\Theta(z)$ has the same zeros and poles as $F(u)$ and with the same multiplicity in each case. Hence, if $\Phi(z) = F(2\omega_1 z/\pi)/\Theta(z)$, the poles of F are canceled by the poles of Θ and the zeros of Θ are canceled by the zeros of F. Thus, $\Phi(z)$ is an elliptic function with no singularites and must, therefore, be a constant A. We have proved

$$F(u) = A \prod_{i=1}^{N} \frac{\theta_1(z - z_i)}{\theta_1(z - p_i)}. \tag{8.9.3}$$

For example, if $F(u) = \text{sn } u$, then $2\omega_1 = 4K$ and $2\omega_3 = 2iK'$, and we can take $u_1 = 0, u_2 = 2K$, for an irreducible pair of zeros and $v_1 = iK', v_2 = 2K - iK'$ for an irreducible pair of poles; clearly, $u_1 + u_2 = v_1 + v_2$. Correspondingly, we have $z_1 = 0, z_2 = \frac{1}{2}\pi, p_1 = \pi iK'/4K = \frac{1}{2}\pi\tau, p_2 = \pi(2K - iK')/4K = \frac{1}{2}\pi - \frac{1}{2}\pi\tau$ (N.B. as explained below, the τ appearing in these equations is different from the τ which enters into equations (2.2.3)). Thus, equation (8.9.3) yields in this case

$$\text{sn } u = A \frac{\theta_1(z)\theta_1(z - \frac{1}{2}\pi)}{\theta_1(z - \frac{1}{2}\pi\tau)\theta_1(z - \frac{1}{2}\pi + \frac{1}{2}\pi\tau)} = B \frac{\theta_1(z)\theta_2(z)}{\theta_3(z)\theta_4(z)}, \tag{8.9.4}$$

after using the identities (1.3.6)–(1.3.9). This result can be simplified using Landen's transformation. Thus, dividing the transformation equations (1.8.8) and (1.8.11), we find that

$$\operatorname{sn} u = B \frac{\theta_1(2z|2\tau)}{\theta_4(2z|2\tau)}. \tag{8.9.5}$$

This is equivalent to our defining identity (2.1.1), since $2z = \pi u/\omega_1 = \pi u/2K$ and $2\tau = 2\omega_3/\omega_1 = iK'/K$.

In view of the close relationship between the functions $\theta_1(z)$ and $\sigma(u)$ as expressed by equation (6.2.1), the result (8.9.3) is equivalent to a similar representation in terms of sigma functions. Thus, substituting for the theta functions from equation (6.2.1), we find that

$$F(u) = A \prod_{i=1}^{N} \frac{\exp\{-a(u-u_i)^2\}\sigma(u-u_i)}{\exp\{-a(u-v_i)^2\}\sigma(u-v_i)}$$

$$= A' \prod_{i=1}^{N} \frac{\sigma(u-u_i)}{\sigma(u-v_i)}, \tag{8.9.6}$$

since the ratio of the exponential functions is a constant by virtue of the condition $\sum u_i = \sum v_i$.

As an example, consider the function $\mathscr{P}(u) - \mathscr{P}(\alpha)$, where α is not a period. This is of the second order and has a double pole at $u = 0$ and zeros at $u = \pm \alpha$. If 2α is also not a period, these zeros are not at congruent points and so provide an irreducible set satisfying the condition $\sum u_i = \sum v_i$ (if 2α is a period, the zeros are also double; we exclude this case for the moment). Hence,

$$\mathscr{P}(u) - \mathscr{P}(\alpha) = A \frac{\sigma(u-\alpha)\sigma(u+\alpha)}{\sigma^2(u)}. \tag{8.9.7}$$

Expanding both sides in powers of u and comparing coefficients of u^{-2}, we find that $A = -1/\sigma^2(\alpha)$. Thus, the identity (6.7.5) has been reestablished. That the identity is also valid when 2α is a period, follows by analytic continuation.

8.10. Representation of a General Elliptic Function in Terms of $\mathscr{P}(u)$

A method similar to that employed in the last section establishes that any elliptic function $F(u)$ can be represented as a rational function of the Weierstrass functions $\mathscr{P}(u)$ and $\mathscr{P}'(u)$. Thus, the characteristic properties of all elliptic functions are essentially encapsulated within those of $\mathscr{P}(u)$.

As before, we shall denote a pair of primitive periods of $F(u)$ by $2\omega_1$ and $2\omega_3$. First, suppose $F(u)$ is even and has no pole or zero at a lattice point.

Choose a cell for which $F(u)$ has no pole or zero on its perimeter. Then, as proved in section 8.2, $F(u)$ is an elliptic function of even order $2N$ whose zeros inside the cell can be arranged in pairs of points congruent to $+\alpha$ and $-\alpha$ or lie at some midlattice point with even multiplicity. Thus, an irreducible set of zeros can be taken to have affixes $\pm u_i$ ($i = 1, 2, \ldots, N$), multiple zeros being repeated the requisite number of times (in the case of a midlattice point α of multiplicity $2r$, we take r zeros at α and the other r at the congruent point $-\alpha$). Similarly, an irreducible set of poles can be taken to have affixes $\pm v_i$.

Now consider the function

$$G(u) = \prod_{i=1}^{N} \frac{\mathscr{P}(u) - \mathscr{P}(u_i)}{\mathscr{P}(u) - \mathscr{P}(v_i)}. \tag{8.10.1}$$

$G(u)$ is periodic with periods $2\omega_1, 2\omega_3$ and has no pole at $u = 0$ (or any other lattice point), since the poles of $\mathscr{P}(u)$ in the numerator and denominator cancel. $\mathscr{P}(u)$ can be expanded in the neighborhood of u_i thus;

$$\mathscr{P}(u) = \mathscr{P}(u_i) + (u - u_i)\mathscr{P}'(u_i) + \cdots. \tag{8.10.2}$$

If u_i is not a midlattice point, $\mathscr{P}'(u_i) \neq 0$ and so $\mathscr{P}(u) - \mathscr{P}(u_i)$ has a simple zero at $u = u_i$. If u_i is a midlattice point, $\mathscr{P}'(u_i) = 0$ and $\mathscr{P}(u) - \mathscr{P}(u_i)$ has a double zero at the point. Since $\mathscr{P}(u)$ is even, this means that $\mathscr{P}(u) - \mathscr{P}(u_i)$ also has a simple zero at $u = -u_i$ and a double zero there if u_i is a midlattice point (in which case $-u_i$ is congruent to u_i). Thus, $G(u)$ has the same zeros and poles as $F(u)$ and with the same multiplicities.

It now follows that $F(u)/G(u)$ is an elliptic function with no poles and, hence, must be a constant A. We have therefore proved that

$$F(u) = A \prod_{i=1}^{N} \frac{\mathscr{P}(u) - \mathscr{P}(u_i)}{\mathscr{P}(u) - \mathscr{P}(v_i)}. \tag{8.10.3}$$

Next, suppose $F(u)$ has a zero at $u = 0$ and hence at all the lattice points. Since $F(u)$ is an even function, the multiplicity of the zero will be even, $2n$ say. The remaining zeros of $F(u)$ forming an irreducible set can be denoted by $\pm u_i$, but i will now run from 1 to $N - n$ and we shall construct the function $G(u)$, thus:

$$G(u) = \prod_{i=1}^{N-n} \{\mathscr{P}(u) - \mathscr{P}(u_i)\} \bigg/ \prod_{j=1}^{N} \{\mathscr{P}(u) - \mathscr{P}(v_i)\}. \tag{8.10.4}$$

The poles at $u = 0$ in the factors of the denominator and numerator no longer cancel and $G(u)$ has a zero of multiplicity $2n$ at $u = 0$. Hence, as before, $F(u)/G(u)$ is regular everywhere and must be constant.

This case can be covered by the formula (8.10.3) provided it is understood that factors corresponding to zeros u_i at the lattice points are always omitted. Similarly, factors in the denominator corresponding to poles v_i at the lattice points must be omitted.

Finally, suppose the restriction that $F(u)$ be even is lifted. Then $F(u)$ can be separated into even and odd components thus,

$$F(u) = F_1(u) + F_2(u) = \tfrac{1}{2}\{F(u) + F(-u)\} + \tfrac{1}{2}\{F(u) - F(-u)\}. \qquad (8.10.5)$$

The last result applies to the component $F_1(u)$, which is even. Since $\mathscr{P}'(u)$ is an odd elliptic function, $F_2(u)/\mathscr{P}'(u)$ is even and the result (8.10.4) can be applied to it also. We conclude that $F(u)$ can be expressed in the form

$$F(u) = P\{\mathscr{P}(u)\} + \mathscr{P}'(u)Q\{\mathscr{P}(u)\}, \qquad (8.10.6)$$

where P and Q are rational functions. $\mathscr{P}'(u)$ can, of course, also be expressed in terms of $\mathscr{P}(u)$ if square roots are acceptable.

As an example, we shall take $F(u) = \operatorname{sn}(u, k)$. Given k, K and K' are fixed and these determine the periods $2\omega_1 = 4K$, $2\omega_3 = 2iK'$ of $\operatorname{sn} u$ and, hence, of $\mathscr{P}(u)$. An irreducible set of zeros of $\operatorname{sn} u$ is $\{0, 2K\}$ and an irreducible set of poles is $\{iK', 2K + iK'\}$, all simple. Now, $\mathscr{P}'(u)$ has a triple pole at $u = 0$ (and at all lattice points) and simple zeros at the midlattice points congruent to $\omega_1 = 2K$, $\omega_3 = iK'$, $\omega_1 + \omega_3 = 2K + iK'$. It follows that the even function $\operatorname{sn} u/\mathscr{P}'(u)$ has an irreducible set of zeros $\{0 \text{ (multiplicity 4)}\}$ and an irreducible set of poles $\{iK' \text{ (double)}, 2K + iK' \text{ (double)}\}$—note that the zeros of $\operatorname{sn} u$ and $\mathscr{P}'(u)$ at $u = 2K$ cancel. The zeros at the lattice points are ignored. The poles at the midlattice points are accounted for by taking $v_1 = iK'$, $v_2 = 2K + iK'$; although these poles are double (necessarily), this is allowed for, since the irreducible set is taken to be $\pm v_1$, $\pm v_2$ and $-v_1$, $-v_2$ are congruent to v_1, v_2, respectively. Application of the result (8.10.4) now gives

$$\frac{\operatorname{sn} u}{\mathscr{P}'(u)} = \frac{A}{\{\mathscr{P}(u) - e_2\}\{\mathscr{P}(u) - e_3\}}, \qquad (8.10.7)$$

where $e_2 = \mathscr{P}(\omega_1 + \omega_3) = \mathscr{P}(2K + iK') = \mathscr{P}(v_2)$, $e_3 = \mathscr{P}(\omega_3) = \mathscr{P}(iK') = \mathscr{P}(v_1)$. Substituting for $\mathscr{P}'(u)$ from equation (6.7.16), we calculate that

$$\operatorname{sn} u = B \sqrt{\frac{\mathscr{P}(u) - e_1}{\{\mathscr{P}(u) - e_2\}\{\mathscr{P}(u) - e_3\}}} = C\frac{\mathscr{P}(u) - e_1}{\mathscr{P}'(u)}, \qquad (8.10.8)$$

where $e_1 = \mathscr{P}(\omega_1) = \mathscr{P}(2K)$. Expanding both sides of this equation in powers of u, and comparing terms of the first degree, we find that $C = -2$. Thus,

$$\operatorname{sn} u = 2\{e_1 - \mathscr{P}(u)\}/\mathscr{P}'(u). \qquad (8.10.9)$$

This formula for $\operatorname{sn} u$ is quite different from the one already found at (6.9.6). The discrepancy is accounted for by the fact that the lattices used to define the Weierstrass functions in the two cases are distinct. Thus, equations (6.9.5) give $\omega_1 = K$, $\omega_3 = iK'$, whereas we have been assuming $\omega_1 = 2K$, $\omega_3 = iK'$. In the former case, a pair of primitive period parallelograms for $\mathscr{P}(u)$ are needed to provide a period parallelogram for $\operatorname{sn} u$; the positive square root in equation (6.9.6) must be taken in one and the negative square root in the other.

If we replace u by $u + iK'$ in equation (8.10.9), we deduce, using equations (2.2.17) and (6.8.13), that

$$\frac{1}{k} \operatorname{ns} u = \operatorname{sn}(u + iK') = 2\{e_1 - \mathscr{P}(u + \omega_3)\}/\mathscr{P}'(u + \omega_3)$$

$$= \frac{2\left[e_1 - e_3 - \dfrac{(e_3 - e_1)(e_3 - e_2)}{\mathscr{P}(u) - e_3} \right]}{-\dfrac{(e_3 - e_1)(e_3 - e_2)}{\{\mathscr{P}(u) - e_3\}^2} \mathscr{P}'(u)}$$

$$= \frac{2\{\mathscr{P}(u) - e_2\}\{\mathscr{P}(u) - e_3\}}{(e_3 - e_2)\mathscr{P}'(u)}$$

$$= \frac{\mathscr{P}'(u)}{2(e_3 - e_2)\{\mathscr{P}(u) - e_1\}}$$

$$= (e_2 - e_3)^{-1} \operatorname{ns} u. \tag{8.10.10}$$

We conclude that, in contrast to equation (6.9.9),

$$k = e_2 - e_3. \tag{8.10.11}$$

8.11. Expansion of an Elliptic Function in Terms of $\zeta(u)$

Given the positions of the zeros and poles of an elliptic function $F(u)$ with periods $2\omega_1$ and $2\omega_3$, it has been shown that the function is completely specified to within a constant factor. Liouville's theorem also permits us to deduce that, if the principal part of the function at each of its poles is known, then the function is determined apart from an additive constant. For, we can construct an elliptic function having the same poles and principal parts at these poles, and the difference between this function and $F(u)$ is regular everywhere and so must be a constant.

Thus, suppose $u = v_i$ $(i = 1, 2, \ldots, N)$ is an irreducible set of poles of $F(u)$ and let the principal part at v_i be represented by

$$\sum_{r=1}^{n} b_{ir}(u - v_i)^{-r}, \tag{8.11.1}$$

where n is the multiplicity of the pole of highest order and some of the coefficients b_{ir} may be zero.

Consider the function

$$g_1(u) = \sum_{i=1}^{N} b_{i1}\zeta(u - v_i), \tag{8.11.2}$$

where ζ is Weierstrass's zeta function (section 6.6). $g_i(u)$ is an elliptic function for, using the identity (6.6.4),

$$g_1(u + 2\omega_1) = \sum_{i=1}^{N} b_{i1}\zeta(u + 2\omega_1 - v_i)$$

$$= \sum_{i=1}^{N} b_{i1}\{\zeta(u - v_i) + 2\eta_1\}$$

$$= g_1(u), \qquad (8.11.3)$$

since $\sum b_{i1}$ is the sum of the residues of $F(u)$ at an irreducible set of singularities and hence vanishes. Similarly, we prove $g_1(u + 2\omega_3) = g_1(u)$. Further, since $\zeta(u)$ has a simple pole at $u = 0$ with residue 1, the principal part of $g_1(u)$ at $u = v_i$ is $b_{i1}/(u - v_i)$.

Further functions $g_r(u)$ $(r = 2, 3, \ldots, n)$ are defined by

$$g_r(u) = \sum_{i=1}^{N} b_{ir}\zeta^{(r-1)}(u - v_i) \qquad (8.11.4)$$

and these are clearly also elliptic functions, since $\zeta'(u) = -\mathscr{P}(u)$. The principal part of $\zeta^{(r-1)}(u)$ at its pole $u = 0$ is $(-1)^{r-1}(r - 1)!/u^r$ and it follows that the principal part of $g_r(u)$ at its pole $u = v_i$ is $(-1)^{r-1}(r - 1)!b_{ir}/(u - v_i)^r$.

Now consider the function $G(u)$ defined by

$$G(u) = \sum_{r=1}^{n} \frac{(-1)^{r-1}}{(r-1)!}g_r(u). \qquad (8.11.5)$$

It is an elliptic function with periods $2\omega_1$ and $2\omega_3$, and its poles at $u = v_i$ form an irreducible set. Its principal part at $u = v_i$ is the expression (8.11.1) and $F(u) - G(u)$ is accordingly regular for all u. Hence, by Liouville's theorem

$$F(u) = G(u) + \text{constant}$$

$$= \sum_{i=1}^{N} \sum_{r=1}^{n} \frac{(-1)^{r-1}b_{ir}}{(r-1)!}\zeta^{(r-1)}(u - v_i) + C. \qquad (8.11.6)$$

This representation of $F(u)$ is particularly useful when it is required to integrate the function. For, clearly, the derivatives of ζ are immediately integrable and ζ itself integrates to $\ln \sigma(u)$ by equation (6.6.1).

As an example, the principal parts of $\operatorname{sn} u$ at its irreducible set of poles iK', $2K + iK'$ are $1/k(u - iK')$, $-1/k(u - 2K - iK')$ (see section 2.8). Hence

$$\operatorname{sn} u = \frac{1}{k}\{\zeta(u - iK') - \zeta(u - 2K - iK')\} + C, \qquad (8.11.7)$$

with $\omega_1 = 2K$, $\omega_3 = iK'$ as the parameters of the ζ-function. Since $\zeta(-iK') = -\zeta(\omega_3) = -\eta_3$ and $\zeta(-2K - iK') = \zeta(\omega_2) = \eta_2$, putting $u = 0$ in equation (8.11.7), we deduce that $C = (\eta_2 + \eta_3)/k = -\eta_1/k$, so that

$$\operatorname{sn} u = \frac{1}{k}\{\zeta(u - iK') - \zeta(u - 2K - iK') - \eta_1\}. \qquad (8.11.8)$$

8.12. Quarter-Periods $K(k)$ and $K'(k)$ as Analytic Functions of k

It has been shown that, for real values of k in the interval $(0, 1)$, $K(k)$ and $K'(k)$ are a pair of independent solutions of the second-order differential equation (3.8.19). However, solutions to this equation when k is complex are readily obtainable and provide analytic continuations of $K(k)$ and $K'(k)$ over the region $|k| < 1$ of the complex k-plane. The analytic functions thus constructed must, necessarily, be identical with the quarter-periods defined by theta functions in Chapter 2, but the advantage of our new approach is that $K(k)$ and $K'(k)$ will be exhibited directly as functions of the modulus k.

Changing the independent variable in equation (3.8.19) by the transformation $c = k^2$, it will be found that it takes the form

$$\frac{d}{dc}\left[c(1 - c)\frac{dw}{dc} \right] = \tfrac{1}{4}w. \tag{8.12.1}$$

This equation has regular singularities at $c = 0, 1$. Assuming a solution in the form

$$w = c^p \sum_{n=0}^{\infty} a_n c^n \tag{8.12.2}$$

and using the method of Frobenius, it will be found that the indicial equation has equal roots $\rho = 0$ and that the general solution takes the form

$$w = AW_1 + B(W_1 \ln c + W_2), \tag{8.12.3}$$

where

$$W_1(c) = 1 + \left(\frac{1}{2}\right)^2 c + \left(\frac{1.3}{2.4}\right)^2 c^2 + \left(\frac{1.3.5}{2.4.6}\right)^2 c^3 + \cdots, \tag{8.12.4}$$

$$W_2(c) = 4\left[\left(\frac{1}{2}\right)^2\left(1 - \frac{1}{2}\right)c + \left(\frac{1.3}{2.4}\right)^2\left(1 - \frac{1}{2} + \frac{1}{3} - \frac{1}{4}\right)c^2 \right.$$
$$\left. + \left(\frac{1.3.5}{2.4.6}\right)^2\left(1 - \frac{1}{2} + \frac{1}{3} - \frac{1}{4} + \frac{1}{5} - \frac{1}{6}\right)c^3 + \cdots \right]. \tag{8.12.5}$$

Both the radii of convergence of these series are 1.

Referring to equation (3.8.5), we note that $W_1(k^2) = (2/\pi)K(k)$. $K'(k)$ must also be a solution. Assume

$$K'(k) = PW_1(k^2) + Q\{2W_1(k^2)\ln k + W_2(k^2)\}. \tag{8.12.6}$$

Equation (3.8.26) shows that

$$K'(k) + \ln(\tfrac{1}{4}k) \to 0, \quad \text{as } k \to 0. \tag{8.12.7}$$

The only values of P and Q consistent with this result are $P = \ln 4$, $Q = -\tfrac{1}{2}$.

Thus,

$$K'(k) = \frac{2}{\pi} K(k) \ln(4/k) - \tfrac{1}{2} W_2(k^2). \tag{8.12.8}$$

Equations (3.8.5) and (8.12.8) now define $K(k)$ and $K'(k)$ for complex values of k such that $|k| < 1$. Evidently, $K'(k)$ is multivalued, having a logarithmic branch point at $k = 0$. If k is taken around a closed contour encircling the branch point in the anticlockwise sense, commencing and terminating with the value k_0, the value of $K'(k)$ will change by $-4iK(k_0)$. Thus, the set of values of K' at any point can represented by $K'(k) - 4inK(k)$ (n an integer). This is in conformity with our earlier analysis since, if $2iK'$ is a period of $\text{sn}(u, k)$, so is $2iK' + 8nK$ (the period $2iK' + 4K$ is obtained by transforming k to $-k$ along a semicircle with centre at $k = 0$ and noting that $\text{sn}(u, -k) = \text{sn}(u, k)$). By cutting the k-plane along its negative real axis from $k = 0$ to $k = -\infty$, the logarithmic function can be restricted to its principal branch and, then, $K'(k)$ is single-valued in the slit plane.

Changing the variable by $c = 1 - c'$ (i.e., $c' = k'^2$, where k' is the complementary modulus) in equation (8.12.1), the form of the equation is unchanged. It follows that this equation has general solutions in the two forms

$$w = AK(k) + BK'(k), \qquad w = PK(k') + QK'(k'). \tag{8.12.9}$$

But, by equation (2.6.5), $K(k') = K'(k)$ and $K'(k') = K(k)$, so these solutions are identical in the region common to $|c| < 1$ and $|c'| = |c - 1| < 1$.

Our next problem is to continue $K(k)$ and $K'(k)$ analytically outside the circle $|k| = 1$.

To perform this service for $K(k)$, we make use of the integral formula (3.1.3), viz.

$$K(k) = \int_0^1 \{(1 - t^2)(1 - k^2 t^2)\}^{-1/2} \, dt, \tag{8.12.10}$$

the integration contour being the real axis from $t = 0$ to $t = 1$ and the branch of the integrand taken being that having value $+1$ at $t = 0$. This has been proved valid for $0 < k < 1$ and so must be identical with the analytic function $K(k)$ already defined for $|k| < 1$. However, the integral exists for all values of k except $+1$ and -1, and therefore provides an analytic continuation of the function $K(k)$ over the whole k-plane except for singularities at $k = \pm 1$. It remains to study the nature of these singularities and the precise definition of the integral when k is real and $k^2 > 1$, in which cases the integrand has a singularity at $t = 1/|k|$ on the line of integration.

Suppose the point k makes one circuit, in the anticlockwise sense, around the circle $|k - 1| = r$, where $0 < r < 1$ (Fig. 8.4). In the t-plane, the point $t = 1/k$ follows a circular path (center $1/(1 - r^2)$, 0; radius $r/(1 - r^2)$) embracing the point $t = 1$, the sense of description also being anticlockwise. Corresponding positions of the points k, $1/k$ in the two planes are indicated in the figure thus: (P, P'), (Q, Q'), and (R, R'). If P, P' represent the initial positions of k, $1/k$ respectively, when k reaches R, $1/k$ reaches R' and the integrand of the integral

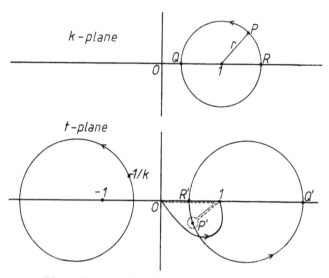

Figure 8.4. Loci: $|k - 1| = r$, $t = 1/k$, $t = -1/k$.

for $K(k)$ now has a singularity on the line of integration. The contour of integration must accordingly be diverted below R' to avoid this singularity. As k returns to P, $1/k$ returns to P' and the contour must be further diverted below P' (the contour cannot be diverted above P', since this would permit the singularity to cross over the contour and thus introduce a discontinuity into the definition of $K(k)$). We conclude that the integral defining $K(k)$ does not return to its initial value after k has made a circuit of the singularity $k = 1$ and hence that $K(k)$ is multivalued with a branch point at $k = 1$. The singularity $-1/k$ describes a circle about $t = -1$ and has no effect on the contour.

Clearly, by making a cut in the k-plane along the real axis from $k = 1$ to $k = +\infty$ and prohibiting excursions across the cut, only one value of K will be accessible for each value of k and this can be taken to be the value given by equation (8.12.10), with the contour along the real axis. In the special case $k > 1$, when k lies on the cut, the integrand's singularity $t = 1/k$ is on the contour, which must accordingly be diverted above or below to avoid it (e.g., by a small semicircle with center at $t = 1/k$). It is conventional to regard such a point k as lying on the upper side of the cut, so that it can be approached from above; $1/k$ can then approach R' from below and the contour must be diverted above the singularity to ensure continuity of $K(k)$ in this neighborhood.

When the point k returns to its starting point P after a circuit of $k = 1$, the contour of integration has been diverted as shown in Fig. 8.4. This contour can be further deformed, without affecting the value of the integral, into the shape indicated by the broken line, viz. (i) straight line from $t = 0$ to $t = 1$, (ii) straight line from $t = 1$ to $t = 1/k$, (iii) small circle about $t = 1/k$ as center, (iv) straight line from $t = 1/k$ to $t = 1$. It follows that the difference between the initial and final values of $K(k)$ is the contribution to the integral of the segments of the contour (ii)–(iv). But the contribution of the small circle vanishes as its radius tends to zero and the circuit of this circle transfers the integrand from one branch to the other, i.e., changes the sign of the square root. Thus, the difference is equal to

$$2 \int_1^{1/k} \{(1 - t^2)(1 - k^2 t^2)\}^{-1/2} \, dt. \tag{8.12.11}$$

Changing to a new complex variable s by the transformation $k^2 t^2 + k'^2 s^2 = 1$, this integral can be expressed in the form

$$- \int_0^1 \{(s^2 - 1)(1 - k'^2 s^2)\}^{-1/2} \, ds = - iK(k'), \tag{8.12.12}$$

by equation (8.12.10) (a direct justification for the minus sign depends on a careful examination of the appropriate branches of the square roots involved, and is tedious; instead an indirect justification will be supplied later). Thus, the increment in $K(k)$ due to a circuit of the branch point $k = 1$ has been found to be $- 2iK(k')$. The possible values for K at a point k are accordingly expressed by $K(k) - 2inK(k')$, where $K(k)$ and $K(k')$ are principal values given by equation (8.12.10).

Equation (8.12.10) shows immediately that $K(-k) = K(k)$ for principal values and hence that the function's behavior near to the singularity $k = -1$ is the same as in the neighborhood of $k = 1$. An anticlockwise circuit of $k = -1$ accordingly increments K by $2iK(k')$. Introducing a cut from $k = -1$ to $k = -\infty$ along the negative real axis we can restrict $K(k)$ to the principal branch given by the formula (8.12.10). In this case, however, points lying on the cut itself must be allocated to the lower edge if the contour is to be indented above the singularity $t = -1/k$.

To continue $K'(k)$ analytically outside the circle $|k| = 1$, we shall make use of the identity

$$K'(k) = K(k'), \tag{8.12.13}$$

where $K(k')$ denotes the multivalued function. Thus, $K'(k)$ is multivalued and we shall need to cut the k-plane if K' is to be confined to its principal value. However, since K is an even function, the ambiguity in the sign of k' does not lead to further values for K'.

The singularities of $K(k')$ at $k' = \pm 1$ correspond to a single singular point for K' at $k = 0$. This is the logarithmic singularity already established. Suppose the point k is taken once around the circle $|k| = \delta$ (where δ is small), embracing the singularity $k = 0$. Then $k = \delta \, e^{i\theta}$ and $k' = 1 - \frac{1}{2}\delta^2 \, e^{2i\theta}$ approximately. Thus, the point k' moves on a circle with center at $k' = 1$, such that one circuit of the circle $|k| = \delta$ corresponds to two circuits of the circle $|k' - 1| = \frac{1}{2}\delta^2$. But one circuit of the point $k' = 1$ increments $K(k')$ by $-2iK(k)$, where k is the starting value and $K(k)$ denotes the principal value. We conclude that one circuit around the point $k = 0$ increments $K'(k)$ by $-4iK(k)$, in agreement with the result derived from equation (8.12.8). To yield the correct sign for this result, it is clearly necessary that a minus sign should be included in the right-hand member of equation (8.12.12); we have now accordingly provided a justification for this sign.

As noted earlier, to confine $K'(k)$ to one of its branches, we can cut the k-plane along the negative real axis. The principal value is taken to be the branch for which $K'(k)$ is real for $0 < k \leqslant 1$. All values of K' are then included in the formula $K'(k) - 4inK(k)$, where $K'(k)$ and $K(k)$ refer to principal values.

Finally, we note that the validity of the identity $K'(k) = K(k')$ depends on an apposite choice of values for the two multivalued functions involved. For the single-valued functions, it is only valid in the half-plane $\mathscr{R}k \geqslant 0$. For $K(k')$ is defined as a single-valued function in the k'-plane with cuts from $k' = +1$ to $+\infty$ and from $k' = -1$ to $-\infty$. Since k cannot cross its imaginary axis unless k' crosses one of these cuts, k is confined to one of the half-planes $\mathscr{R}k \leqslant 0$ or $\mathscr{R}k \geqslant 0$. But the half-plane $\mathscr{R}k \leqslant 0$ contains the cut from $k = 0$ to $k = -\infty$ used in defining the single-valued function $K'(k)$. Thus, $K'(k)$ and $K(k')$ are only defined together as single-valued analytic functions in the half-plane $\mathscr{R}k \geqslant 0$. In this region, they are certainly identical on the real axis from $k = 0$ to $k = 1$ and, therefore, by the principle of analytic continuation, throughout the region. In the second and third quadrants the principal value $K'(k) = K'(-k) \mp 2iK(-k)$.

8.13. The Complete Integrals $E(k)$, $E'(k)$ as Analytic Functions of k

The complete integrals of the second kind $E(k)$ and $E'(k)$, defined by equations (3.8.3) and (3.8.4) respectively, can be studied as analytic functions of k in a manner very similar to that followed for K and K' in the previous section. However, it is more convenient first to study the functions E and $J' = K' - E'$.

Referring to equation (3.8.23), we note that this has independent solutions $E(k)$ and $J'(k)$. Changing the independent variable to $c = k^2$, this equation assumes the form

$$4c(1 - c)\frac{d^2w}{dc^2} + 4(1 - c)\frac{dw}{dc} + w = 0. \tag{8.13.1}$$

Using the method of Frobenius, we calculate the general solution in the form

$$w = AW_1 + B(W_1 \ln c + W_2), \tag{8.13.2}$$

where

$$W_1 = 1 - \left(\frac{1}{2}\right)^2 c - \frac{1}{3}\left(\frac{1.3}{2.4}\right)^2 c^2 - \frac{1}{5}\left(\frac{1.3.5}{2.4.6}\right)^2 c^3 - \cdots, \tag{8.13.3}$$

$$W_2 = 4\left[\left(\frac{1}{2}\right)^2\left(\frac{1}{1} - 1 + \frac{1}{2}\right)c + \frac{1}{3}\left(\frac{1.3}{2.4}\right)^2\left(\frac{2}{3} - 1 + \frac{1}{2} - \frac{1}{3} + \frac{1}{4}\right)c^2\right.$$
$$\left. + \frac{1}{5}\left(\frac{1.3.5}{2.4.6}\right)^2\left(\frac{3}{5} - 1 + \frac{1}{2} - \frac{1}{3} + \frac{1}{4} - \frac{1}{5} + \frac{1}{6}\right)c^3 + \cdots\right]. \tag{8.13.4}$$

Equation (3.8.6) shows that

$$E = \tfrac{1}{2}\pi W_1. \tag{8.13.5}$$

To obtain a formula for J', we note that

$$J' + \ln(\tfrac{1}{4}k) = K' + \ln(\tfrac{1}{4}k) - E' \to -1 \tag{8.13.6}$$

as $k \to 0$ (since $E'(0) = 1$). Hence, if the solution (8.13.2) is to represent J', we must take $A = \ln 4 - 1$, $B = -\tfrac{1}{2}$. Thus,

$$J' = W_1\{\ln(4/k) - 1\} - \tfrac{1}{2}W_2. \tag{8.13.7}$$

We see that, inside the circle $|k| = 1$, E is regular and J' possesses a logarithmic branch point at $k = 0$.

$E(k)$ can be continued outside the circle $|k| = 1$ by use of the integral (3.8.3). Putting $t = \sin\theta$, we have

$$E(k) = \int_0^1 \left[\frac{1 - k^2t^2}{1 - t^2}\right]^{1/2} dt, \tag{8.13.8}$$

where it is understood that the contour lies along the positive real axis, and both branches of $(1 - t^2)^{1/2}$ and $(1 - k^2t^2)^{1/2}$ are taken to be $+1$ at $t = 0$. Like $K(k)$, $E(k)$ has branch points at $k = \pm 1$, which can be investigated by the method explained in the previous section. It will be found that a circuit of the branch point $k = +1$ increments $E(k)$ by the quantity

$$2\int_1^{1/k} \left[\frac{1 - k^2t^2}{1 - t^2}\right]^{1/2} dt. \tag{8.13.9}$$

The transformation $k^2t^2 + k'^2s^2 = 1$ brings this integral to the form

$$-\int_0^1 k'^2s^2\{(s^2 - 1)(1 - k'^2s^2)\}^{-1/2}\, ds = \int_0^1 \sqrt{\left[\frac{1 - k'^2s^2}{s^2 - 1}\right]}\, ds$$

$$-\int_0^1 \{(s^2 - 1)(1 - k'^2s^2)\}^{-1/2}\, ds = i[E(k') - K(k')] = -iJ(k'), \tag{8.13.10}$$

where E and K are assumed to take principal values (the principal value of E is given by the formula (8.13.8), when the path of integration is directly along the real axis from 0 to 1). Thus, the multivalued function is represented by $E(k)$ $- 2inJ(k')$ in the vicinity of the branch point $k = +1$. By cutting the k-plane along the real axis from $k = +1$ to $+\infty$ and from $k = -1$ to $k = -\infty$, the function $E(k)$ can be restricted to its principal branch.

To continue $J'(k)$ analytically outside the circle $|k| = 1$, we make use of the identity

$$J'(k) = K'(k) - E'(k) = K(k') - E(k'), \tag{8.13.11}$$

where $K(k')$, $E(k')$ denote the multivalued functions. The singularities of $K(k')$ and $E(k')$ at $k' = \pm 1$ correspond to a single singular point for $J'(k)$ at $k = 0$. This is the logarithmic singularity derived at (8.13.7).

As in the previous section, we can show that one circuit of $k = 0$ induces two circuits of $k' = 1$, resulting in increments for $K(k')$ and $E(k')$ of $-4iK(k)$ and $-4iJ(k)$ respectively (principal values of K and J). Thus, $J'(k)$ increments by $-4iK(k) + 4iJ(k) = -4iE(k)$, in perfect conformity with equation (8.13.7). A cut made along the negative real axis from $k = 0$ to $k = -\infty$ restricts J' to its principal branch, which takes real values for $0 < k \leqslant 1$. All values of J' are comprehended by the formula $J'(k) - 4niE(k)$, where $J'(k)$ and $E(k)$ both refer to principal values.

Since $E' = K' - J'$, we now deduce that E' is a multivalued function with a branch point at $k = 0$ and associated general value $E' - 4niJ$. E' is confined to its principal branch by cutting the k-plane from $k = 0$ to $k = -\infty$.

Remarks made at the end of the previous section relating to the validity of the identity $K'(k) = K(k')$ for principal values apply equally to the identities $E'(k) = E(k')$ and $J'(k) = J(k')$.

8.14. Elliptic Integrals in the Complex Plane

First, we shall generalize the elliptic integral (3.1.2) to complex values by studying the integral

$$w = \int_0^z \{(1 - t^2)(1 - k^2 t^2)\}^{-1/2} dt \qquad (8.14.1)$$

calculated along a contour in the complex t-plane connecting the origin to the point $t = z$. It will be shown that it is multivalued, its value depending on the position of the contour in relation to the singularities $t = \pm 1, \pm 1/k$ of the integrand. It will be understood that, at the lower limit $t = 0$, the integrand assumes the value $+1$ and that it is then varied continuously along the contour to the upper limit $t = z$. k will be permitted to be complex, but we shall exclude the values $k = 0, \pm 1$.

If the contour is confined to a simply connected domain D which excludes the integrand's singularities, then the integrand is regular and single-valued in D and, by Cauchy's theorem, around a closed contour lying in D, the integral will be zero. Hence, the value of w is independent of the contour and the integral is single-valued. In particular, when $z = 0$, then $w = 0$. Differentiating equation (8.14.1), we find that

$$\frac{dz}{dw} = \{(1 - z^2)(1 - k^2 z^2)\}^{1/2}, \qquad (8.14.2)$$

a differential equation having in D a unique solution $z = \operatorname{sn} w$ such that $z = 0$ when $w = 0$. Hence,

$$w = \operatorname{sn}^{-1} z, \qquad (8.14.3)$$

for one of the values of the right-hand member. Since any contour which does not encircle a singularity can be embedded in a suitable domain D, this last result is valid in all such cases.

In the particular case when the contour is chosen to be the straight line joining the origin to the point z, the value of w will be termed the *principal value* of $sn^{-1}z$. All other values of w will be denoted by $Sn^{-1}z$. Clearly, it must be assumed that the straight contour does not pass through a singularity; if it does, then we replace z by $z + \delta z$ so that the line passes to one side or other of the singularity and then let $\delta z \to 0$; two limits are possible, depending on whether the contour approaches the singularity from one side or the other, and it is necessary to adopt some convention regarding which shall be accepted as the principal value (e.g., require that the line rotates clockwise about O as $\delta z \to 0$).

Now consider the integral (8.14.1) taken along any contour connecting O and z. Any such contour can always be deformed in the t-plane, without passing through a branch point ± 1, $\pm 1/k$, so that it comprises none or one or more of the closed contours C_i ($i = 1, 2, 3, 4$) indicated in Fig. 8.5 and the straight line from O to z (assumed not to pass through a singularity). The sense of description of a contour C_i will depend on the form of the original contour. By Cauchy's theorem, the deformation will have no effect on the value of w.

For each contour C_i, the contribution of the circle around the singularity becomes vanishing as its radius tends to zero. However, a circuit of the circle transfers the integrand from one branch to the other and accordingly reverses its sign; thus, the contributions of the two straight segments of the contour augment one another. In particular, an integral taken around a contour C_1 has value

$$2 \int_0^1 \{(1 - t^2)(1 - k^2 t^2)\}^{-1/2} \, dt = \pm 2K, \qquad (8.14.4)$$

using equation (8.12.10); the sign to be chosen depends on whether the integrand leaves $t = 0$ with the value $+1$ or -1.

Similarly, an integral around C_2 has value

$$2 \int_0^{-1} \{(1 - t^2)(1 - k^2 t^2)\}^{-1/2} \, dt = \pm 2K, \qquad (8.14.5)$$

after changing the variable of integration to $-t$.

The contribution of a contour of type C_3 is

$$2 \int_0^{1/k} \{(1 - t^2)(1 - k^2 t^2)\}^{-1/2} \, dt = 2 \int_0^1 + 2 \int_1^{1/k} \{(1 - t^2)(1 - k^2 t^2)\}^{-1/2} \, dt$$

$$= \pm 2K \pm 2iK', \qquad (8.14.6)$$

where we have replaced the straight contour from O to $1/k$ by (i) a straight line from O to 1, (ii) a small circular arc with center at $t = 1$, and (iii) a straight line from 1 to $1/k$ (and made use of the result (8.12.12)). The sign ambiguities are not related and depend on the integrand's value upon leaving $t = 0$ and the position of the end point $1/k$.

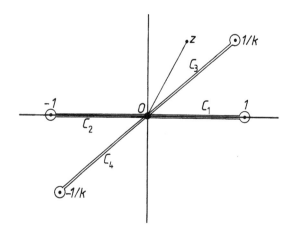

Figure 8.5. Contours C_1, C_2, C_3, and C_4.

A contour of type C_4 also makes a contribution in the form (8.14.6).

Now suppose that the contour from O to z has been deformed into n_1 contours of type C_1 or C_2 each contributing $+2K$, n_2 contours of type C_1 or C_2 each contributing $-2K$, n_3 contours of type C_3 or C_4 each contributing $+2K \pm 2iK'$, and n_4 contours of type C_3 or C_4 each contributing $-2K \pm 2iK'$. Each circuit of a contour C_i brings the integrand back to $t = 0$ with its sign reversed. Hence, on returning to the origin for the last time, the integrand's value will be $(-1)^{n_1+n_2+n_3+n_4}$ and the contribution of the straight contour from O to z will accordingly be

$$(-1)^{n_1+n_2+n_3+n_4} \operatorname{sn}^{-1} z. \tag{8.14.7}$$

Thus, the net contribution of the complete contour must be

$$2K(n_1 - n_2 + n_3 - n_4) + 2inK' + (-1)^{n_1+n_2+n_3+n_4} \operatorname{sn}^{-1} z$$
$$= 2mK + 2inK' + (-1)^m \operatorname{sn}^{-1} z, \tag{8.14.8}$$

where m and n are integers. We conclude that, for a general contour connecting O and z,

$$\operatorname{Sn}^{-1} z = \int_0^z \{(1 - t^2)(1 - k^2 t^2)\}^{-1/2}\, dt = 2mK + 2inK' + (-1)^m \operatorname{sn}^{-1} z.$$

$$\tag{8.14.9}$$

Writing $w = \operatorname{sn}^{-1} z$, it follows immediately that

$$\operatorname{sn}\{2mK + 2inK' + (-1)^m w\} = z = \operatorname{sn} w \tag{8.14.10}$$

and, hence, that $\operatorname{sn} w$ is doubly periodic with periods $4K$ and $2iK'$. Also, $\operatorname{sn}(K - w) = \operatorname{sn} w$.

If the straight line joining O and z passes through a singularity, then we replace z by $z + \delta z$ and let $\delta z \to 0$ as necessary to ensure that $\operatorname{sn}^{-1}(z + \delta z)$ tends to the principal value of $\operatorname{sn}^{-1} z$.

The integral of the second kind

$$w = \int_0^z \left[\frac{1 - k^2 t^2}{1 - t^2} \right]^{1/2} dt \tag{8.14.11}$$

can be treated similarly. If w_0 denotes the principal value, with the aid of equations (8.13.8) and (8.13.10) we prove that the general value is

$$w = 2mE + 2inJ' + (-1)^m w_0. \tag{8.14.12}$$

Alternatively, we note that, if $u = \alpha$ is a pole of dn u, its Laurent expansion in a neighborhood of this point (see section 2.8) is

$$\operatorname{dn} u = \pm \frac{i}{u - \alpha} + A(u - \alpha) + B(u - \alpha)^2 + \cdots. \tag{8.14.13}$$

Hence,

$$\operatorname{dn}^2 u = -\frac{1}{(u - \alpha)^2} \pm 2iA + O(u - \alpha), \tag{8.14.14}$$

showing that $\operatorname{dn}^2 u$ has a double pole at $u = \alpha$ with zero residue. It follows that Jacobi's E-function

$$E(u) = \int_0^u \operatorname{dn}^2 v \, dv \tag{8.14.15}$$

is independent of the contour and, therefore, single-valued. Further, we find

$$E(u) = \frac{1}{u - \alpha} + \phi(u), \tag{8.14.16}$$

where $\phi(u)$ is regular near $u = \alpha$; i.e., $E(u)$ has simple poles with unit residues at the poles of dn u.

Now

$$w = \int_0^z \left[\frac{1 - k^2 t^2}{1 - t^2} \right]^{1/2} dt = E(\operatorname{Sn}^{-1} z) \tag{8.14.17}$$

and this integral is accordingly multivalued with $\operatorname{Sn}^{-1} z$. Using the result (8.14.9), it now follows that

$$w = E(2mK + 2inK' + (-1)^m \operatorname{sn}^{-1} z)$$
$$= 2mE + 2inJ' + (-1)^m E(\operatorname{sn}^{-1} z), \tag{8.14.18}$$

where we have made use of the identities (3.5.8), (3.6.14), and the fact that $E(u)$ is odd. This confirms the result (8.14.12).

Elliptic integrals of the third kind have been made to depend (section 3.7) on the function $\Pi(u, a, k)$. For example, putting $t = \operatorname{sn} v$ in equation (3.7.5), we have

$$w = \Pi(\operatorname{Sn}^{-1} z, a, k) = \int_0^z \frac{k^2 t^2 \operatorname{sn} a \operatorname{cn} a \operatorname{dn} a}{(1 - k^2 t^2 \operatorname{sn}^2 a) \sqrt{\{(1 - t^2)(1 - k^2 t^2)\}}} dt. \tag{8.14.19}$$

This integral can now be seen to be a multivalued function of z on two accounts. First, $\text{Sn}^{-1}z$ is multivalued, and secondly, $\Pi(u, a, k)$ is a multivalued function of u, as is indicated by the identity (3.7.8) (Π is arbitrary to the extent of the addition of a multiple of $i\pi$). The multivalued nature of Π also follows immediately from equation (3.7.5), since the integrand has simple poles with nonvanishing residues and its value accordingly depends on the contour chosen. Of course, we expect the integral (8.14.19) to be multivalued, its value depending on the position of the integration contour in relation to the branch points $t = \pm 1$, $\pm 1/k$ and poles $t = \pm 1/(k \, \text{sn} \, a)$ of its integrand.

The general value of the integral (8.14.19) is given by

$$w = \Pi(2mK + 2inK' + (-1)^m \, \text{sn}^{-1}z) + iN\pi$$

$$= (2mK + 2inK')Z(a) + \frac{in\pi a}{K} + iN\pi + (-1)^m(\text{sn}^{-1}z), \qquad (8.14.20)$$

after making use of Exercises 27 and 28 in Chapter 3 and the circumstance that $\Pi(u)$ is odd.

EXERCISES

1. If 3α is a period of $\mathscr{P}(u)$, show that

$$F(u) = \{\mathscr{P}(u) - \mathscr{P}(\alpha)\}\{\mathscr{P}(u + \alpha) - \mathscr{P}(\alpha)\}\{\mathscr{P}(u + 2\alpha) - \mathscr{P}(\alpha)\} \to -\mathscr{P}'^2(\alpha)$$

as $u \to 0$. Deduce that $F(u)$ is an elliptic function without singularities and hence that $F(u) = -\mathscr{P}'^2(\alpha)$.

2. Show that $\mathscr{P}'''(u) = 12\mathscr{P}(u)\mathscr{P}'(u)$. If

$$f(u) = \begin{vmatrix} \mathscr{P}(u) & \mathscr{P}'(u) & 1 \\ \mathscr{P}(\alpha) & \mathscr{P}'(\alpha) & 1 \\ \mathscr{P}(u + \alpha) & -\mathscr{P}'(u + \alpha) & 1 \end{vmatrix},$$

prove that $f(u) \to 0$ as $u \to 0$. Deduce that $f(u)$ is an elliptic function without singularities and hence obtain an addition theorem for $\mathscr{P}$.

3. If

$$g(u) = \begin{vmatrix} 1 & \mathscr{P}(u) & \mathscr{P}'(u) \\ 1 & \mathscr{P}(a) & \mathscr{P}'(a) \\ 1 & \mathscr{P}(b) & \mathscr{P}'(b) \end{vmatrix},$$

where $a \neq b$ and none of $a, b, a \pm b$ is a period, show that $g(u)$ is an elliptic function of the third order, with a pole at $u = 0$ and zeros at $u = a, b, -a - b$. Deduce that

$$g(u) = A\sigma(u + a + b)\sigma(u - a)\sigma(u - b)/\sigma^3(u),$$

where A is constant. As $u \to 0$, show that

$$u^3 g(u) \to 2\{\mathscr{P}(a) - \mathscr{P}(b)\}.$$

Hence prove that $A = -2\sigma(a - b)/\{\sigma^3(a)\sigma^3(b)\}$. (Hint: Refer to (8.9.6) and (8.9.7).)

4. Express the Fourier expansions of $\operatorname{sn} u$, $\operatorname{cn} u$, and $\operatorname{dn} u$ in the following forms:

$$\operatorname{sn} u = \frac{\pi}{kK} \sum_{n=0}^{\infty} \frac{\sin\{(n+\tfrac{1}{2})\pi u/K\}}{\sinh\{(n+\tfrac{1}{2})\pi K'/K\}},$$

$$\operatorname{cn} u = \frac{\pi}{kK} \sum_{n=0}^{\infty} \frac{\cos\{(n+\tfrac{1}{2})\pi u/K\}}{\cosh\{(n+\tfrac{1}{2})\pi K'/K\}},$$

$$\operatorname{dn} u = \frac{\pi}{2K} + \frac{\pi}{K} \sum_{n=1}^{\infty} \frac{\cos(n\pi u/K)}{\cosh(n\pi K'/K)}.$$

5. By integrating the Fourier series for $\operatorname{dn} u$, obtain the expansion

$$\operatorname{am} u = z + \sum_{n=1}^{\infty} \frac{2q^n}{n(1+q^{2n})} \sin 2nz,$$

where $z = \pi u/2K$.

6. Prove the identity

$$\operatorname{sn} u \operatorname{dc} u = \operatorname{ns} 2u - \operatorname{cs} 2u.$$

Deduce the Fourier expansion

$$\operatorname{sn} u \operatorname{dc} u = \frac{\pi}{2K} \tan z + \frac{2\pi}{K} \sum_{n=1}^{\infty} \frac{q^n}{1+(-q)^n} \sin 2nz,$$

where $z = \pi u/2K$, valid in the strip $|\mathscr{I}z| < \pi\mathscr{I}\tau$, except at the infinities of $\tan z$. By integration with respect to u, show further that

$$\ln(\operatorname{cn} u) = \ln(\cos z) - 4 \sum_{n=1}^{\infty} \frac{q^n}{n(1+(-q)^n)} \sin^2 nz.$$

7. Using equation (3.6.1), obtain the following Fourier expansion for Jacobi's zeta function:

$$Z(u) = \frac{2\pi}{K} \sum_{n=1}^{\infty} \frac{q^n}{1-q^{2n}} \sin(n\pi u/K).$$

8. Prove that

$$\mathscr{P}'(u) = 2\frac{\sigma(u-\omega_1)\sigma(u-\omega_2)\sigma(u-\omega_3)}{\sigma^3(u)\sigma(\omega_1)\sigma(\omega_2)\sigma(\omega_3)}.$$

Deduce that

$$\sigma(2u) = -2\{\sigma(u)\sigma(u-\omega_1)\sigma(u-\omega_2)\sigma(u-\omega_3)\}/\{\sigma(\omega_1)\sigma(\omega_2)\sigma(\omega_3)\}.$$

(Hint: Use the identity (6.7.8).)

9. Show that $\{\mathscr{P}(u-\alpha)-\mathscr{P}(u+\alpha)\}/\mathscr{P}'(u)$ has a zero of order 4 at $u=0$ and double poles at $u=\alpha$, $-\alpha$ (there are no poles at the midlattice points; why?). Deduce that

$$\mathscr{P}(u-\alpha) - \mathscr{P}(u+\alpha) = \mathscr{P}'(u)\mathscr{P}'(\alpha)\{\mathscr{P}(u)-\mathscr{P}(\alpha)\}^{-2}.$$

10. Show that $\{\mathscr{P}'(u)-\mathscr{P}'(\alpha)\}/\{\mathscr{P}(u)-\mathscr{P}(\alpha)\}$ has an irreducible set of simple poles given by $u=0$, $-\alpha$ and that the residues at these poles are $-2, 2$ respectively. Using

the formula (8.11.6), deduce the identity (6.8.4). Hence prove that, if $u + v + w = 0$, then

$$\{\zeta(u) + \zeta(v) + \zeta(w)\}^2 + \zeta'(u) + \zeta'(v) + \zeta'(w) = 0.$$

11. If

$$F(u) = \zeta(u - \alpha) - \zeta(u - \beta) - \zeta(\alpha - \beta) + \zeta(2\alpha - 2\beta),$$

show that $F(u)$ is an elliptic function with periods $2\omega_1$ and $2\omega_3$. Show, further, that $F(u)$ has an irreducible set of simple poles at $u = \alpha, \beta$ and an irreducible set of simple zeros at $u = 2\alpha - \beta, 2\beta - \alpha$. Hence prove that

$$F(u) = \frac{\sigma(u - 2\alpha + \beta)\sigma(u - 2\beta + \alpha)}{\sigma(2\beta - 2\alpha)\sigma(u - \alpha)\sigma(u - \beta)}.$$

12. Obtain the Fourier expansions

$$\mathscr{P}(u + \omega_1) = -\frac{\eta_1}{\omega_1} + \left(\frac{\pi}{\omega_1}\right)^2 \left[\tfrac{1}{4}\sec^2 z - 2 \sum_{n=1}^{\infty} \frac{n(-1)^n q^{2n}}{1 - q^{2n}} \cos 2nz \right],$$

$$\mathscr{P}(u + \omega_2) = -\frac{\eta_1}{\omega_1} - 2\left(\frac{\pi}{\omega_1}\right)^2 \sum_{n=1}^{\infty} \frac{n(-1)^n q^n}{1 - q^{2n}} \cos 2nz,$$

$$\mathscr{P}(u + \omega_3) = -\frac{\eta_1}{\omega_1} - 2\left(\frac{\pi}{\omega_1}\right)^2 \sum_{n=1}^{\infty} \frac{n q^n}{1 - q^{2n}} \cos 2nz.$$

Deduce the identity given in Exercise 13, Chapter 6.

13. Show that, in the neighborhood of $k = 1$,

$$K(k) = \frac{1}{\pi} K'(k) \ln(8/h) - \tfrac{1}{4} h - \tfrac{7}{32} h^2 + O(h^3),$$

$$K'(k) = \tfrac{1}{2}\pi(1 + \tfrac{1}{2} h + \tfrac{5}{16} h^2 + O(h^3)),$$

$$E(k) = \frac{1}{\pi} J'(k) \ln(8/h) + 1 - \tfrac{1}{2} h - \tfrac{5}{16} h^2 + O(h^3),$$

$$J'(k) = \tfrac{1}{2}\pi(h + \tfrac{1}{4} h^2 + O(h^3)),$$

$$E'(k) = \tfrac{1}{2}\pi(1 - \tfrac{1}{2} h + \tfrac{1}{16} h^2 + O(h^3)),$$

where $h = 1 - k$.

CHAPTER 9

Modular Transformations

9.1. The Modular Group

In sections 1.7 and 1.8, we studied the effect on the theta functions of transforming from a parameter τ to a parameter τ', where $\tau' = -1/\tau$ and $\tau' = 2\tau$ respectively. Then, in sections 2.6 and 3.9, it was shown that these transformations change the modulus of the Jacobian elliptic functions from k to $k' = \sqrt{(1 - k^2)}$ and from k to $k_1 = (1 - k')/(1 + k')$ respectively; the modulus transformation $\kappa = 1/k$ was also examined.

We now generalize these ideas by considering transformations of the type

$$\tau' = \frac{c + d\tau}{a + b\tau}, \tag{9.1.1}$$

where a, b, c, d, are positive or negative integers or zero such that $(ad - bc)$ is positive. The latter restriction ensures that $\mathscr{I}\tau' > 0$, whenever $\mathscr{I}\tau > 0$ (as required for the definition of the theta functions). Without loss of generality, we shall assume that the integers a, b, c, d have no common factor.

It is helpful to construct the transformation (9.1.1) as follows: First transform the pair of complex numbers (ω_1, ω_3) into the pair (ω_1', ω_3') by the equations

$$\omega_1' = a\omega_1 + b\omega_3, \qquad \omega' = c\omega_1 + d\omega_3. \tag{9.1.2}$$

Then, defining $\tau = \omega_3/\omega_1, \tau' = \omega_3'/\omega_1'$, we find that τ and τ' are related by the transformation (9.1.1).

Using a matrix notation, the transformation (9.1.2) can be written

$$\Omega' = M\Omega, \tag{9.1.3}$$

where

$$\Omega = \begin{bmatrix} \omega_1 \\ \omega_3 \end{bmatrix}, \qquad \Omega' = \begin{bmatrix} \omega_1' \\ \omega_3' \end{bmatrix}, \qquad M = \begin{bmatrix} a & b \\ c & d \end{bmatrix}. \tag{9.1.4}$$

M will be termed the matrix representing the transformation (9.1.1); note that it is arbitrary to the extent of a factor -1, but is otherwise uniquely determined. If Δ is the determinant of M, then $\Delta = |M| = ad - bc > 0$ and,

$$\Omega = M^{-1}\Omega' = \frac{1}{\Delta}(\text{adj } M)\Omega', \tag{9.1.5}$$

where adj M is the adjoint matrix given by

$$\text{adj } M = \begin{bmatrix} d & -b \\ -c & a \end{bmatrix}. \tag{9.1.6}$$

It follows from the last pair of equations that

$$\tau = \frac{-c + a\tau'}{d - b\tau'}, \tag{9.1.7}$$

as is easily verified by solving (9.1.1) for τ. This is the inverse transformation and is clearly of the same type as the original transformation (9.1.1), with the same value of Δ and with matrix adj M.

The resultant (or product) of the transformations $\Omega' = M\Omega, \Omega'' = N\Omega'$ is $\Omega'' = NM\Omega$. After removing any common factor from the elements of NM, we obtain the matrix representing the resultant of two transformations of the type (9.1.1). Since $|NM| = |N||M| > 0$, the product of any pair of transformations (9.1.1) must be a transformation of the same type. The inverse (9.1.7) is also of this type, so that the transformations (9.1.1) form a group, called the *extended modular group*.

We shall write $\Delta = ad - bc$ and term Δ the *order* of the transformation (9.1.1). We note that the order of an inverse transformation is the same as that of the direct transformation. However, it may be possible to remove a factor from the product matrix NM, so that we can only assert that the order of the product of two transformations is some factor of the product of their orders.

If we confine our attention to transformations of unit order, the inverses and products will all be of unit order and we have therefore identified a subgroup of the extended modular group called, simply, the *modular group*. If $2\omega_1, 2\omega_3$ is a pair of primitive periods of an elliptic function and $2\omega_1', 2\omega_3'$ is another pair of such periods for the same function, then equations (2.3.4) show that $\tau = \omega_3/\omega_1$ and $\tau' = \omega_3'/\omega_1'$ are related by the transformation (9.1.1) with $\Delta = 1$. Thus, the modular group is a homomorphic image of the group of primitive period transformations (two period transformations, differing only in sign, correspond to each modular transformation).

9.2. Generators of the Modular Group

In this section, we shall prove that any transformation from the modular group can be expressed as a product of transformations of the two fundamental types (i) $\tau' = 1 + \tau$, and (ii) $\tau' = -1/\tau$. It will then follow that we need only calculate the effects of each of these transformations on the theta and elliptic functions, to enable us to determine the effects on these functions of the general modular transformation.

The matrices representing the fundamental transformations are

$$\text{(i)} \quad P = \begin{bmatrix} 1 & 0 \\ 1 & 1 \end{bmatrix}, \qquad \text{(ii)} \quad Q = \begin{bmatrix} 0 & 1 \\ -1 & 0 \end{bmatrix}, \tag{9.2.1}$$

respectively (or their negatives). We have to prove that any matrix M (equation (9.1.4)) with unit determinant can be expressed as a product of factors P and Q. P and Q are then said to be *generators* of the modular group.

We start by considering some special cases:

Note that

$$Q^2 = \begin{bmatrix} -1 & 0 \\ 0 & -1 \end{bmatrix} = -I \tag{9.2.2}$$

is a matrix representing the identical transformation $\tau' = \tau$.

Also

$$QPQPQ = -\begin{bmatrix} 1 & 0 \\ -1 & 1 \end{bmatrix} = -R. \tag{9.2.3}$$

R represents the transformation $\tau' = -1 + \tau$, which has accordingly been shown to be the resultant of five transformations, viz.

$$\tau_1 = -1/\tau, \qquad \tau_2 = 1 + \tau_1, \qquad \tau_3 = -1/\tau_2,$$
$$\tau_4 = 1 + \tau_3, \qquad \tau' = -1/\tau_4. \tag{9.2.4}$$

Note that $R = P^{-1}$, as follows from the circumstance that $\tau' = -1 + \tau$ is the transformation inverse to $\tau = 1 + \tau'$.

Clearly, any (positive or negative) integral power of P can now be generated; we find that

$$P^n = \begin{bmatrix} 1 & 0 \\ n & 1 \end{bmatrix}. \tag{9.2.5}$$

Other group elements which can easily be expressed as products of the generators are

$$PQP = \begin{bmatrix} 1 & 1 \\ 0 & 1 \end{bmatrix} = S, \qquad QPQ = -\begin{bmatrix} 1 & -1 \\ 0 & 1 \end{bmatrix} = -S^{-1}. \tag{9.2.6}$$

Then,

$$\begin{bmatrix} 1 & n \\ 0 & 1 \end{bmatrix} = S^n, \qquad \begin{bmatrix} n & 1 \\ -1 & 0 \end{bmatrix} = QP^n, \qquad \begin{bmatrix} 0 & -1 \\ 1 & n \end{bmatrix} = -QS^n, \tag{9.2.7}$$

for positive or negative n.

Premultiplying M by P^n, we find

$$P^n M = \begin{bmatrix} 1 & 0 \\ n & 1 \end{bmatrix} \begin{bmatrix} a & b \\ c & d \end{bmatrix} = \begin{bmatrix} a & b \\ c + na & d + na \end{bmatrix}. \tag{9.2.8}$$

Note that the effect is to add or subtract a multiple of the first row to or from the second row of M. Similarly, premultiplication by S^n adds or subtracts a multiple of the second row to or from the first.

Since $ad - bc = 1$, if a is a multiple of c, then c divides 1 and, hence $|c| = 1$. Similarly, if c is a multiple of a, then $|a| = 1$. The same observations relate to b and d. We will dispose of these special cases first.

If $|a| = 1$, we can reduce the element c of M to zero (if it is not already zero) by addition or subtraction of a multiple of the first row to or from the second row. The transformed matrix must be $\pm S^n$ for some value of n. Thus, $P^m M = \pm S^n$ and, therefore, $M = \pm P^{-m} S^n$. M has been expressed as a product of the generators.

The cases $|b| = 1, |c| = 1, |d| = 1$ can be treated similarly, making use of the matrices QP^n, QS^n, and P^n respectively.

We now transfer attention to the general case, when neither a nor c is a multiple of the other and b, d satisfy the same condition.

Then, we cannot have $|a| > |c|$ and $|d| > |b|$, except in the special case $a = d = 1, b = c = 0$ (the identity transformation); for, otherwise, $|ad|$ would be greater than $|bc|$ by more than unity. Neither can we have $|a| < |c|, |d| < |b|$, except in the special case $a = d = 0, b = 1, c = -1$ (i.e., Q), for, otherwise, $|ad|$ would be less than $|bc|$ by more than unity. Hence, setting aside the two special cases (which require no transformation) the only remaining possibilities for M to be studied are when $|a| > |c|$ and $|b| > |d|$ or when the reverse inequalities hold.

If $|a| > |c|$ and $|b| > |d|$, we can add or subtract a multiple of the second row to or from the first, so that the new element a' satisfies $|a'| < |c|$. The new matrix must now be one of the special cases already reduced to a product of generators or, if not, we must have $|b'| < |d|$. In the latter case, having reduced the magnitudes of the elements in the first row to be smaller than the magnitudes of the corresponding elements in the second row, we then apply a similar transformation to reduce the magnitudes of the elements in the second row to be smaller than those in the first row. Ultimately, this process must terminate when one of the special cases is reached. The reverse inequalities are treated similarly.

Hence, we shall arrive at an equation of the type

$$P^{m_1} S^{n_1} P^{m_2} S^{n_2} \cdots M = X, \tag{9.2.9}$$

where X is one of the special matrices already reduced to a product of generators. Then,

$$M = \cdots S^{-n_2} P^{-m_2} S^{-n_1} P^{-m_1} X \qquad (9.2.10)$$

expresses M as a product of generators.

Since $PQPQPQ = -I$, this representation of M is not unique.

$\tau' = -1/\tau$ is Jacobi's transformation and its effect upon the theta and elliptic functions has already been determined in sections 1.7 and 2.6 respectively. In the next section, the effect of the other fundamental transformation $\tau' = 1 + \tau$ will be investigated.

9.3. The Transformation $\tau' = 1 + \tau$

Since the nome $q = e^{i\pi\tau}$, the parameter transformation $\tau_1 = 1 + \tau$ is equivalent to the nome transformation

$$q_1 = e^{i\pi(1 + \tau)} = -q. \qquad (9.3.1)$$

Equations (1.2.11)–(1.2.14) now show that the associated theta function transformations are

$$\begin{aligned}
\theta_1(z|\tau_1) &= i^{1/2}\theta_1(z|\tau), \\
\theta_2(z|\tau_1) &= i^{1/2}\theta_2(z|\tau), \\
\theta_3(z|\tau_1) &= \theta_4(z|\tau), \\
\theta_4(z|\tau_1) &= \theta_3(z|\tau),
\end{aligned} \right\} \qquad (9.3.2)$$

where, unambiguously, $i^{1/2} = e^{(1/4)i\pi}$.

Turning to the elliptic functions, let k_1, k_1' denote the modulus and complementary modulus associated with the parameter τ_1. Then, the first of equations (2.1.7) coupled with the transformation equations (9.3.2) yield the following result:

$$k_1 = \theta_2^2(0|\tau_1)/\theta_3^2(0|\tau_1) = i\theta_2^2(0|\tau)/\theta_4^2(0|\tau) = ik/k'. \qquad (9.3.3)$$

Also, using the second of equations (2.1.7),

$$\theta_3^2(0|\tau)/\theta_3^2(0|\tau_1) = \theta_3^2(0|\tau)/\theta_4^2(0|\tau) = 1/k'. \qquad (9.3.4)$$

Thus, if $z = u/\theta_3^2(0|\tau)$, $z_1 = u/\theta_3^2(0|\tau_1)$, then

$$z_1 = z/k'. \qquad (9.3.5)$$

The transformation equations for the Jacobi functions now follows from equations (2.1.1)–(2.1.3), thus:

$$\begin{aligned}
\operatorname{sn}(u, k_1) &= \frac{\theta_3(0|\tau_1)}{\theta_2(0|\tau_1)} \cdot \frac{\theta_1(z_1|\tau_1)}{\theta_4(z_1|\tau_1)} = \frac{\theta_4(0|\tau)}{\theta_2(0|\tau)} \cdot \frac{\theta_1(z/k'|\tau)}{\theta_3(z/k'|\tau)} \\
&= k' \operatorname{sd}(u/k', k),
\end{aligned} \qquad (9.3.6)$$

$$\operatorname{cn}(u, k_1) = \frac{\theta_4(0|\tau_1)}{\theta_2(0|\tau_1)} \cdot \frac{\theta_2(z_1|\tau_1)}{\theta_4(z_1|\tau_1)} = \frac{\theta_3(0|\tau)}{\theta_2(0|\tau)} \cdot \frac{\theta_2(z/k'|\tau)}{\theta_3(z/k'|\tau)}$$

$$= \operatorname{cd}(u/k', k),$$ (9.3.7)

$$\operatorname{dn}(u, k_1) = \frac{\theta_4(0|\tau_1)}{\theta_3(0|\tau_1)} \cdot \frac{\theta_3(z_1|\tau_1)}{\theta_4(z_1|\tau_1)} = \frac{\theta_3(0|\tau)}{\theta_4(0|\tau)} \cdot \frac{\theta_4(z/k'|\tau)}{\theta_3(z/k'|\tau)}$$

$$= \operatorname{nd}(u/k', k).$$ (9.3.8)

Replacing u by $k'u$, these transformations can be presented in their more usual form, as follows:

$$\left. \begin{array}{l} \operatorname{sn}(k'u, ik/k') = k'\operatorname{sd}(u, k), \\[4pt] \operatorname{cn}(k'u, ik/k') = \operatorname{cd}(u, k), \\[4pt] \operatorname{dn}(k'u, ik/k') = \operatorname{nd}(u, k). \end{array} \right\}$$ (9.3.9)

If K_1, iK_1' are the transformed quarter-periods, then using equations (2.2.3), we calculate that

$$\left. \begin{array}{l} K_1 = \tfrac{1}{2}\pi\theta_3^2(0|\tau_1) = \tfrac{1}{2}\pi\theta_4^2(0|\tau) = \tfrac{1}{2}\pi k'\theta_3^2(0|\tau) = k'K, \\[4pt] iK_1' = \tau_1 K_1 = (1 + \tau)k'K = k'(K + iK'). \end{array} \right\}$$ (9.3.10)

These are quarter-periods for the functions $\operatorname{sn}(u, k_1)$, etc. If we replace u by $k'u$, to give the transformation equations (9.3.9), the quarter-periods of $\operatorname{sn}(k'u, k_1)$, etc. are K and $K + iK'$; this is in accord with these transformation equations, since $\operatorname{sd}(u, k)$, etc. have alternative primitive periods $4K$, $4(K + iK')$.

9.4. General Modular Transformation of the Elliptic Functions

The other transformation generating the modular group is $\tau_1 = -1/\tau$ and has been treated in sections 1.7 and 2.6 (Jacobi's transformation). The results we have obtained are summarized below:

$$\left. \begin{array}{l} \theta_1(z|\tau_1) = -i(-i\tau)^{1/2} e^{i\tau z^2/\pi}\theta_1(\tau z|\tau), \\[4pt] \theta_2(z|\tau_1) = (-i\tau)^{1/2} e^{i\tau z^2/\pi}\theta_4(\tau z|\tau), \\[4pt] \theta_3(z|\tau_1) = (-i\tau)^{1/2} e^{i\tau z^2/\pi}\theta_3(\tau z|\tau), \\[4pt] \theta_4(z|\tau_1) = (-i\tau)^{1/2} e^{i\tau z^2/\pi}\theta_2(\tau z|\tau), \end{array} \right\}$$ (9.4.1)

where the real part of $(-i\tau)^{1/2}$ is to be taken positively. Also

$$\left. \begin{array}{l} \operatorname{sn}(u, k_1) = -i\operatorname{sc}(iu, k), \\[4pt] \operatorname{cn}(u, k_1) = \operatorname{nc}(iu, k), \\[4pt] \operatorname{dn}(u, k_1) = \operatorname{dc}(iu, k), \end{array} \right\}$$ (9.4.2)

where $k_1 = k'$.

These transformation equations, together with the equations (9.3.2), (9.3.9), permit us to calculate the effect of any transformation belonging to the modular group.

As an example, consider the modular transformation $\tau_3 = \tau/(1-\tau)$. This is the resultant of the successive transformations

$$\tau_3 = -1/\tau_2, \qquad \tau_2 = 1 + \tau_1, \qquad \tau_1 = -1/\tau. \tag{9.4.4}$$

We treat these transformations in turn from the left.

Referring to equations (9.4.2), we see that the transformation $\tau_3 = -1/\tau_2$ induces the elliptic function transformations

$$\left. \begin{array}{l} \mathrm{sn}(u, k_3) = -i\,\mathrm{sc}(iu, k_2), \qquad \mathrm{cn}(u, k_3) = \mathrm{nc}(iu, k_2), \\ \mathrm{dn}(u, k_3) = \mathrm{dc}(iu, k_2), \end{array} \right\} \tag{9.4.5}$$

where

$$k_3 = k_2'. \tag{9.4.6}$$

Then, equations (9.3.6)–(9.3.8) show that the transformation $\tau_2 = 1 + \tau_1$ leads to the relationships

$$\left. \begin{array}{l} \mathrm{sn}(u, k_2) = k_1'\,\mathrm{sd}(u/k_1', k_1), \qquad \mathrm{cn}(u, k_2) = \mathrm{cd}(u/k_1', k_1), \\ \mathrm{dn}(u, k_2) = \mathrm{nd}(u/k_1', k_1), \end{array} \right\} \tag{9.4.7}$$

where

$$k_2 = ik_1/k_1'. \tag{9.4.8}$$

Finally, the transformation $\tau_1 = -1/\tau$ yields the equations

$$\left. \begin{array}{l} \mathrm{sn}(u, k_1) = -i\,\mathrm{sc}(iu, k), \qquad \mathrm{cn}(u, k_1) = \mathrm{nc}(iu, k), \\ \mathrm{dn}(u, k_1) = \mathrm{dc}(iu, k), \end{array} \right\} \tag{9.4.9}$$

where

$$k_1 = k'. \tag{9.4.10}$$

It now follows from these successive sets of transformation equations that

$$\left. \begin{array}{l} \mathrm{sn}(u, k_3) = k\,\mathrm{sn}(u/k, k), \qquad \mathrm{cn}(u, k_3) = \mathrm{dn}(u/k, k), \\ \mathrm{dn}(u, k_3) = \mathrm{cn}(u/k, k), \end{array} \right\} \tag{9.4.11}$$

where $k_3 = 1/k$. These are the earlier equations (3.9.4)–(3.9.6).

It is left as an exercise for the reader to show that the corresponding transformation equations for the theta functions are

$$\left. \begin{array}{l} \theta_1(z|\tau_3) = i^{1/2}F\theta_1[(1-\tau)z|\tau], \\ \theta_2(z|\tau_3) = F\theta_3[(1-\tau)z|\tau], \\ \theta_3(z|\tau_3) = F\theta_2[(1-\tau)z|\tau], \\ \theta_4(z|\tau_3) = i^{1/2}F\theta_4[(1-\tau)z|\tau], \end{array} \right\} \tag{9.4.12}$$

where

$$F = (1-\tau)^{1/2}\,e^{iz^2(\tau-1)/\pi}. \tag{9.4.13}$$

9.5. Transformations of Higher Order

We shall show that any transformation of order $\Delta > 1$ can be represented as a product of transformations of the first order and of transformations of higher order with matrix

$$N = \begin{bmatrix} 1 & 0 \\ 0 & n \end{bmatrix}. \tag{9.5.1}$$

Hence, the only transformations of order greater than unity we need study are those of the type $\tau' = n\tau \ (n > 1)$.

Suppose the matrix M (equation (9.1.4)) is such that $ad - bc = \Delta > 1$. First note that, if the elements c, d, have a common factor $f \, (> 0)$, we can separate M into factors as follows:

$$M = \begin{bmatrix} 1 & 0 \\ 0 & f \end{bmatrix} \begin{bmatrix} a & b \\ c' & d' \end{bmatrix}, \tag{9.5.2}$$

where $c = fc'$, $d = fd'$. The first matrix factor is of the form (9.5.1) and constitutes the first step in the reduction of M.

We can now assume that c', d' have no common factor and hence that integers λ and μ can be found such that

$$\lambda c' + \mu d' = 1. \tag{9.5.3}$$

Then, writing

$$ad' - bc' = \Delta/f = n, \qquad \lambda a + \mu b = \alpha, \tag{9.5.4}$$

we have

$$\begin{bmatrix} a & b \\ c' & d' \end{bmatrix} \begin{bmatrix} \lambda & -d' \\ \mu & c' \end{bmatrix} = \begin{bmatrix} \alpha & -n \\ 1 & 0 \end{bmatrix} = \begin{bmatrix} \alpha & -1 \\ 1 & 0 \end{bmatrix} \begin{bmatrix} 1 & 0 \\ 0 & n \end{bmatrix}. \tag{9.5.5}$$

In view of equation (9.5.3), we calculate that

$$\begin{bmatrix} \lambda & -d' \\ \mu & c' \end{bmatrix}^{-1} = \begin{bmatrix} c' & d' \\ -\mu & \lambda \end{bmatrix}, \tag{9.5.6}$$

from which it follows that

$$\begin{bmatrix} a & b \\ c' & d' \end{bmatrix} = \begin{bmatrix} \alpha & -1 \\ 1 & 0 \end{bmatrix} \begin{bmatrix} 1 & 0 \\ 0 & n \end{bmatrix} \begin{bmatrix} c' & d' \\ -\mu & \lambda \end{bmatrix}. \tag{9.5.7}$$

The two nondiagonal matrices on the right-hand side of this equation both have unit determinants and M has accordingly been represented as a product of matrices of the type (9.5.1) and of the first order.

Clearly, any transformation $\tau' = n\tau$ can be separated into a product of transformations $\tau_1 = p\tau$, $\tau_2 = q\tau_1$, etc., where $p, q, \dots$ are the prime factors of n. It follows that we need only study the transformation $\tau' = n\tau$ for the cases where n is a prime.

9.6. Third-Order Transformation of Theta Functions

In this and the following section, we shall illustrate a general method of treating the nth-order transformation $\tau' = n\tau$, by detailed consideration of the case $n = 3$. The second-order transformation $\tau' = 2\tau$ (Landen's transformation) has already been analyzed in sections 1.8 and 3.9, and we leave it as an exercise for the reader to derive our earlier results by application of the general method described below (Exercise 2).

In the case where n is odd, the general method is to consider the function

$$F(z) = \frac{\theta_3(nz \mid n\tau)}{\theta_3(z) \prod \theta_3\left(\dfrac{r\pi}{n} + z\right) \prod \theta_3\left(\dfrac{r\pi}{n} - z\right)} \tag{9.6.1}$$

where, in both products, r ranges over the integers from 1 to $\tfrac{1}{2}(n-1)$. (Note: In this and the following section, wherever the parameter of a theta function is not explicitly indicated, it is assumed to be τ; further, as in the past, we shall often shorten $\theta_2(0 \mid \tau)$, $\theta_3(0 \mid \tau)$, $\theta_4(0 \mid \tau)$, to θ_2, θ_3, θ_4, respectively.) It is easily verified that $F(z)$ is a doubly periodic function with periods π/n and $\pi\tau$. But the numerator of $F(z)$ possesses the same zeros (all simple) as the denominator, so that $F(z)$ has no singularities and accordingly must be a constant. It remains only to identify this constant by putting $z = 0$.

Incrementation of z by $\tfrac{1}{2}\pi$, $\tfrac{1}{2}\pi\tau$, and $\tfrac{1}{2}\pi + \tfrac{1}{2}\pi\tau$, successively, now leads to identities for $\theta_4(nz \mid n\tau)$, $\theta_2(nz \mid n\tau)$, and $\theta_1(nz \mid n\tau)$, respectively.

If n is even, the function we examine is

$$G(z) = \frac{\theta_3(nz \mid n\tau)}{\prod \theta_3\left(\dfrac{r\pi}{2n} + z\right) \prod \theta_3\left(\dfrac{r\pi}{2n} - z\right)} \tag{9.6.2}$$

r ranging over the odd integers from 1 to $n-1$. Similar reasoning to before shows that $G(z)$ is constant.

In the case $n = 3$, we have shown that

$$\theta_3(3z \mid 3\tau) = A\theta_3(z)\theta_3(\tfrac{1}{3}\pi + z)\theta_3(\tfrac{1}{3}\pi - z), \tag{9.6.3}$$

where A is constant. Incrementation of z yields the additional identities

$$\theta_4(3z \mid 3\tau) = A\theta_4(z)\theta_4(\tfrac{1}{3}\pi + z)\theta_4(\tfrac{1}{3}\pi - z), \tag{9.6.4}$$

$$\theta_2(3z \mid 3\tau) = A\theta_2(z)\theta_2(\tfrac{1}{3}\pi + z)\theta_2(\tfrac{1}{3}\pi - z), \tag{9.6.5}$$

$$\theta_1(3z \mid 3\tau) = A\theta_1(z)\theta_1(\tfrac{1}{3}\pi + z)\theta_1(\tfrac{1}{3}\pi - z). \tag{9.6.6}$$

To compute the factor A, first note that

$$\lim_{z \to 0} \frac{\theta_1(3z \mid 3\tau)}{\theta_1(z \mid \tau)} = \lim_{z \to 0} \frac{3\theta_1'(3z \mid 3\tau)}{\theta_1'(z \mid \tau)} = \frac{3\theta_2(0 \mid 3\tau)\theta_3(0 \mid 3\tau)\theta_4(0 \mid 3\tau)}{\theta_2(0 \mid \tau)\theta_3(0 \mid \tau)\theta_4(0 \mid \tau)}, \tag{9.6.7}$$

where we have referred to the result (1.5.11). Then, putting $z = 0$ in the identities (9.6.3)–(9.6.5), we calculate that

$$\lim_{z \to 0} \frac{\theta_1(3z|3\tau)}{\theta_1(z|\tau)} = 3A^3\theta_2^2(\tfrac{1}{3}\pi)\theta_3^2(\tfrac{1}{3}\pi)\theta_4^2(\tfrac{1}{3}\pi). \tag{9.6.8}$$

But this limit can also be found from equation (9.6.6) directly, its value being $A\theta_1^2(\tfrac{1}{3}\pi)$. Since A is clearly positive, it now follows that

$$1/A = \sqrt{3}\theta_2(\tfrac{1}{3}\pi)\theta_3(\tfrac{1}{3}\pi)\theta_4(\tfrac{1}{3}\pi)/\theta_1(\tfrac{1}{3}\pi). \tag{9.6.9}$$

Putting $z = 0$ in equations (9.6.3)–(9.6.5), we can now show that

$$\theta_3(0|3\tau) = \frac{1}{\sqrt{3}}\theta_3(0)\frac{\theta_1(\tfrac{1}{3}\pi)\theta_3(\tfrac{1}{3}\pi)}{\theta_2(\tfrac{1}{3}\pi)\theta_4(\tfrac{1}{3}\pi)}, \tag{9.6.10}$$

$$\theta_4(0|3\tau) = \frac{1}{\sqrt{3}}\theta_4(0)\frac{\theta_1(\tfrac{1}{3}\pi)\theta_4(\tfrac{1}{3}\pi)}{\theta_2(\tfrac{1}{3}\pi)\theta_3(\tfrac{1}{3}\pi)}, \tag{9.6.11}$$

$$\theta_2(0|3\tau) = \frac{1}{\sqrt{3}}\theta_2(0)\frac{\theta_1(\tfrac{1}{3}\pi)\theta_2(\tfrac{1}{3}\pi)}{\theta_3(\tfrac{1}{3}\pi)\theta_4(\tfrac{1}{3}\pi)}. \tag{9.6.12}$$

9.7. Third-Order Transformation of Jacobi's Functions

These transformations now follow from those for the theta functions by use of the equations (2.1.1)–(2.1.3).

First note that, if k is the modulus corresponding to the parameter τ and k_1 is the modulus for parameter 3τ, then the first of equations (2.1.7) gives

$$k_1 = \frac{\theta_2^2(0|3\tau)}{\theta_3^2(0|3\tau)} = \frac{\theta_2^2}{\theta_3^2}\frac{\theta_2^4(\tfrac{1}{3}\pi)}{\theta_3^4(\tfrac{1}{3}\pi)} = k\frac{\theta_2^4(\tfrac{1}{3}\pi)}{\theta_3^4(\tfrac{1}{3}\pi)} = k\frac{\theta_1^4(\tfrac{1}{6}\pi)}{\theta_4^4(\tfrac{1}{6}\pi)}, \tag{9.7.1}$$

having referred to equations (9.6.10) and (9.6.12). Putting $z = \tfrac{1}{6}\pi$, $u = \tfrac{1}{6}\pi\theta_3^2(0) = \tfrac{1}{3}K$ (*vide* equations (2.2.3)), in equation (2.1.1), we find that

$$\theta_1(\tfrac{1}{6}\pi)/\theta_4(\tfrac{1}{6}\pi) = k^{1/2} \operatorname{sn} \tfrac{1}{3}K, \tag{9.7.2}$$

where, throughout this section, elliptic functions are understood to be computed to modulus k and theta functions to parameter τ, unless otherwise indicated. It follows from the last two equations that

$$k_1 = k^3 \operatorname{sn}^4(\tfrac{1}{3}K). \tag{9.7.3}$$

Similarly, from the second of equations (2.1.7), we calculate, using equations (9.6.10) and (9.6.11), that

$$k_1' = \frac{\theta_4^2(0|3\tau)}{\theta_3^2(0|3\tau)} = \frac{\theta_4^2}{\theta_3^2}\frac{\theta_4^4(\tfrac{1}{3}\pi)}{\theta_3^4(\tfrac{1}{3}\pi)} = k'\frac{\theta_3^4(\tfrac{1}{6}\pi)}{\theta_4^4(\tfrac{1}{6}\pi)}, \tag{9.7.4}$$

and then from equation (2.1.3),

$$\theta_3(\tfrac{1}{6}\pi)/\theta_4(\tfrac{1}{6}\pi) = k'^{-1/2}\,\mathrm{dn}\,\tfrac{1}{3}K, \tag{9.7.5}$$

showing that

$$k_1' = \frac{1}{k'}\,\mathrm{dn}^4(\tfrac{1}{3}K). \tag{9.7.6}$$

Clearly, $\mathrm{sn}\,\tfrac{1}{3}K$ is a function of k and, inversely therefore, k can be expressed as a function of $s = \mathrm{sn}\,\tfrac{1}{3}K$. s proves to be the more convenient parameter and we shall accordingly find k in terms of s. To do this, we note from equations (2.2.17) that

$$\mathrm{cd}\,\tfrac{1}{3}K = \mathrm{sn}(K - \tfrac{1}{3}K) = \mathrm{sn}_3^2 K. \tag{9.7.7}$$

Then, using the identity (2.4.4) and writing $s = \mathrm{sn}\,\tfrac{1}{3}K, c = \mathrm{cn}\,\tfrac{1}{3}K, d = \mathrm{dn}\,\tfrac{1}{3}K$, the last equation yields

$$\frac{c}{d} = \frac{2scd}{1 - k^2 s^4}. \tag{9.7.8}$$

Thus,

$$k^2 = \frac{2s - 1}{s^3(2 - s)}, \qquad k'^2 = \frac{(1 + s)(1 - s)^3}{s^3(2 - s)}. \tag{9.7.9}$$

The equations (9.7.3) and (9.7.6) can now be expressed in the forms

$$k_1^2 = \frac{(2s - 1)^3}{s(2 - s)^3}, \tag{9.7.10}$$

$$k_1'^2 = \frac{(1 - s)(1 + s)^3}{s(2 - s)^3}. \tag{9.7.11}$$

Working with the parameter 3τ, equation (2.1.1) yields

$$\mathrm{sn}(\lambda u, k_1) = k_1^{-1/2}\frac{\theta_1(\lambda z_1|3\tau)}{\theta_4(\lambda z_1|3\tau)}, \tag{9.7.12}$$

where $z_1 = u/\theta_3^2(0|3\tau)$. We now choose λ such that $\lambda z_1 = 3z = 3u/\theta_3^2(0)$, i.e.,

$$\lambda = \frac{3\theta_3^2(0|3\tau)}{\theta_3^2(0)} = \frac{\theta_1^2(\tfrac{1}{3}\pi)\theta_3^2(\tfrac{1}{3}\pi)}{\theta_2^2(\tfrac{1}{3}\pi)\theta_4^2(\tfrac{1}{3}\pi)} = \frac{\theta_2^2(\tfrac{1}{6}\pi)\theta_4^2(\tfrac{1}{6}\pi)}{\theta_1^2(\tfrac{1}{6}\pi)\theta_3^2(\tfrac{1}{6}\pi)}, \tag{9.7.13}$$

by equation (9.6.10). Putting $z = \tfrac{1}{6}\pi$ in equations (2.1.2) and (2.1.3), we deduce that

$$\theta_2(\tfrac{1}{6}\pi)/\theta_4(\tfrac{1}{6}\pi) = (k/k')^{1/2}\,\mathrm{cn}\,\tfrac{1}{3}K, \tag{9.7.14}$$

$$\theta_3(\tfrac{1}{6}\pi)/\theta_4(\tfrac{1}{6}\pi) = k'^{-1/2}\,\mathrm{dn}\,\tfrac{1}{3}K. \tag{9.7.15}$$

It now follows from this pair of equations and equation (9.7.2) that

$$\lambda = \frac{\mathrm{cn}^2(\tfrac{1}{3}K)}{\mathrm{sn}^2(\tfrac{1}{3}K)\mathrm{dn}^2(\tfrac{1}{3}K)} = \frac{\mathrm{sn}^2(\tfrac{2}{3}K)}{\mathrm{sn}^2(\tfrac{1}{3}K)} = \frac{2 - s}{s}. \tag{9.7.16}$$

Thus, with this value of λ, equation (9.7.12) gives

$$\mathrm{sn}(\lambda u, k_1) = k_1^{-1/2}\frac{\theta_1(3z|3\tau)}{\theta_4(3z|3\tau)} = k_1^{-1/2}\frac{\theta_1(z)\theta_1(\frac{1}{3}\pi + z)\theta_1(\frac{1}{3}\pi - z)}{\theta_4(z)\theta_4(\frac{1}{3}\pi + z)\theta_4(\frac{1}{3}\pi - z)}$$

$$= k^{3/2}k_1^{-1/2}\,\mathrm{sn}\,u\,\mathrm{sn}(\tfrac{2}{3}K + u)\,\mathrm{sn}(\tfrac{2}{3}K - u)$$

$$= \mathrm{ns}^2(\tfrac{1}{3}K)\,\mathrm{sn}\,u\,\mathrm{sn}(\tfrac{2}{3}K + u)\,\mathrm{sn}(\tfrac{2}{3}K - u). \tag{9.7.17}$$

This is the first of the transformation equations for Jacobi's elliptic functions.

Equations (2.1.2) and (2.1.3) now yield, in a similar manner, transformation equations for cn and dn. Thus,

$$\mathrm{cn}(\lambda u, k_1) = (k_1'/k_1)^{1/2}\frac{\theta_2(3z|3\tau)}{\theta_4(3z|3\tau)} = (k_1'/k_1)^{1/2}\frac{\theta_2(z)\theta_2(\frac{1}{3}\pi + z)\theta_2(\frac{1}{3}\pi - z)}{\theta_4(z)\theta_4(\frac{1}{3}\pi + z)\theta_4(\frac{1}{3}\pi - z)}$$

$$= (k^3k_1'/k'^3k_1)^{1/2}\,\mathrm{cn}\,u\,\mathrm{cn}(\tfrac{2}{3}K + u)\,\mathrm{cn}(\tfrac{2}{3}K - u)$$

$$= (1 - s)^{-2}\,\mathrm{cn}\,u\,\mathrm{cn}(\tfrac{2}{3}K + u)\,\mathrm{cn}(\tfrac{2}{3}K - u)$$

$$= \mathrm{nc}^2(\tfrac{1}{3}K)\,\mathrm{cn}\,u\,\mathrm{cn}(\tfrac{2}{3}K + u)\,\mathrm{cn}(\tfrac{2}{3}K - u), \tag{9.7.18}$$

since $\mathrm{sn}\,\tfrac{1}{3}K + \mathrm{cn}\,\tfrac{2}{3}K = 1$ (Exercise 22 in Chapter 2). Also

$$\mathrm{dn}(\lambda u, k_1) = k_1'^{1/2}\frac{\theta_3(3z|3\tau)}{\theta_4(3z|3\tau)} = k_1'^{1/2}\frac{\theta_3(z)\theta_3(\frac{1}{3}\pi + z)\theta_3(\frac{1}{3}\pi - z)}{\theta_4(z)\theta_4(\frac{1}{3}\pi + z)\theta_4(\frac{1}{3}\pi - z)}$$

$$= (k_1'/k'^3)^{1/2}\,\mathrm{dn}\,u\,\mathrm{dn}(\tfrac{2}{3}K + u)\,\mathrm{dn}(\tfrac{2}{3}K - u)$$

$$= k'^{-2}\,\mathrm{dn}^2(\tfrac{1}{3}K)\,\mathrm{dn}\,u\,\mathrm{dn}(\tfrac{2}{3}K + u)\,\mathrm{dn}(\tfrac{2}{3}K - u)$$

$$= \mathrm{nd}^2(\tfrac{2}{3}K)\,\mathrm{dn}\,u\,\mathrm{dn}(\tfrac{2}{3}K + u)\,\mathrm{dn}(\tfrac{2}{3}K - u), \tag{9.7.19}$$

having used equations (9.7.6) and (2.2.19) in the last steps of the calculation.

Instead of s, λ can be very conveniently adopted as the parameter of transformation. Then, substituting from equation (9.7.16) into equations (9.7.9)–(9.7.11), we calculate that

$$\left.\begin{array}{ll} k^2 = \dfrac{1}{16\lambda}(\lambda + 1)^3(3 - \lambda), & k'^2 = \dfrac{1}{16\lambda}(\lambda - 1)^3(3 + \lambda), \\[3mm] k_1^2 = \dfrac{1}{16\lambda^3}(\lambda + 1)(3 - \lambda)^3, & k_1'^2 = \dfrac{1}{16\lambda^3}(\lambda - 1)(3 + \lambda)^3. \end{array}\right\} \tag{9.7.20}$$

If k lies between 0 and 1, then λ is confined to the interval $(1, 3)$.

Denoting the quarter-periods of the transformed functions by Ω and $i\Omega'$, using equations (2.2.3), we calculate that

$$\Omega = \tfrac{1}{2}\pi\theta_3^2(0|3\tau) = K\frac{\theta_3^2(0|3\tau)}{\theta_3^2(0)} = \tfrac{1}{3}\lambda K, \tag{9.7.21}$$

$$i\Omega' = 3\tau\Omega = \lambda\tau K = i\lambda K', \tag{9.7.22}$$

where reference has been made to equation (9.7.13).

The inverse third-order transformation $\tau_3 = \tau/3$ can be constructed as a product of transformations $\tau_3 = -1/\tau_2, \tau_2 = 3\tau_1, \tau_1 = -1/\tau$; the derivation is left as an exercise for the reader (the result will be found in Exercise 4).

9.8. Transformation of Weierstrass's Function

The modular transformation (9.1.1) is less convenient than the linear transformation (9.1.2) when the Weierstrass function $\mathscr{P}(u, \omega_1, \omega_3)$ is under consideration and we shall accordingly confine our attention to the latter. This linear transformation may be regarded as replacing a lattice generated by periods $2\omega_1, 2\omega_3$ by another lattice generated by periods $2\omega'_1, 2\omega'_3$. Associated with each of these lattices is a Weierstrass function whose primitive periods are determined by the pair of generators. We recall that the Weierstrass function is completely defined when the lattice of vertices of its primitive period parallelograms is known. Thus, if $(2\omega_1, 2\omega_3)$, $(2\omega'_1, 2\omega'_3)$ generate identical lattices, then $\mathscr{P}(u, \omega_1, \omega_3) \equiv \mathscr{P}(u, \omega'_1, \omega'_3)$. But, if $ad - bc = 1$, then it may be proved (see, e.g., G.H. Hardy and E.M. Wright, *Introduction to the Theory of Numbers*, Oxford University Press, 1945, p. 28) that the lattice generated by $2\omega'_1, 2\omega'_3$ is identical with that generated by $2\omega_1, 2\omega_3$. We conclude that, in this case, the transformation (9.1.2) has no effect on the Weierstrass function. This disposes of transformations of the first order.

It was shown in section 9.5 that any transformation of higher order than the first can be represented as a product of first-order transformations and transformations with matrix $\begin{bmatrix} 1 & 0 \\ 0 & n \end{bmatrix}$. Thus, we need only study the effect on the Weierstrass function of transformations having this type of matrix.

Suppose $n = 2$, so that

$$\omega'_1 = \omega_1, \qquad \omega'_3 = 2\omega_3. \tag{9.8.1}$$

Consider the function

$$f(u) = \mathscr{P}(\tfrac{1}{2}u, \omega_1, \omega_3) + \mathscr{P}(\tfrac{1}{2}u + \omega_3, \omega_1, \omega_3). \tag{9.8.2}$$

$f(u)$ is a doubly periodic function with periods $4\omega_1, 2\omega_3$, having double poles at the origin and all points congruent to the origin mod$(4\omega_1, 2\omega_3)$. The principal part of $f(u)$ at $u = 0$ is $4/u^2$ and it follows, therefore, using Liouville's theorem, that $f(u)$ differs from $4\mathscr{P}(u, 2\omega_1, \omega_3)$ by at most a constant. Thus,

$$4\mathscr{P}(u, 2\omega_1, \omega_3) = \mathscr{P}(\tfrac{1}{2}u, \omega_1, \omega_3) + \mathscr{P}(\tfrac{1}{2}u + \omega_3, \omega_1, \omega_3) + C. \tag{9.8.3}$$

Letting $u \to 0$, we find $C = -\mathscr{P}(\omega_3, \omega_1, \omega_3) = -e_3$. Hence,

$$\mathscr{P}(u, 2\omega_1, \omega_3) = \tfrac{1}{4}[\mathscr{P}(\tfrac{1}{2}u, \omega_1, \omega_3) + \mathscr{P}(\tfrac{1}{2}u + \omega_3, \omega_1, \omega_3) - e_3]. \tag{9.8.4}$$

Similarly, we can prove that

$$\mathscr{P}(u, \omega_1, 2\omega_3) = \tfrac{1}{4}[\mathscr{P}(\tfrac{1}{2}u, \omega_1, \omega_3) + \mathscr{P}(\tfrac{1}{2}u + \omega_1, \omega_1, \omega_3) - e_1]. \tag{9.8.5}$$

Using the identity (6.6.17), we can now prove that

$$\mathscr{P}(u, \tfrac{1}{2}\omega_1, \omega_3) = 4\mathscr{P}(2u, \omega_1, 2\omega_3)$$
$$= \mathscr{P}(u, \omega_1, \omega_3) + \mathscr{P}(u + \omega_1, \omega_1, \omega_3) - e_1, \tag{9.8.6}$$

as also follows directly by application of Liouville's theorem. Similarly, we show that

$$\mathscr{P}(u, \omega_1, \tfrac{1}{2}\omega_3) = \mathscr{P}(u, \omega_1, \omega_3) + \mathscr{P}(u + \omega_3, \omega_1, \omega_3) - e_3. \tag{9.8.7}$$

Alternative forms for these transformations can be found by using the identities (6.8.11) and (6.8.13). Thus, (9.8.7) can be written

$$\mathscr{P}(u, \omega_1, \tfrac{1}{2}\omega_3) = \mathscr{P}(u, \omega_1, \omega_3) + \frac{(e_3 - e_1)(e_3 - e_2)}{\mathscr{P}(u, \omega_1, \omega_3) - e_3}. \tag{9.8.8}$$

To calculate the invariants of the transformed functions, we need to find their stationary values. For example, the function $\mathscr{P}(u, \tfrac{1}{2}\omega_1, \omega_3)$ is stationary at $u = \omega_1' = \tfrac{1}{2}\omega_1$, $u = \omega_3' = \omega_3$, and $u = \omega_2' = -\tfrac{1}{2}\omega_1 - \omega_3$. Its stationary values are therefore

$$e_1' = \mathscr{P}(\tfrac{1}{2}\omega_1, \tfrac{1}{2}\omega_1, \omega_3) = \mathscr{P}(\tfrac{1}{2}\omega_1, \omega_1, \omega_3) + \mathscr{P}(\tfrac{3}{2}\omega_1, \omega_1, \omega_3) - e_1$$
$$= 2\mathscr{P}(\tfrac{1}{2}\omega_1, \omega_1, \omega_3) - e_1 = e_1 + 2\{(e_1 - e_2)(e_1 - e_3)\}^{1/2}, \tag{9.8.9}$$

$$e_3' = \mathscr{P}(\omega_3, \tfrac{1}{2}\omega_1, \omega_3) = \mathscr{P}(\omega_3, \omega_1, \omega_3) + \mathscr{P}(\omega_1 + \omega_3, \omega_1, \omega_3) - e_1$$
$$= e_3 + e_2 - e_1 = -2e_1, \tag{9.8.10}$$

after referring to Exercise 8 in Chapter 6. Since $e_1' + e_2' + e_3' = 0$, we deduce that

$$e_2' = e_1 - 2\{(e_1 - e_2)(e_1 - e_3)\}^{1/2}. \tag{9.8.11}$$

Using the definitions (6.7.19) of the invariants, we now calculate that

$$g_2' = 4(11e_1^2 + 4e_2e_3) = 60e_1^2 - 4g_2, \tag{9.8.12}$$

$$g_3' = 8e_1(7e_1^2 + 4e_2e_3) = 56e_1^3 + 8g_3. \tag{9.8.13}$$

Transformations of higher order can be constructed in a similar way. Thus, a third-order transformation is

$$\mathscr{P}(u, 3\omega_1, \omega_3) = \tfrac{1}{9}\{\mathscr{P}(\tfrac{1}{3}u, \omega_1, \omega_3) + \mathscr{P}(\tfrac{1}{3}u + \tfrac{2}{3}\omega_3, \omega_1, \omega_3)$$
$$+ \mathscr{P}(\tfrac{1}{3}u - \tfrac{2}{3}\omega_3, \omega_1, \omega_3) - 2\mathscr{P}(\tfrac{2}{3}\omega_3, \omega_1, \omega_3)\}. \tag{9.8.14}$$

EXERCISES

1. Show that the transformation $\tau_1 = \tau + 2$ generates the following transformation equations:

$$k_1 = -k, \quad k_1' = k', \quad K_1 = K, \quad K_1' = K' - 2iK,$$
$$\text{sn}(u, k_1) = \text{sn}(u, k), \quad \text{cn}(u, k_1) = \text{cn}(u, k), \quad \text{dn}(u, k_1) = \text{dn}(u, k).$$

2. Use the method of section (9.6) to derive the identities

$$\theta_1(2z|2\tau) = A\theta_1(z)\theta_2(z),$$
$$\theta_2(2z|2\tau) = A\theta_2(\tfrac{1}{4}\pi + z)\theta_2(\tfrac{1}{4}\pi - z),$$
$$\theta_3(2z|2\tau) = A\theta_3(\tfrac{1}{4}\pi + z)\theta_3(\tfrac{1}{4}\pi - z),$$
$$\theta_4(2z|2\tau) = A\theta_3(z)\theta_4(z),$$

where

$$A = \theta_2/[\sqrt{2\theta_2(\tfrac{1}{4}\pi)\theta_3(\tfrac{1}{4}\pi)}].$$

Verify that these transformation equations are equivalent to those already established in section 1.8. (Hint: Use the identity (1.4.32).)

3. Writing $\theta_i(0|\tau) = \theta_i, \theta_i(0|3\tau) = \Theta_i$ $(i = 1, 2, 3, 4)$, establish the following relationships:

(i) $\theta_2\Theta_2 + \theta_4\Theta_4 = \theta_3\Theta_3$,

(ii) $\theta_2^3\Theta_4 - \theta_4^3\Theta_2 = 3\theta_3\Theta_2\Theta_3\Theta_4$,

(iii) $\theta_3^2(\theta_2^2 - \Theta_2^2) = \Theta_3^2(\theta_2^2 + 3\Theta_2^2)$,

(iv) $\theta_2^2 + 3\Theta_2^2 = 2\sqrt{(\theta_2\theta_3^3\Theta_2/\Theta_3)}$,

(v) $\theta_2^2 - 3\Theta_2^2 = 2\sqrt{(\theta_2\theta_4^3\Theta_2/\Theta_4)}$,

(vi) $\dfrac{\theta_2^2 + \Theta_2^2}{2\sqrt{(\theta_2\Theta_2)}} = \sqrt{\dfrac{\Theta_4^3}{\theta_4}}$,

(vii) $\dfrac{\theta_2^2 - \Theta_2^2}{2\sqrt{(\theta_2\Theta_2)}} = \sqrt{\dfrac{\Theta_3^3}{\theta_3}}$,

(viii) $\theta_2^2\Theta_3^2 - \theta_3^2\Theta_2^2 = 2\theta_4\Theta_4\sqrt{(\theta_2\theta_3\Theta_2\Theta_3)}$,

(ix) $\sqrt{\dfrac{\theta_3^3}{\Theta_3}} + \sqrt{\dfrac{\theta_4^3}{\Theta_4}} = \sqrt{\dfrac{\theta_2^3}{\Theta_2}} = \sqrt{\dfrac{\Theta_3^3}{\theta_3}} + \sqrt{\dfrac{\Theta_4^3}{\theta_4}}.$

4. Treating the transformation $\tau_3 = \tau/3$ as the product of the transformations $\tau_3 = -1/\tau_2$, $\tau_2 = 3\tau_1$, $\tau_1 = -1/\tau$, obtain the transformation equations

$$sn(\lambda u, k_3) = dc^2(\tfrac{2}{3}iK')\,sn\,u\,sn(\tfrac{2}{3}iK' + u)\,sn(\tfrac{2}{3}iK' - u),$$
$$cn(\lambda u, k_3) = nc^2(\tfrac{2}{3}iK')\,cn\,u\,cn(\tfrac{2}{3}iK' + u)\,cn(\tfrac{2}{3}iK' - u),$$
$$dn(\lambda u, k_3) = nd^2(\tfrac{2}{3}iK')\,dn\,u\,dn(\tfrac{2}{3}iK' + u)\,dn(\tfrac{2}{3}iK' - u),$$

where

$$\lambda = (2-s)/s, \qquad k^2 = \frac{(1+s)(1-s)^3}{s^3(2-s)}, \qquad k_3^2 = \frac{(1-s)(1+s)^3}{s(2-s)^3},$$

and $s = sn(\tfrac{1}{3}K', k')$, the modulus being k unless otherwise indicated.

5. Applying the method of section 9.6, obtain the fourth-order transformation equations for the theta functions given in Exercise 21 in Chapter 1. Deduce the corresponding transformation equations for the Jacobi functions in the form

$$sn(\lambda u, k_1) = (1 + k')^2(1 + \sqrt{k'})^2 \frac{sn\,u\,cn\,u(sn^2\tfrac{1}{2}K - sn^2 u)}{dn\,u(dn^2 u + dn^2\tfrac{1}{2}K)},$$

$$cn(\lambda u, k_1) = (1 + k')^2(1 + \sqrt{k'})^2 \frac{(cn^2 u - cn^2\tfrac{1}{4}K)(cn^2 u - cn^2\tfrac{3}{4}K)}{dn\,u(dn^2 u + dn^2\tfrac{1}{2}K)},$$

$$dn(\lambda u, k_1) = (1 + \sqrt{k'})^{-2} \frac{[dn^2 u - dn^2(\tfrac{1}{4}K + iK')][dn^2 u - dn^2(\tfrac{3}{4}K + iK')]}{dn\,u(dn^2 u + dn^2\tfrac{1}{2}K)},$$

where $\lambda = (1 + \sqrt{k'})^2$, $k_1 = ((1 - \sqrt{k'})/(1 + \sqrt{k'}))^2$. Show also that $K_1 = \tfrac{1}{4}K$, $K'_1 = \lambda K'$. (Hint: The identities (1.4.23), (1.4.32), and (1.4.33) should prove helpful.)

6. Follow through the calculations of sections 9.6 and 9.7 for the case of the fifth-order modular transformation $\tau_1 = 5\tau$ to obtain the following transformations for the Jacobi functions:

$$sn(\lambda u, k_1) = ns^2(\tfrac{1}{5}K)\,ns^2(\tfrac{2}{5}K)\,sn\,u\,sn(\tfrac{2}{5}K + u)\,sn(\tfrac{2}{5}K - u)\,sn(\tfrac{4}{5}K + u)\,sn(\tfrac{4}{5}K - u),$$

$$cn(\lambda u, k_1) = nc^2(\tfrac{2}{5}K)\,nc^2(\tfrac{4}{5}K)\,cn\,u\,cn(\tfrac{2}{5}K + u)\,cn(\tfrac{2}{5}K - u)\,cn(\tfrac{4}{5}K + u)\,cn(\tfrac{4}{5}K - u),$$

$$dn(\lambda u, k_1) = nd^2(\tfrac{2}{5}K)\,nd^2(\tfrac{4}{5}K)\,dn\,u\,dn(\tfrac{2}{5}K + u)\,dn(\tfrac{2}{5}K - u)\,dn(\tfrac{4}{5}K + u)\,dn(\tfrac{4}{5}K - u),$$

where

$$k_1 = k^5\,sn^4(\tfrac{1}{5}K)\,sn^4(\tfrac{3}{5}K), \qquad k'_1 = k'^{-3}\,dn^4(\tfrac{1}{5}K)\,dn^4(\tfrac{3}{5}K),$$

$$\lambda = \frac{cn^2(\tfrac{1}{5}K)\,cn^2(\tfrac{3}{5}K)}{sn^2(\tfrac{2}{5}K)\,sn^2(\tfrac{3}{5}K)\,dn^2(\tfrac{1}{5}K)\,dn^2(\tfrac{3}{5}K)}.$$

7. With the notation used in section 9.7, prove that

$$\sqrt{(kk_1)} + \sqrt{(k'k_1)} = 1.$$

8. Prove the third-order transformation equation

$$\mathscr{P}(u, \tfrac{1}{3}\omega_1, \omega_3) = \mathscr{P}(u, \omega_1, \omega_3) + \mathscr{P}(u + \tfrac{2}{3}\omega_1, \omega_1, \omega_3)$$
$$+ \mathscr{P}(u - \tfrac{2}{3}\omega_1, \omega_1, \omega_3) - 2\mathscr{P}(\tfrac{2}{3}\omega_1, \omega_1, \omega_3)$$

and reduce this to the equivalent form

$$\mathscr{P}(u, \tfrac{1}{3}\omega_1, \omega_3) = (x - a)^{-2}\{x^3 - 2ax^2 + (7a^2 - \tfrac{1}{2}g_2)x - 2a^3 - \tfrac{1}{2}g_2a - g_3\},$$

where $x = \mathscr{P}(u, \omega_1, \omega_3)$, $a = \mathscr{P}(\tfrac{2}{3}\omega_1, \omega_1, \omega_3)$.

9. If $a = \mathscr{P}(\tfrac{2}{3}\omega_1, \omega_1, \omega_3)$ and g'_2, g'_3 are the invariants of the Weierstrass function $\mathscr{P}(u, \tfrac{1}{3}\omega_1, \omega_3)$, show that

$$g'_2 = 120a^2 - 9g_2, \qquad g'_3 = 280a^3 - 42g_2a - 27g_3.$$

(Hint: Calculate $\mathscr{P}'(u, \tfrac{1}{3}\omega_1, \omega_3)$ from the previous exercise and use (6.7.26).)

10. Putting $sn(u, k) = x$, $sn(\lambda u, k_1) = y$ in equation (9.7.17), show that the change of variable

$$y = \frac{x(x^2 + s^2 - 2s)}{(2s - 1)x^2 - s^2}$$

generates the transformation

$$[(1 - y^2)(1 - k_1^2 y^2)]^{-1/2}\,dy = \lambda[(1 - x^2)(1 - k^2 x^2)]^{-1/2}\,dx.$$

Fourier Series for a Periodic Analytic Function

Suppose $f(z)$ $(z = x + iy)$ is periodic, with period 2π, and is regular in the strip $a < y < b$ of complex z-plane. Consider the regular transformation $w = e^{iz}$. Noting that

$$|w| = e^{-y}, \qquad \arg w = x, \tag{A1}$$

it follows that a rectangular region $c \leqslant x \leqslant c + 2\pi$, $a < y < b$ in the z-plane transforms into the annular region $e^{-b} < |w| < e^{-a}$, $c \leqslant \arg w \leqslant c + 2\pi$ in the w-plane. If $f(z) = F(w)$, $F(w)$ will be regular over the annular region and single-valued in consequence of the periodicity of $f(z)$. It follows that $F(w)$ is expansible in a Laurent series, thus

$$F(w) = \sum_{r=-\infty}^{\infty} c_r w^r, \tag{A2}$$

where

$$c_r = \frac{1}{2\pi i} \oint F(w) w^{-r-1} \, dw, \tag{A3}$$

the contour integral being taken around any circle $|w| = R$ lying inside the annular region (see, e.g., E.T. Copson, *Theory of Functions of a Complex Variable*, pp. 75–77, Oxford University Press, 1935).

Transforming back to the variable z, the equations (A2) and (A3) take the forms

$$f(z) = \sum_{r=-\infty}^{\infty} c_r e^{irz}, \tag{A4}$$

$$c_r = \frac{1}{2\pi} \int f(z) e^{-irz} \, dz, \tag{A5}$$

the integral being evaluated along a line $y = \alpha, c \leqslant x \leqslant c + 2\pi$ lying in the strip $a < y < b$. In particular, if the real axis lies in this strip, we can take $\alpha = 0, c = 0$ and then

$$c_r = \frac{1}{2\pi} \int_0^{2\pi} f(x) e^{-irx} dx. \tag{A6}$$

Equation (A4) provides a Fourier expansion for $f(z)$.

In the special case when $f(z)$ is odd, since $f(-z) = -f(z)$, we must have $c_{-r} = -c_r$ and, hence,

$$f(z) = \sum_{r=1}^{\infty} c_r(e^{irz} - e^{-irz}) = \sum_{r=1}^{\infty} b_r \sin rz, \tag{A7}$$

where

$$b_r = 2ic_r = \frac{i}{\pi} \int_0^{2\pi} f(x) e^{-irx} dx = \frac{i}{\pi} \int_{-\pi}^{\pi} f(x) e^{-irx} dx. \tag{A8}$$

If desired, this formula for the coefficients of the sine expansion can be further reduced to the form

$$b_r = \frac{2}{\pi} \int_0^{\pi} f(x) \sin rx \, dx, \tag{A9}$$

since $f(x) \cos rx$ is odd and its integral vanishes.

Similarly, if $f(z)$ is even, we must have $c_{-r} = c_r$ and, therefore,

$$f(z) = \tfrac{1}{2}a_0 + \sum_{r=1}^{\infty} a_r \cos rz, \tag{A10}$$

where

$$a_r = \frac{2}{\pi} \int_0^{\pi} f(x) \cos rx \, dx. \tag{A11}$$

APPENDIX B

Calculation of a Definite Integral

Consider the contour integral

$$\int e^{-z^2}\, dz \tag{B1}$$

calculated along the arc of a circle $z = R e^{i\theta}$ from $\theta = 0$ to $\theta = \alpha$, where $0 < \alpha < \frac{1}{4}\pi$. We shall show this tends to zero as the radius R tends to infinity. Taking the modulus, we calculate that

$$\left| \int e^{-z^2}\, dz \right| = \left| \int_0^\alpha \exp\{- R^2(\cos 2\theta + i \sin 2\theta)\} iR\, e^{i\theta}\, d\theta \right|$$

$$< R \int_0^{\pi/4} \exp(- R^2 \cos 2\theta)\, d\theta. \tag{B2}$$

But, inspection of the graph of $\cos 2\theta$ shows that

$$\cos 2\theta \geqslant 1 - 2\theta/\pi \tag{B3}$$

for $0 \leqslant \theta \leqslant \frac{1}{4}\pi$. We deduce that

$$\left| \int e^{-z^2}\, dz \right| < R \int_0^{\pi/4} \exp\{- R^2(1 - 2\theta/\pi)\}\, d\theta = \frac{\pi}{2R}(e^{-(1/2)R^2} - e^{-R^2}). \tag{B4}$$

Hence, as $R \to \infty$, the contour integral approaches zero.

Now consider the integral of e^{-z^2} around the closed contour $ABCDA$ (Fig. B1), where AB is a straight line (L) inclined at an acute angle β to the real axis (with $|\beta| < \frac{1}{4}\pi$), BC and AD are arcs of a circle of radius R and center O, and CD is the real axis. For sufficiently large R, the circular arcs will subtend angles at O of magnitudes less than $\frac{1}{4}\pi$. It follows from our earlier result that the contributions of these arcs approach 0 as $R \to \infty$. But, e^{-z^2} is regular

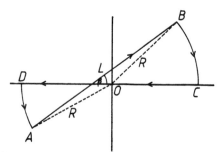

Figure B.1. Integration contour for $\exp(-z^2)$.

everywhere, and so by Cauchy's theorem the integral taken around the closed contour vanishes. We conclude that, in the limit,

$$\int_L e^{-z^2}\,dz = \int_{-\infty}^{\infty} e^{-x^2}\,dx = \sqrt{\pi}, \tag{B5}$$

the real integral being well known.

Suppose L has equation $z = au + c$, where u is a real variable and $|\arg a| < \tfrac{1}{4}\pi$. Then

$$\int_L e^{-z^2}\,dz = a\int_{-\infty}^{\infty} e^{-(au+c)^2}\,du. \tag{B6}$$

Thus

$$\int_{-\infty}^{\infty} e^{-(au+c)^2}\,du = \sqrt{\pi}/a. \tag{B7}$$

Putting $c = b/2a$, we obtain finally that

$$\int_{-\infty}^{\infty} e^{-a^2u^2 - bu}\,du = \frac{\sqrt{\pi}}{a}e^{b^2/4a^2}, \tag{B8}$$

where a, b can be any complex numbers subject to the single restriction $|\arg a| < \tfrac{1}{4}\pi$.

BASIC Program for Reduction of Elliptic Integral to Standard Form

```
10  INPUT "Enter coefficients a, b, c, d, e.   " a, b, c, d, e
20  P = b↑3 + 8∗a∗a∗d − 4∗a∗b∗c
30  Q = 2∗(16∗a∗a∗e − 4∗a∗c∗c + b∗b∗c + 2∗a∗b∗d)
40  R = 5∗(b∗b∗d + 8∗a∗b∗e − 4∗a∗c∗d)
50  S = 20∗(b∗b∗e − a∗d∗d)
60  T = 5∗(4∗b∗c∗e − 8∗a∗d∗e − b∗d∗d)
70  U = 2∗(4∗c∗c∗e − 16∗a∗e∗e − c∗d∗d − 2∗b∗d∗e)
80  V = 4∗c∗d∗e − d↑3 − 8∗b∗e∗e
90  INPUT "Enter trial value of the sextic root.   " L
100 r = FNr(L)
110 IF r = 0 GOTO 230
120 r1 = r
130 L = L + 1
140 r = FNr(L)
150 IF r = 0 GOTO 230
160 IF r∗r1 < 0 THEN s% = 1: GOTO 190
170 IF ABSr > ABSr1 THEN PROCdown(0) ELSE PROCup(0)
180 IF r = 0 GOTO 230
190 FOR n% = 1 TO 8
200 IF s% = 1 THEN PROCdown(n%) ELSE PROCup(n%)
210 NEXT n%
220 IF ABSr > ABSr1 THEN L = L1:r = r1
230 PRINT "r = "; r
240 X = L∗(L∗(4∗a∗L + 3∗b) + 2∗c) + d
250 Y = L∗(L∗(b∗L + 2∗c) + 3∗d) + 4∗e
260 IF Y = 0 THEN PRINT "Value of mu is infinite for this sextic root. Try
    new trial value": GOTO 90
```

```
270 M = − X/Y
280 s = FNr(1/M)
290 PRINT "s = "; s
300 A = a + M*(b + M*(c + M*(d + e*M)))
310 B = 6*a*L*L + 6*e*M*M + (L*M + 1)*(L*b + M*d)*3
     + c*(L*L*M*M + 4*L*M + 1)
320 C = L*(L*(L*(a*L + b) + c) + d) + e
330 D = B*B − 4*A*C
340 IF D < 0 THEN PRINT "Imaginary factors. Try new sextic root.":
     GOTO 90
350 R1 = (B − SQRD)/2/A
360 R2 = (B + SQRD)/2/A
370 PRINT A; "(t2 + "; R1;")(t2 + "; R2;")"
380 PRINT "x = (t + "; L;")/("; M;" t + 1)"
390 PRINT "Do you wish to try another trial value of the sextic root?"
400 A$ = GET$
410 IF A$ = "Y" OR A$ = "y" GOTO 90
420 END
430 DEF FNr(L)
440    = L*(L*(L*(L*(L*(P*L + Q) + R) + S) + T) + U) + V
450 DEF PROCdown(n%)
460 s% = − 1
470 i = .1↑n%
480 REPEAT
490 r1 = r
500 L1 = L
510 L = L − i
520 r = FNr(L)
530 UNTIL r*r1 = 0 OR r*r1 < 0
540 ENDPROC
550 DEF PROCup(n%)
560 s% = 1
570 i = .1↑n%
580 REPEAT
590 r1 = r
600 L1 = L
610 L = L + i
620 r = FNr(L)
630 UNTIL r*r1 = 0 OR r*r1 < 0
640 ENDPROC
```

Note: r and s are residues after the calculated values of the roots λ and $1/\mu$ have been substituted in the sextic.

APPENDIX D

Computation of Tables

Equation (1.2.11) defines

$$\theta_1(z, q) = 2q^{1/4} \sum_{n=0}^{\infty} (-1)^n q^{n(n+1)} \sin(2n + 1)z.$$

The following program, based on this formula, expressed in BBC BASIC, displays the first half of Table A:

```
 10 MODE 7
 20 VDU 14
 30 PRINT TAB(7); "Function Theta 1"
 40 PRINT: PRINT
 50 @% = &2050C
 60 FOR q = .1 TO .81 STEP .1
 70 PRINT "q = "; q: PRINT
 80 PRINT TAB(6); "x deg"; TAB(17); "Theta 1": PRINT
 90 r = SQR(SQRq)
100 FOR x% = 0 TO 90
110 T = 0:n% = 0:s% = 1
120 z = RAD(x%)
130 REPEAT
140 p = q↑(n%*(n% + 1))
150 T = T + s%*p*SIN((2*n% + 1)*z)
160 n% = n% + 1: s% = - s%
170 UNTIL p < .0000001
180 PRINT x%, 2*r*T
190 NEXT x%
200 PRINT: PRINT
```

```
210 NEXT q
220 END
```

The second half of the table is computed from the formula (1.2.13) for $\theta_3(z, q)$ by a very similar program.

To construct Table B, we first compute K as a function of k^2 by making use of the result (3.9.39). A BASIC program for this purpose is given below:

```
10  MODE 7:VDU 14:@% = &2050B
20  PRINT TAB(2) "Quarter period K as function of k2":PRINT
30  PRINT TAB(7) "k2"; SPC(9); "K"
40  PRINT: PRINT 0,PI/2
50  FOR k2 = .01 TO .991 STEP .01
60  a = 1:b = SQR(1 – k2)
70  REPEAT
80  a1 = 1
90  a = (a + b)/2: b = SQR(a1*b)
100 UNTIL a – b < .0000001
110 PRINT k2, PI/2/a
120 NEXT k2
130 END
```

q as a function of k^2 can now be derived from the formula $q = \exp(-\pi K'/K)$, recalling that $K'(k) = K(k')$.

Since we have derived formulae for all other functions (including $E(k)$) in terms of theta functions, it is now straightforward to construct programs which will display these in the manner indicated in the remaining tables. We have preferred the modulus k rather than the nome q as the tabulated parameter in Tables B and C, but q can be calculated for each value of k and used when computing the theta functions.

The integral of the first kind $F(\phi, k)$ (Table D) can be computed as $\mathrm{sn}^{-1}(\sin \phi, k)$, but a computational procedure for inverting the elliptic function is slow in execution and it may be considered preferable, therefore, to obtain F directly from the integral (3.1.8), using one of the many procedures available for numerical calculation of an integral.

Equation (3.4.27) shows that $D(\phi, k) = E[\mathrm{sn}^{-1}(\sin \phi), k] = E[F(\phi, k), k]$. It follows that, when F has been computed, the integral of the second kind, $D(\phi, k)$, can be found by substituting $u = F(\phi, k)$ into equation (3.5.3).

Table J may be computed from the formula given in Exercise 22 in Chapter 6. The sign factor s is best controlled within a REPEAT loop by the instruction: IF n% MOD 2 = 1 THEN s% = – s%. This will reverse the sign of s% in alternate passages round the loop.

Table K was computed using equations (6.7.29)–(6.7.31) and (6.7.19).

To construct Table M, it is necessary to invert the functions $G(\kappa)$ and $G(v)$, in the two cases $g_3 > 0$ and $g_3 < 0$. Table L then follows by calculation of the functions $\omega_1^2 \mathcal{P}(u, \omega_1, i\kappa\omega_1)$ and $\omega_1^2 \mathcal{P}(u, \omega_1, \frac{1}{2}\omega_1(1 + iv))$ using values of κ and v from Table M.

Tables

Table A. Theta Functions $\theta_1(x)$ and $\theta_2(x)$

			$q \rightarrow$		
x^0	0.1	0.2	0.3	0.4	x^0
0	0.00000	0.00000	0.00000	0.00000	90
1	0.01904	0.02055	0.01895	0.01500	89
2	0.03808	0.04110	0.03792	0.03003	88
3	0.05710	0.06165	0.05691	0.04511	87
4	0.07612	0.08220	0.07592	0.06026	86
5	0.09511	0.10276	0.09498	0.07550	85
6	0.11409	0.12332	0.11409	0.09086	84
7	0.13303	0.14387	0.13327	0.10636	83
8	0.15195	0.16444	0.15251	0.12202	82
9	0.17083	0.18500	0.17183	0.13786	81
10	0.18968	0.20557	0.19125	0.15392	80
11	0.20848	0.22614	0.21076	0.17020	79
12	0.22722	0.24671	0.23038	0.18672	78
13	0.24592	0.26728	0.25011	0.20352	77
14	0.26456	0.28785	0.26996	0.22060	76
15	0.28314	0.30842	0.28994	0.23798	75
16	0.30165	0.32899	0.31005	0.25568	74
17	0.32009	0.34955	0.33031	0.27372	73
18	0.33845	0.37011	0.35070	0.29211	72
19	0.35673	0.39066	0.37125	0.31087	71
20	0.37493	0.41120	0.39194	0.33000	70
21	0.39303	0.43172	0.41279	0.34953	69
22	0.41104	0.45224	0.43380	0.36945	68
23	0.42895	0.47273	0.45496	0.38979	67
24	0.44675	0.49320	0.47628	0.41054	66
25	0.46445	0.51364	0.49775	0.43171	65
26	0.48203	0.53405	0.51938	0.45331	64
27	0.49949	0.55442	0.54117	0.47535	63
28	0.51682	0.57476	0.56310	0.49782	62
29	0.53403	0.59505	0.58519	0.52072	61
30	0.55110	0.61528	0.60741	0.54405	60
31	0.56802	0.63546	0.62977	0.56782	59
32	0.58481	0.65558	0.65225	0.59201	58
33	0.60144	0.67563	0.67486	0.61662	57
34	0.61791	0.69559	0.69758	0.64165	56
35	0.63423	0.71548	0.72041	0.66707	55
36	0.65038	0.73527	0.74333	0.69289	54
37	0.66635	0.75496	0.76633	0.71909	53
38	0.68215	0.77455	0.78940	0.74565	52
39	0.69776	0.79401	0.81252	0.77255	51
40	0.71319	0.81336	0.83570	0.79979	50
41	0.72843	0.83256	0.85890	0.82733	49
42	0.74346	0.85162	0.88211	0.85516	48
43	0.75829	0.87053	0.90532	0.88326	47
44	0.77291	0.88928	0.92852	0.91160	46
45	0.78732	0.90785	0.95168	0.94015	45

$\theta_1(x)$ (left column, arrow pointing down) $\theta_2(x)$ (right column, arrow pointing up)

Table A. (*Continued*)

$$q \rightarrow$$

	x^0	0.1	0.2	0.3	0.4	x^0	
	46	0.80150	0.92624	0.97478	0.96888	44	
	47	0.81546	0.94443	0.99781	0.99777	43	
	48	0.82919	0.96242	1.02074	1.02679	42	
	49	0.84268	0.98019	1.04356	1.05590	41	
	50	0.85593	0.99774	1.06625	1.08506	40	
	51	0.86894	1.01505	1.08878	1.11426	39	
	52	0.88169	1.03210	1.11114	1.14344	38	
	53	0.89418	1.04890	1.13330	1.17257	37	
	54	0.90641	1.06543	1.15523	1.20161	36	
	55	0.91837	1.08167	1.17693	1.23053	35	
	56	0.93006	1.09761	1.19835	1.25928	34	
	57	0.94148	1.11325	1.21949	1.28782	33	
	58	0.95261	1.12858	1.24031	1.31611	32	
	59	0.96345	1.14357	1.26080	1.34412	31	
	60	0.97400	1.15822	1.28093	1.37178	30	
	61	0.98426	1.17252	1.30067	1.39908	29	
	62	0.99421	1.18645	1.32001	1.42595	28	
	63	1.00386	1.20001	1.33891	1.45236	27	
	64	1.01320	1.21319	1.35737	1.47827	26	
	65	1.02222	1.22597	1.37534	1.50362	25	
$\theta_1(x)$	66	1.03092	1.23834	1.39282	1.52839	24	$\theta_2(x)$
	67	1.03931	1.25029	1.40978	1.55252	23	
	68	1.04736	1.26182	1.42620	1.57598	22	
	69	1.05509	1.27291	1.44205	1.59872	21	
	70	1.06248	1.28356	1.45732	1.62071	20	
	71	1.06953	1.29374	1.47198	1.64190	19	
	72	1.07625	1.30347	1.48602	1.66226	18	
	73	1.08262	1.31271	1.49942	1.68175	17	
	74	1.08864	1.32148	1.51215	1.70033	16	
	75	1.09431	1.32976	1.52421	1.71797	15	
	76	1.09963	1.33754	1.53557	1.73464	14	
	77	1.10460	1.34481	1.54621	1.75030	13	
	78	1.10921	1.35158	1.55613	1.76493	12	
	79	1.11345	1.35782	1.56531	1.77849	11	
	80	1.11734	1.36355	1.57374	1.79097	10	
	81	1.12086	1.36874	1.58140	1.80233	9	
	82	1.12401	1.37340	1.58829	1.81255	8	
	83	1.12680	1.37753	1.59438	1.82162	7	
	84	1.12922	1.38111	1.59969	1.82952	6	
	85	1.13127	1.38414	1.60419	1.83623	5	
	86	1.13295	1.38663	1.60788	1.84174	4	
	87	1.13425	1.38857	1.61075	1.84603	3	
	88	1.13518	1.38996	1.61281	1.84911	2	
	89	1.13574	1.39079	1.61405	1.85095	1	
	90	1.13593	1.39107	1.61446	1.85157	0	

Table A. (*Continued*)

$q \rightarrow$

	x^0	0.5	0.6	0.7	0.8	x^0	
	0	0.00000	0.00000	0.00000	0.00000	90	
	1	0.00959	0.00426	0.00091	0.00003	89	
	2	0.01921	0.00855	0.00183	0.00006	88	
	3	0.02890	0.01290	0.00279	0.00009	87	
	4	0.03870	0.01736	0.00379	0.00013	86	
	5	0.04863	0.02196	0.00487	0.00018	85	
	6	0.05874	0.02673	0.00603	0.00023	84	
	7	0.06905	0.03170	0.00730	0.00030	83	
	8	0.07960	0.03691	0.00870	0.00038	82	
	9	0.09043	0.04240	0.01025	0.00048	81	
	10	0.10156	0.04819	0.01197	0.00060	80	
	11	0.11304	0.05434	0.01389	0.00075	79	
	12	0.12488	0.06088	0.01603	0.00092	78	
	13	0.13713	0.06783	0.01842	0.00114	77	
	14	0.14982	0.07525	0.02109	0.00141	76	
	15	0.16297	0.08317	0.02408	0.00173	75	
	16	0.17661	0.09163	0.02743	0.00213	74	
	17	0.19078	0.10067	0.03116	0.00260	73	
	18	0.20550	0.11033	0.03531	0.00317	72	
	19	0.22080	0.12064	0.03994	0.00385	71	
$\theta_1(x)$	20	0.23670	0.13166	0.04509	0.00467	70	$\theta_2(x)$
	21	0.25324	0.14342	0.05080	0.00564	69	
	22	0.27042	0.15597	0.05713	0.00681	68	
	23	0.28829	0.16934	0.06413	0.00818	67	
	24	0.30685	0.18357	0.07186	0.00981	66	
	25	0.32612	0.19871	0.08038	0.01173	65	
	26	0.34613	0.21479	0.08975	0.01399	64	
	27	0.36689	0.23186	0.10005	0.01664	63	
	28	0.38841	0.24995	0.11132	0.01974	62	
	29	0.41071	0.26910	0.12366	0.02335	61	
	30	0.43379	0.28935	0.13713	0.02754	60	
	31	0.45767	0.31072	0.15180	0.03240	59	
	32	0.48234	0.33326	0.16776	0.03801	58	
	33	0.50782	0.35698	0.18507	0.04447	57	
	34	0.53410	0.38193	0.20383	0.05189	56	
	35	0.56118	0.40812	0.22409	0.06038	55	
	36	0.58905	0.43557	0.24596	0.07006	54	
	37	0.61772	0.46431	0.26950	0.08108	53	
	38	0.64716	0.49435	0.29478	0.09358	52	
	39	0.67737	0.52569	0.32189	0.10771	51	
	40	0.70833	0.55835	0.35089	0.12363	50	
	41	0.74003	0.59233	0.38184	0.14152	49	
	42	0.77243	0.62762	0.41483	0.16156	48	
	43	0.80551	0.66422	0.44989	0.18393	47	
	44	0.83925	0.70211	0.48708	0.20883	46	
	45	0.87362	0.74127	0.52644	0.23645	45	

Table A. (*Continued*)

$$q \rightarrow$$

	x^0	0.5	0.6	0.7	0.8	x^0	
	46	0.90857	0.78168	0.56802	0.26699	44	
	47	0.94408	0.82332	0.61183	0.30066	43	
	48	0.98009	0.86613	0.65789	0.33765	42	
	49	1.01658	0.91008	0.70622	0.37816	41	
	50	1.05348	0.95513	0.75681	0.42238	40	
	51	1.09076	1.00120	0.80963	0.47048	39	
	52	1.12835	1.04825	0.86466	0.52262	38	
	53	1.16621	1.09620	0.92186	0.57897	37	
	54	1.20427	1.14498	0.98116	0.63963	36	
	55	1.24247	1.19450	1.04250	0.70473	35	
	56	1.28076	1.24467	1.10578	0.77434	34	
	57	1.31906	1.29541	1.17090	0.84851	33	
	58	1.35730	1.34661	1.23774	0.92724	32	
	59	1.39543	1.39816	1.30616	1.01051	31	
	60	1.43336	1.44996	1.37601	1.09826	30	
	61	1.47103	1.50188	1.44712	1.19038	29	
	62	1.50836	1.55381	1.51931	1.28670	28	
	63	1.54527	1.60562	1.59238	1.38703	27	
	64	1.58170	1.65717	1.66611	1.49110	26	
	65	1.61756	1.70835	1.74029	1.59861	25	
$\theta_1(x)$	66	1.65278	1.75900	1.81466	1.70920	24	$\theta_2(x)$
	67	1.68728	1.80900	1.88898	1.82246	23	
	68	1.72098	1.85820	1.96299	1.93792	22	
	69	1.75382	1.90646	2.03642	2.05509	21	
	70	1.78571	1.95365	2.10899	2.17339	20	
	71	1.81659	1.99962	2.18042	2.29223	19	
	72	1.84637	2.04423	2.25042	2.41099	18	
	73	1.87499	2.08734	2.31871	2.52898	17	
	74	1.90239	2.12882	2.38499	2.64551	16	
	75	1.92848	2.16854	2.44897	2.75987	15	
	76	1.95322	2.20637	2.51039	2.87132	14	
	77	1.97654	2.24218	2.56895	2.97913	13	
	78	1.99837	2.27586	2.62439	3.08255	12	
	79	2.01868	2.30729	2.67645	3.18087	11	
	80	2.03739	2.33636	2.72488	3.27338	10	
	81	2.05448	2.36299	2.76946	3.35939	9	
	82	2.06988	2.38706	2.80996	3.43827	8	
	83	2.08358	2.40851	2.84619	3.50940	7	
	84	2.09551	2.42725	2.87797	3.57223	6	
	85	2.10567	2.44323	2.90513	3.62628	5	
	86	2.11401	2.45638	2.92755	3.67111	4	
	87	2.12053	2.46665	2.94510	3.70635	3	
	88	2.12519	2.47402	2.95771	3.73174	2	
	89	2.12800	2.47845	2.96529	3.74705	1	
	90	2.12893	2.47993	2.96783	3.75217	0	

Theta Functions $\theta_3(x)$ and $\theta_4(x)$

$q \rightarrow$

x^0	0.1	0.2	0.3	0.4	x^0
0	1.20020	1.40320	1.61624	1.85173	90
1	1.20008	1.40295	1.61583	1.85111	89
2	1.19971	1.40220	1.61462	1.84927	88
3	1.19910	1.40094	1.61260	1.84620	87
4	1.19825	1.39918	1.60977	1.84191	86
5	1.19715	1.39693	1.60614	1.83641	85
6	1.19581	1.39418	1.60172	1.82972	84
7	1.19424	1.39094	1.59651	1.82183	83
8	1.19242	1.38722	1.59052	1.81278	82
9	1.19037	1.38301	1.58376	1.80258	81
10	1.18809	1.37833	1.57625	1.79124	80
11	1.18558	1.37318	1.56798	1.77879	79
12	1.18284	1.36756	1.55898	1.76526	78
13	1.17988	1.36149	1.54926	1.75067	77
14	1.17670	1.35497	1.53883	1.73504	76
15	1.17331	1.34801	1.52772	1.71842	75
16	1.16970	1.34062	1.51593	1.70083	74
17	1.16588	1.33281	1.50348	1.68230	73
18	1.16187	1.32460	1.49040	1.66287	72
19	1.15765	1.31598	1.47671	1.64258	71
20	1.15324	1.30697	1.46242	1.62146	70
21	1.14865	1.29759	1.44756	1.59956	69
22	1.14387	1.28785	1.43214	1.57691	68
23	1.13892	1.27775	1.41620	1.55355	67
24	1.13381	1.26732	1.39975	1.52953	66
25	1.12852	1.25656	1.38283	1.50488	65
26	1.12308	1.24549	1.36544	1.47966	64
27	1.11750	1.23412	1.34763	1.45391	63
28	1.11176	1.22248	1.32941	1.42766	62
29	1.10590	1.21056	1.31081	1.40097	61
30	1.09990	1.19840	1.29186	1.37388	60
31	1.09378	1.18600	1.27258	1.34642	59
32	1.08755	1.17338	1.25301	1.31866	58
33	1.08121	1.16055	1.23316	1.29063	57
34	1.07478	1.14754	1.21307	1.26238	56
35	1.06825	1.13436	1.19277	1.23394	55
36	1.06164	1.12102	1.17227	1.20537	54
37	1.05496	1.10754	1.15161	1.17670	53
38	1.04821	1.09394	1.13082	1.14798	52
39	1.04140	1.08024	1.10992	1.11925	51
40	1.03454	1.06645	1.08895	1.09054	50
41	1.02764	1.05259	1.06792	1.06191	49
42	1.02071	1.03868	1.04686	1.03338	48
43	1.01375	1.02473	1.02580	1.00500	47
44	1.00678	1.01077	1.00478	0.97679	46
45	0.99980	0.99680	0.98380	0.94880	45

$\theta_3(x)$

$\theta_4(x)$

Table A. (*Continued*)

$$q \rightarrow$$

	x^0	0.1	0.2	0.3	0.4	x^0	
	46	0.99282	0.98285	0.96290	0.92106	44	
	47	0.98585	0.96893	0.94211	0.89360	43	
	48	0.97890	0.95506	0.92145	0.86646	42	
	49	0.97197	0.94126	0.90094	0.83966	41	
	50	0.96508	0.92753	0.88061	0.81323	40	
	51	0.95823	0.91391	0.86048	0.78721	39	
	52	0.95144	0.90041	0.84057	0.76161	38	
	53	0.94470	0.88703	0.82091	0.73646	37	
	54	0.93803	0.87381	0.80152	0.71179	36	
	55	0.93144	0.86074	0.78241	0.68762	35	
	56	0.92493	0.84786	0.76362	0.66396	34	
	57	0.91852	0.83517	0.74516	0.64085	33	
	58	0.91220	0.82268	0.72704	0.61829	32	
	59	0.90599	0.81042	0.70930	0.59631	31	
	60	0.89990	0.79840	0.69194	0.57492	30	
	61	0.89393	0.78663	0.67499	0.55414	29	
	62	0.88809	0.77513	0.65845	0.53398	28	
	63	0.88238	0.76390	0.64236	0.51445	27	
	64	0.87682	0.75296	0.62672	0.49556	26	
	65	0.87141	0.74233	0.61155	0.47733	25	
	66	0.86615	0.73201	0.59686	0.45977	24	
	67	0.86106	0.72203	0.58267	0.44288	23	
$\theta_3(x)$	68	0.85614	0.71238	0.56899	0.42666	22	$\theta_4(x)$
	69	0.85139	0.70308	0.55583	0.41114	21	
	70	0.84683	0.69414	0.54321	0.39632	20	
	71	0.84245	0.68557	0.53113	0.38219	19	
	72	0.83826	0.67738	0.51961	0.36877	18	
	73	0.83427	0.66958	0.50865	0.35606	17	
	74	0.83048	0.66218	0.49828	0.34406	16	
	75	0.82689	0.65519	0.48848	0.33278	15	
	76	0.82352	0.64861	0.47929	0.32222	14	
	77	0.82036	0.64245	0.47069	0.31238	13	
	78	0.81742	0.63672	0.46270	0.30326	12	
	79	0.81471	0.63143	0.45533	0.29487	11	
	80	0.81221	0.62657	0.44857	0.28721	10	
	81	0.80995	0.62217	0.44245	0.28027	9	
	82	0.80792	0.61821	0.43696	0.27406	8	
	83	0.80612	0.61471	0.43210	0.26858	7	
	84	0.80455	0.61166	0.42788	0.26383	6	
	85	0.80323	0.60908	0.42430	0.25981	5	
	86	0.80214	0.60697	0.42138	0.25652	4	
	87	0.80129	0.60532	0.41910	0.25397	3	
	88	0.80069	0.60414	0.41747	0.25214	2	
	89	0.80032	0.60343	0.41649	0.25104	1	
	90	0.80020	0.60320	0.41616	0.25068	0	

Table A. (*Continued*)

$q \rightarrow$

$\theta_3(x)$ ↓ $\theta_4(x)$ ↑

x^0	0.5	0.6	0.7	0.8	x^0
0	2.12894	2.47993	2.96783	3.75217	90
1	2.12800	2.47845	2.96529	3.74705	89
2	2.12520	2.47402	2.95771	3.73174	88
3	2.12053	2.46665	2.94510	3.70635	87
4	2.11402	2.45638	2.92755	3.67111	86
5	2.10568	2.44323	2.90513	3.62628	85
6	2.09552	2.42725	2.87797	3.57223	84
7	2.08358	2.40851	2.84619	3.50940	83
8	2.06989	2.38706	2.80996	3.43827	82
9	2.05449	2.36299	2.76946	3.35939	81
10	2.03741	2.33636	2.72488	3.27338	80
11	2.01869	2.30729	2.67645	3.18087	79
12	1.99839	2.27586	2.62439	3.08255	78
13	1.97656	2.24218	2.56895	2.97913	77
14	1.95324	2.20637	2.51039	2.87132	76
15	1.92851	2.16854	2.44897	2.75987	75
16	1.90242	2.12882	2.38499	2.64551	74
17	1.87503	2.08734	2.31871	2.52898	73
18	1.84641	2.04423	2.25042	2.41099	72
19	1.81664	1.99962	2.18042	2.29223	71
20	1.78577	1.95365	2.10899	2.17339	70
21	1.75388	1.90646	2.03642	2.05509	69
22	1.72106	1.85820	1.96299	1.93792	68
23	1.68736	1.80900	1.88898	1.82246	67
24	1.65287	1.75900	1.81466	1.70920	66
25	1.61767	1.70835	1.74029	1.59861	65
26	1.58183	1.65718	1.66611	1.49110	64
27	1.54542	1.60562	1.59238	1.38703	63
28	1.50852	1.55381	1.51931	1.28670	62
29	1.47122	1.50189	1.44712	1.19038	61
30	1.43358	1.44997	1.37601	1.09826	60
31	1.39568	1.39817	1.30616	1.01051	59
32	1.35759	1.34662	1.23774	0.92724	58
33	1.31938	1.29542	1.17090	0.84851	57
34	1.28112	1.24469	1.10578	0.77434	56
35	1.24289	1.19451	1.04250	0.70473	55
36	1.20474	1.14500	0.98116	0.63963	54
37	1.16674	1.09623	0.92186	0.57897	53
38	1.12896	1.04828	0.86466	0.52262	52
39	1.09144	1.00124	0.80963	0.47048	51
40	1.05426	0.95517	0.75681	0.42238	50
41	1.01745	0.91013	0.70622	0.37816	49
42	0.98108	0.86619	0.65790	0.33765	48
43	0.94519	0.82338	0.61183	0.30066	47
44	0.90983	0.78176	0.56802	0.26699	46
45	0.87503	0.74136	0.52644	0.23645	45

Table A. (*Continued*)

$$q \rightarrow$$

x^0	0.5	0.6	0.7	0.8	x^0
46	0.84084	0.70222	0.48708	0.20883	44
47	0.80730	0.66435	0.44989	0.18393	43
48	0.77444	0.62777	0.41483	0.16156	42
49	0.74228	0.59251	0.38185	0.14152	41
50	0.71087	0.55856	0.35089	0.12363	40
51	0.68021	0.52593	0.32189	0.10771	39
52	0.65034	0.49463	0.29479	0.09358	38
53	0.62127	0.46464	0.26950	0.08108	37
54	0.59303	0.43596	0.24597	0.07006	36
55	0.56561	0.40856	0.22410	0.06038	35
56	0.53905	0.38245	0.20384	0.05189	34
57	0.51333	0.35758	0.18509	0.04447	33
58	0.48848	0.33395	0.16777	0.03801	32
59	0.46450	0.31152	0.15182	0.03240	31
60	0.44139	0.29027	0.13716	0.02754	30
61	0.41915	0.27017	0.12369	0.02335	29
62	0.39778	0.25118	0.11136	0.01974	28
63	0.37728	0.23327	0.10009	0.01664	27
64	0.35764	0.21642	0.08981	0.01399	26
65	0.33886	0.20057	0.08045	0.01173	25
66	0.32093	0.18571	0.07195	0.00981	24
67	0.30385	0.17178	0.06424	0.00818	23
68	0.28761	0.15877	0.05726	0.00681	22
69	0.27219	0.14662	0.05096	0.00565	21
70	0.25759	0.13531	0.04528	0.00467	20
71	0.24379	0.12480	0.04017	0.00385	19
72	0.23079	0.11505	0.03559	0.00317	18
73	0.21858	0.10604	0.03149	0.00260	17
74	0.20714	0.09773	0.02783	0.00213	16
75	0.19646	0.09009	0.02457	0.00174	15
76	0.18653	0.08309	0.02167	0.00141	14
77	0.17734	0.07670	0.01911	0.00115	13
78	0.16889	0.07090	0.01685	0.00093	12
79	0.16115	0.06566	0.01486	0.00075	11
80	0.15412	0.06095	0.01313	0.00061	10
81	0.14778	0.05676	0.01162	0.00049	9
82	0.14214	0.05306	0.01033	0.00039	8
83	0.13719	0.04984	0.00923	0.00032	7
84	0.13291	0.04708	0.00830	0.00026	6
85	0.12929	0.04477	0.00754	0.00021	5
86	0.12635	0.04290	0.00693	0.00018	4
87	0.12406	0.04145	0.00646	0.00015	3
88	0.12243	0.04042	0.00614	0.00013	2
89	0.12145	0.03981	0.00594	0.00012	1
90	0.12112	0.03960	0.00588	0.00012	0

$\theta_3(x)$

$\theta_4(x)$

Table B. Nome and Complete
Integrals of the First and Second
Kinds as Functions of the
Squared Modulus

k^2	q	K	E	k'^2
0.00	0.00000	1.57080	1.57080	1.00
0.01	0.00063	1.57475	1.56686	0.99
0.02	0.00126	1.57874	1.56291	0.98
0.03	0.00190	1.58278	1.55895	0.97
0.04	0.00255	1.58687	1.55497	0.96
0.05	0.00321	1.59100	1.55097	0.95
0.06	0.00387	1.59519	1.54696	0.94
0.07	0.00454	1.59942	1.54294	0.93
0.08	0.00521	1.60371	1.53889	0.92
0.09	0.00589	1.60805	1.53483	0.91
0.10	0.00658	1.61244	1.53076	0.90
0.11	0.00728	1.61689	1.52667	0.89
0.12	0.00799	1.62139	1.52256	0.88
0.13	0.00870	1.62595	1.51843	0.87
0.14	0.00943	1.63058	1.51428	0.86
0.15	0.01016	1.63526	1.51012	0.85
0.16	0.01090	1.64000	1.50594	0.84
0.17	0.01164	1.64481	1.50174	0.83
0.18	0.01240	1.64968	1.49753	0.82
0.19	0.01317	1.65462	1.49329	0.81
0.20	0.01394	1.65962	1.48904	0.80
0.21	0.01473	1.66470	1.48476	0.79
0.22	0.01552	1.66985	1.48047	0.78
0.23	0.01633	1.67507	1.47615	0.77
0.24	0.01715	1.68037	1.47182	0.76
0.25	0.01797	1.68575	1.46746	0.75
0.26	0.01881	1.69121	1.46309	0.74
0.27	0.01966	1.69675	1.45869	0.73
0.28	0.02052	1.70237	1.45427	0.72
0.29	0.02139	1.70809	1.44983	0.71
0.30	0.02228	1.71389	1.44536	0.70
0.31	0.02317	1.71978	1.44088	0.69
0.32	0.02409	1.72578	1.43637	0.68
0.33	0.02501	1.73186	1.43183	0.67
0.34	0.02595	1.73806	1.42727	0.66
0.35	0.02690	1.74435	1.42269	0.65
0.36	0.02786	1.75075	1.41808	0.64
0.37	0.02885	1.75727	1.41345	0.63
0.38	0.02984	1.76390	1.40879	0.62
0.39	0.03085	1.77065	1.40411	0.61
0.40	0.03188	1.77752	1.39939	0.60
0.41	0.03293	1.78452	1.39465	0.59
0.42	0.03399	1.79165	1.38988	0.58
0.43	0.03507	1.79892	1.38509	0.57
0.44	0.03618	1.80633	1.38026	0.56
0.45	0.03730	1.81388	1.37540	0.55
0.46	0.03844	1.82159	1.37051	0.54
0.47	0.03960	1.82946	1.36560	0.53
0.48	0.04078	1.83749	1.36064	0.52
0.49	0.04199	1.84569	1.35566	0.51
0.50	0.04321	1.85407	1.35064	0.50

Table B. (*Continued*)

k^2	q	K	E	k'^2
0.51	0.04447	1.86264	1.34559	0.49
0.52	0.04575	1.87140	1.34051	0.48
0.53	0.04705	1.88036	1.33538	0.47
0.54	0.04838	1.88953	1.33022	0.46
0.55	0.04974	1.89892	1.32502	0.45
0.56	0.05113	1.90855	1.31979	0.44
0.57	0.05255	1.91841	1.31451	0.43
0.58	0.05401	1.92853	1.30919	0.42
0.59	0.05550	1.93891	1.30383	0.41
0.60	0.05702	1.94957	1.29843	0.40
0.61	0.05858	1.96052	1.29298	0.39
0.62	0.06018	1.97178	1.28748	0.38
0.63	0.06182	1.98337	1.28194	0.37
0.64	0.06351	1.99530	1.27635	0.36
0.65	0.06524	2.00760	1.27071	0.35
0.66	0.06702	2.02028	1.26501	0.34
0.67	0.06885	2.03337	1.25926	0.33
0.68	0.07074	2.04689	1.25346	0.32
0.69	0.07268	2.06088	1.24759	0.31
0.70	0.07469	2.07536	1.24167	0.30
0.71	0.07676	2.09037	1.23568	0.29
0.72	0.07890	2.10595	1.22963	0.28
0.73	0.08112	2.12213	1.22351	0.27
0.74	0.08341	2.13897	1.21732	0.26
0.75	0.08580	2.15652	1.21106	0.25
0.76	0.08827	2.17483	1.20471	0.24
0.77	0.09085	2.19397	1.19829	0.23
0.78	0.09353	2.21402	1.19178	0.22
0.79	0.09634	2.23507	1.18518	0.21
0.80	0.09927	2.25721	1.17849	0.20
0.81	0.10235	2.28055	1.17170	0.19
0.82	0.10559	2.30523	1.16480	0.18
0.83	0.10900	2.33141	1.15779	0.17
0.84	0.11261	2.35926	1.15066	0.16
0.85	0.11644	2.38902	1.14340	0.15
0.86	0.12052	2.42093	1.13600	0.14
0.87	0.12488	2.45534	1.12845	0.13
0.88	0.12957	2.49264	1.12074	0.12
0.89	0.13465	2.53333	1.11286	0.11
0.90	0.14017	2.57809	1.10477	0.10
0.91	0.14624	2.62777	1.09648	0.09
0.92	0.15298	2.68355	1.08794	0.08
0.93	0.16055	2.74707	1.07912	0.07
0.94	0.16921	2.82075	1.06999	0.06
0.95	0.17932	2.90834	1.06047	0.05
0.96	0.19150	3.01611	1.05050	0.04
0.97	0.20688	3.15587	1.03995	0.03
0.98	0.22793	3.35414	1.02859	0.02
0.99	0.26220	3.69564	1.01599	0.01
1.00	1.00000	∞	1.00000	0.00

Table C. Jacobi's Functions sn x,
cn x, and dn x for a Range of
Values of the Modulus

$$k = 0.1$$

x	sn x	cn x	dn x
0.0	0.00000	1.00000	1.00000
0.1	0.09983	0.99500	0.99995
0.2	0.19866	0.98007	0.99980
0.3	0.29548	0.95535	0.99956
0.4	0.38932	0.92110	0.99924
0.5	0.47925	0.87768	0.99885
0.6	0.56437	0.82552	0.99841
0.7	0.64382	0.76518	0.99793
0.8	0.71683	0.69724	0.99743
0.9	0.78268	0.62242	0.99693
1.0	0.84073	0.54145	0.99646
1.1	0.89042	0.45515	0.99603
1.2	0.93126	0.36437	0.99565
1.3	0.96286	0.27001	0.99535
1.4	0.98492	0.17300	0.99514
1.5	0.99724	0.07430	0.99502

$$K = 1.57475$$

$$k = 0.2$$

x	sn x	cn x	dn x
0.0	0.00000	1.00000	1.00000
0.1	0.09983	0.99500	0.99980
0.2	0.19862	0.98008	0.99921
0.3	0.29535	0.95539	0.99825
0.4	0.38904	0.92122	0.99697
0.5	0.47873	0.87796	0.99541
0.6	0.56354	0.82609	0.99363
0.7	0.64263	0.76618	0.99171
0.8	0.71526	0.69886	0.98972
0.9	0.78075	0.62484	0.98773
1.0	0.83851	0.54488	0.98584
1.1	0.88803	0.45978	0.98410
1.2	0.92888	0.37038	0.98259
1.3	0.96072	0.27753	0.98137
1.4	0.98328	0.18211	0.98047
1.5	0.99638	0.08501	0.97994

$$K = 1.58687$$

Table C. (*Continued*)

$$k = 0.3$$

x	sn x	cn x	dn x
0.0	0.00000	1.00000	1.00000
0.1	0.09982	0.99501	0.99955
0.2	0.19855	0.98009	0.99822
0.3	0.29514	0.95545	0.99607
0.4	0.38856	0.92142	0.99318
0.5	0.47786	0.87844	0.98967
0.6	0.56215	0.82703	0.98568
0.7	0.64065	0.76783	0.98136
0.8	0.71264	0.70153	0.97688
0.9	0.77752	0.62885	0.97242
1.0	0.83479	0.55057	0.96813
1.1	0.88400	0.46748	0.96419
1.2	0.92483	0.38037	0.96074
1.3	0.95701	0.29004	0.95790
1.4	0.98035	0.19729	0.95577
1.5	0.99469	0.10291	0.95444
1.6	0.99997	0.00768	0.95394

$$K = 1.60805$$

$$k = 0.4$$

x	sn x	cn x	dn x
0.0	0.00000	1.00000	1.00000
0.1	0.09981	0.99501	0.99920
0.2	0.19846	0.98011	0.99684
0.3	0.29485	0.95555	0.99302
0.4	0.38790	0.92170	0.98789
0.5	0.47665	0.87910	0.98166
0.6	0.56022	0.82834	0.97457
0.7	0.63787	0.77014	0.96690
0.8	0.70896	0.70525	0.95895
0.9	0.77298	0.63444	0.95100
1.0	0.82951	0.55849	0.94335
1.1	0.87826	0.47819	0.93626
1.2	0.91899	0.39428	0.92999
1.3	0.95155	0.30749	0.92473
1.4	0.97583	0.21852	0.92067
1.5	0.99177	0.12803	0.91794
1.6	0.99933	0.03665	0.91663

$$K = 1.64000$$

Table C. (*Continued*)

$$k = 0.5$$

x	sn x	cn x	dn x
0.0	0.00000	1.00000	1.00000
0.1	0.09979	0.99501	0.99875
0.2	0.19835	0.98013	0.99507
0.3	0.29447	0.95566	0.98910
0.4	0.38704	0.92206	0.98110
0.5	0.47508	0.87994	0.97138
0.6	0.55773	0.83002	0.96033
0.7	0.63429	0.77309	0.94838
0.8	0.70421	0.70999	0.93596
0.9	0.76709	0.64155	0.92352
1.0	0.82264	0.56857	0.91149
1.1	0.87069	0.49183	0.90026
1.2	0.91117	0.41202	0.89019
1.3	0.94405	0.32981	0.88159
1.4	0.96933	0.24576	0.87470
1.5	0.98705	0.16040	0.86973
1.6	0.99724	0.07422	0.86682

$$K = 1.68575$$

$$k = 0.6$$

x	sn x	cn x	dn x
0.0	0.00000	1.00000	1.00000
0.1	0.09977	0.99501	0.99821
0.2	0.19820	0.98016	0.99290
0.3	0.29400	0.95580	0.98432
0.4	0.38600	0.92250	0.97281
0.5	0.47318	0.88097	0.95885
0.6	0.55470	0.83205	0.94299
0.7	0.62992	0.77666	0.92583
0.8	0.69839	0.71572	0.90797
0.9	0.75982	0.65013	0.89003
1.0	0.81409	0.58074	0.87259
1.1	0.86118	0.50830	0.85616
1.2	0.90116	0.43348	0.84122
1.3	0.93416	0.35686	0.82816
1.4	0.96032	0.27890	0.81731
1.5	0.97980	0.20000	0.80895
1.6	0.99272	0.12047	0.80326
1.7	0.99918	0.04060	0.80037

$$K = 1.75075$$

Table C. (*Continued*)

k = 0.7

x	sn x	cn x	dn x
0.0	0.00000	1.00000	1.00000
0.1	0.09975	0.99501	0.99756
0.2	0.19803	0.98020	0.99035
0.3	0.29345	0.95597	0.97867
0.4	0.38477	0.92301	0.96305
0.5	0.47092	0.88217	0.94410
0.6	0.55111	0.83443	0.92259
0.7	0.62474	0.78083	0.89931
0.8	0.69147	0.72241	0.87505
0.9	0.75115	0.66013	0.85061
1.0	0.80380	0.59490	0.82669
1.1	0.84957	0.52747	0.80395
1.2	0.88870	0.45849	0.78294
1.3	0.92147	0.38846	0.76416
1.4	0.94817	0.31777	0.74798
1.5	0.96909	0.24669	0.73472
1.6	0.98450	0.17541	0.72462
1.7	0.99457	0.10404	0.71785
1.8	0.99947	0.03263	0.71451

K = 1.84569

k = 0.8

x	sn x	cn x	dn x
0.0	0.00000	1.00000	1.00000
0.1	0.09973	0.99501	0.99681
0.2	0.19784	0.98023	0.98740
0.3	0.29282	0.95617	0.97217
0.4	0.38335	0.92360	0.95181
0.5	0.46833	0.88355	0.92716
0.6	0.54697	0.83715	0.89918
0.7	0.61876	0.78558	0.86889
0.8	0.68345	0.73000	0.83729
0.9	0.74103	0.67147	0.80533
1.0	0.79168	0.61093	0.77387
1.1	0.83571	0.54918	0.74365
1.2	0.87350	0.48682	0.71532
1.3	0.90550	0.42434	0.68938
1.4	0.93215	0.36206	0.66626
1.5	0.95388	0.30018	0.64627
1.6	0.97107	0.23878	0.62967
1.7	0.98405	0.17787	0.61664
1.8	0.99309	0.11739	0.60730
1.9	0.99836	0.05721	0.60174

K = 1.99530

Table C. (*Continued*)

$$k = 0.9$$

x	sn x	cn x	dn x
0.0	0.00000	1.00000	1.00000
0.1	0.09970	0.99502	0.99597
0.2	0.19762	0.98028	0.98406
0.3	0.29211	0.95638	0.96482
0.4	0.38174	0.92427	0.93913
0.5	0.46539	0.88510	0.90805
0.6	0.54228	0.84020	0.87281
0.7	0.61197	0.79088	0.83466
0.8	0.67431	0.73845	0.79480
0.9	0.72943	0.68406	0.75434
1.0	0.77764	0.62871	0.71426
1.1	0.81941	0.57321	0.67538
1.2	0.85526	0.51820	0.63837
1.3	0.88576	0.46414	0.60374
1.4	0.91149	0.41133	0.57188
1.5	0.93298	0.35993	0.54308
1.6	0.95074	0.30999	0.51753
1.7	0.96520	0.26149	0.49537
1.8	0.97676	0.21432	0.47666
1.9	0.98573	0.16832	0.46146
2.0	0.99237	0.12327	0.44979
2.1	0.99688	0.07896	0.44165
2.2	0.99938	0.03513	0.43704

$$K = 2.28055$$

Table D. Legendre's Incomplete Integrals of First
and Second Kinds

First Kind, $F(\phi, k)$

$k \rightarrow$

ϕ^0	0.1	0.2	0.3	0.4	0.5
0	0.00000	0.00000	0.00000	0.00000	0.00000
1	0.01745	0.01745	0.01745	0.01745	0.01745
2	0.03491	0.03491	0.03491	0.03491	0.03491
3	0.05236	0.05236	0.05236	0.05236	0.05237
4	0.06981	0.06982	0.06982	0.06982	0.06983
5	0.08727	0.08727	0.08728	0.08728	0.08729
6	0.10472	0.10473	0.10474	0.10475	0.10477
7	0.12218	0.12219	0.12220	0.12222	0.12225
8	0.13963	0.13964	0.13967	0.13970	0.13974
9	0.15709	0.15711	0.15714	0.15718	0.15724
10	0.17454	0.17457	0.17461	0.17467	0.17475
11	0.19200	0.19203	0.19209	0.19217	0.19228
12	0.20945	0.20950	0.20958	0.20968	0.20982
13	0.22691	0.22697	0.22707	0.22720	0.22738
14	0.24437	0.24444	0.24456	0.24473	0.24495
15	0.26183	0.26192	0.26207	0.26227	0.26254
16	0.27929	0.27940	0.27958	0.27983	0.28015
17	0.29675	0.29688	0.29709	0.29739	0.29779
18	0.31421	0.31436	0.31462	0.31498	0.31544
19	0.33167	0.33185	0.33215	0.33257	0.33312
20	0.34914	0.34934	0.34969	0.35018	0.35082
21	0.36660	0.36684	0.36724	0.36781	0.36855
22	0.38406	0.38434	0.38480	0.38545	0.38630
23	0.40153	0.40184	0.40237	0.40311	0.40408
24	0.41900	0.41935	0.41995	0.42079	0.42189
25	0.43647	0.43687	0.43754	0.43849	0.43973
26	0.45394	0.45439	0.45514	0.45621	0.45761
27	0.47141	0.47191	0.47275	0.47395	0.47551
28	0.48888	0.48944	0.49038	0.49171	0.49345
29	0.50635	0.50697	0.50801	0.50949	0.51142
30	0.52383	0.52451	0.52566	0.52729	0.52943
31	0.54130	0.54205	0.54332	0.54511	0.54747
32	0.55878	0.55960	0.56099	0.56296	0.56555
33	0.57626	0.57716	0.57867	0.58083	0.58367
34	0.59374	0.59472	0.59637	0.59873	0.60183
35	0.61122	0.61228	0.61408	0.61665	0.62003
36	0.62870	0.62986	0.63181	0.63459	0.63827
37	0.64619	0.64743	0.64954	0.65256	0.65655
38	0.66367	0.66502	0.66730	0.67056	0.67487
39	0.68116	0.68261	0.68506	0.68858	0.69324
40	0.69865	0.70020	0.70284	0.70662	0.71165
41	0.71614	0.71781	0.72064	0.72470	0.73010
42	0.73363	0.73542	0.73845	0.74280	0.74860
43	0.75112	0.75303	0.75627	0.76093	0.76714
44	0.76862	0.77065	0.77411	0.77908	0.78573

Table D. (*Continued*)

ϕ^0	0.1	0.2	0.3	0.4	0.5
45	0.78611	0.78828	0.79196	0.79726	0.80437
46	0.80361	0.80591	0.80983	0.81547	0.82305
47	0.82111	0.82355	0.82771	0.83371	0.84178
48	0.83861	0.84120	0.84560	0.85198	0.86055
49	0.85611	0.85885	0.86352	0.87027	0.87937
50	0.87362	0.87651	0.88144	0.88859	0.89825
51	0.89112	0.89417	0.89938	0.90694	0.91716
52	0.90863	0.91185	0.91734	0.92532	0.93613
53	0.92614	0.92952	0.93531	0.94372	0.95514
54	0.94365	0.94721	0.95329	0.96215	0.97420
55	0.96116	0.96490	0.97129	0.98061	0.99331
56	0.97867	0.98259	0.98930	0.99910	1.01247
57	0.99619	1.00029	1.00733	1.01761	1.03167
58	1.01370	1.01800	1.02537	1.03615	1.05092
59	1.03122	1.03571	1.04342	1.05472	1.07021
60	1.04874	1.05343	1.06149	1.07331	1.08955
61	1.06626	1.07115	1.07957	1.09193	1.10894
62	1.08378	1.08888	1.09766	1.11057	1.12837
63	1.10130	1.10662	1.11577	1.12924	1.14784
64	1.11883	1.12436	1.13389	1.14793	1.16735
65	1.13635	1.14210	1.15202	1.16665	1.18691
66	1.15388	1.15985	1.17016	1.18539	1.20651
67	1.17140	1.17761	1.18832	1.20415	1.22615
68	1.18893	1.19537	1.20648	1.22293	1.24583
69	1.20646	1.21313	1.22466	1.24174	1.26555
70	1.22399	1.23090	1.24284	1.26056	1.28530
71	1.24152	1.24867	1.26104	1.27941	1.30509
72	1.25906	1.26645	1.27924	1.29827	1.32491
73	1.27659	1.28423	1.29746	1.31715	1.34477
74	1.29412	1.30201	1.31568	1.33605	1.36466
75	1.31166	1.31980	1.33391	1.35496	1.38457
76	1.32919	1.33759	1.35215	1.37389	1.40452
77	1.34673	1.35538	1.37040	1.39284	1.42449
78	1.36427	1.37317	1.38865	1.41180	1.44449
79	1.38180	1.39097	1.40691	1.43077	1.46451
80	1.39934	1.40877	1.42518	1.44975	1.48455
81	1.41688	1.42658	1.44345	1.46875	1.50462
82	1.43442	1.44438	1.46173	1.48775	1.52470
83	1.45196	1.46219	1.48001	1.50676	1.54479
84	1.46950	1.48000	1.49829	1.52578	1.56490
85	1.48704	1.49781	1.51658	1.54481	1.58503
86	1.50458	1.51562	1.53487	1.56384	1.60516
87	1.52212	1.53343	1.55316	1.58288	1.62530
88	1.53966	1.55124	1.57146	1.60192	1.64545
89	1.55720	1.56905	1.58975	1.62096	1.66560
90	1.57475	1.58687	1.60805	1.64000	1.68575

Table D. (*Continued*)

First Kind, $F(\phi, k)$

$k \rightarrow$

ϕ^0	0.6	0.7	0.8	0.9
0	0.00000	0.00000	0.00000	0.00000
1	0.01745	0.01745	0.01745	0.01745
2	0.03491	0.03491	0.03491	0.03491
3	0.05237	0.05237	0.05238	0.05238
4	0.06983	0.06984	0.06985	0.06986
5	0.08731	0.08732	0.08734	0.08736
6	0.10479	0.10481	0.10484	0.10488
7	0.12228	0.12232	0.12237	0.12242
8	0.13979	0.13985	0.13992	0.14000
9	0.15731	0.15740	0.15749	0.15760
10	0.17485	0.17497	0.17510	0.17525
11	0.19241	0.19256	0.19274	0.19295
12	0.20999	0.21019	0.21042	0.21069
13	0.22759	0.22785	0.22814	0.22848
14	0.24522	0.24554	0.24591	0.24634
15	0.26287	0.26327	0.26372	0.26425
16	0.28056	0.28103	0.28159	0.28223
17	0.29827	0.29884	0.29951	0.30028
18	0.31601	0.31670	0.31750	0.31841
19	0.33379	0.33460	0.33554	0.33663
20	0.35161	0.35255	0.35365	0.35492
21	0.36946	0.37055	0.37183	0.37332
22	0.38735	0.38861	0.39009	0.39180
23	0.40528	0.40672	0.40842	0.41039
24	0.42326	0.42490	0.42684	0.42909
25	0.44128	0.44314	0.44534	0.44791
26	0.45934	0.46144	0.46393	0.46684
27	0.47746	0.47981	0.48261	0.48590
28	0.49562	0.49826	0.50140	0.50509
29	0.51384	0.51677	0.52028	0.52441
30	0.53211	0.53537	0.53927	0.54388
31	0.55043	0.55404	0.55837	0.56350
32	0.56881	0.57279	0.57758	0.58328
33	0.58725	0.59163	0.59691	0.60322
34	0.60574	0.61055	0.61636	0.62333
35	0.62430	0.62956	0.63594	0.64362
36	0.64292	0.64867	0.65565	0.66409
37	0.66161	0.66786	0.67549	0.68475
38	0.68036	0.68715	0.69548	0.70562
39	0.69917	0.70655	0.71560	0.72669
40	0.71805	0.72604	0.73588	0.74797
41	0.73700	0.74563	0.75631	0.76948
42	0.75603	0.76533	0.77689	0.79123
43	0.77512	0.78514	0.79763	0.81321
44	0.79428	0.80506	0.81854	0.83545

Table D. (*Continued*)

ϕ^0	0.6	0.7	0.8	0.9
45	0.81352	0.82509	0.83962	0.85794
46	0.83283	0.84523	0.86088	0.88070
47	0.85222	0.86549	0.88231	0.90374
48	0.87168	0.88587	0.90392	0.92707
49	0.89121	0.90637	0.92572	0.95070
50	0.91083	0.92698	0.94771	0.97464
51	0.93052	0.94772	0.96989	0.99890
52	0.95029	0.96858	0.99228	1.02348
53	0.97013	0.98957	1.01486	1.04841
54	0.99006	1.01068	1.03765	1.07369
55	1.01006	1.03192	1.06065	1.09934
56	1.03014	1.05329	1.08387	1.12536
57	1.05030	1.07479	1.10730	1.15177
58	1.07053	1.09641	1.13094	1.17858
59	1.09084	1.11816	1.15481	1.20580
60	1.11123	1.14004	1.17890	1.23345
61	1.13170	1.16205	1.20322	1.26152
62	1.15224	1.18419	1.22776	1.29005
63	1.17286	1.20646	1.25253	1.31903
64	1.19354	1.22885	1.27753	1.34848
65	1.21431	1.25137	1.30276	1.37840
66	1.23514	1.27401	1.32822	1.40882
67	1.25604	1.29677	1.35390	1.43973
68	1.27701	1.31965	1.37981	1.47115
69	1.29805	1.34265	1.40595	1.50308
70	1.31915	1.36577	1.43230	1.53552
71	1.34031	1.38899	1.45888	1.56849
72	1.36153	1.41233	1.48567	1.60199
73	1.38282	1.43577	1.51267	1.63600
74	1.40415	1.45932	1.53987	1.67054
75	1.42554	1.48296	1.56727	1.70560
76	1.44699	1.50669	1.59486	1.74117
77	1.46848	1.53052	1.62264	1.77724
78	1.49001	1.55442	1.65058	1.81379
79	1.51159	1.57841	1.67870	1.85082
80	1.53320	1.60247	1.70696	1.88829
81	1.55485	1.62659	1.73537	1.92619
82	1.57654	1.65078	1.76391	1.96449
83	1.59825	1.67502	1.79257	2.00315
84	1.61999	1.69931	1.82133	2.04214
85	1.64175	1.72364	1.85018	2.08141
86	1.66353	1.74801	1.87911	2.12094
87	1.68532	1.77241	1.90811	2.16066
88	1.70713	1.79682	1.93715	2.20054
89	1.72894	1.82126	1.96622	2.24052
90	1.75075	1.84569	1.99530	2.28055

Table D. (*Continued*)

Second Kind, $D(\phi, k)$

$k \rightarrow$

ϕ^0	0.1	0.2	0.3	0.4	0.5
0	0.00000	0.00000	0.00000	0.00000	0.000000
1	0.01745	0.01745	0.01745	0.01745	0.01745
2	0.03491	0.03491	0.03491	0.03491	0.03490
3	0.05236	0.05236	0.05236	0.05236	0.05235
4	0.06981	0.06981	0.06981	0.06980	0.06980
5	0.08727	0.08726	0.08726	0.08725	0.08724
6	0.10472	0.10471	0.10470	0.10469	0.10467
7	0.12217	0.12216	0.12215	0.12212	0.12210
8	0.13962	0.13961	0.13959	0.13955	0.13951
9	0.15707	0.15705	0.15702	0.15698	0.15692
10	0.17452	0.17450	0.17445	0.17439	0.17431
11	0.19197	0.19194	0.19188	0.19180	0.19169
12	0.20942	0.20938	0.20930	0.20920	0.20906
13	0.22687	0.22682	0.22672	0.22658	0.22641
14	0.24432	0.24425	0.24413	0.24396	0.24374
15	0.26177	0.26168	0.26153	0.26133	0.26106
16	0.27922	0.27911	0.27893	0.27868	0.27836
17	0.29666	0.29653	0.29632	0.29602	0.29563
18	0.31411	0.31396	0.31370	0.31335	0.31289
19	0.33155	0.33137	0.33108	0.33066	0.33012
20	0.34900	0.34879	0.34844	0.34796	0.34733
21	0.36644	0.36620	0.36580	0.36524	0.36451
22	0.38388	0.38361	0.38315	0.38250	0.38167
23	0.40132	0.40101	0.40048	0.39975	0.39880
24	0.41876	0.41841	0.41781	0.41698	0.41590
25	0.43620	0.43580	0.43513	0.43419	0.43298
26	0.45364	0.45319	0.45244	0.45138	0.45002
27	0.47107	0.47057	0.46973	0.46856	0.46703
28	0.48851	0.48795	0.48702	0.48571	0.48402
29	0.50594	0.50532	0.50429	0.50284	0.50097
30	0.52337	0.52269	0.52155	0.51995	0.51788
31	0.54080	0.54005	0.53880	0.53704	0.53476
32	0.55823	0.55741	0.55604	0.55411	0.55161
33	0.57566	0.57476	0.57327	0.57116	0.56842
34	0.59309	0.59211	0.59048	0.58818	0.58520
35	0.61051	0.60945	0.60768	0.60518	0.60194
36	0.62794	0.62679	0.62486	0.62215	0.61864
37	0.64536	0.64412	0.64204	0.63911	0.63530
38	0.66278	0.66144	0.65920	0.65603	0.65193
39	0.68020	0.67876	0.67634	0.67294	0.66851
40	0.69762	0.69607	0.69348	0.68982	0.68506
41	0.71503	0.71337	0.71059	0.70667	0.70157
42	0.73245	0.73067	0.72770	0.72350	0.71804
43	0.74986	0.74797	0.74479	0.74030	0.73446
44	0.76727	0.76525	0.76187	0.75708	0.75085

Table D. (*Continued*)

ϕ^0	0.1	0.2	0.3	0.4	0.5
45	0.78468	0.78254	0.77893	0.77384	0.76720
46	0.80209	0.79981	0.79598	0.79056	0.78350
47	0.81950	0.81708	0.81301	0.80727	0.79977
48	0.83691	0.83434	0.83004	0.82394	0.81599
49	0.85431	0.85160	0.84704	0.84059	0.83217
50	0.87171	0.86885	0.86404	0.85722	0.84832
51	0.88911	0.88609	0.88102	0.87382	0.86442
52	0.90651	0.90333	0.89798	0.89040	0.88048
53	0.92391	0.92056	0.91493	0.90695	0.89650
54	0.94131	0.93779	0.93187	0.92348	0.91248
55	0.95870	0.95501	0.94880	0.93998	0.92843
56	0.97610	0.97222	0.96571	0.95646	0.94433
57	0.99349	0.98943	0.98261	0.97291	0.96019
58	1.01088	1.00664	0.99949	0.98934	0.97602
59	1.02827	1.02383	1.01636	1.00575	0.99180
60	1.04566	1.04103	1.03322	1.02213	1.00756
61	1.06305	1.05821	1.05007	1.03849	1.02327
62	1.08043	1.07539	1.06691	1.05483	1.03895
63	1.09782	1.09257	1.08373	1.07115	1.05459
64	1.11520	1.10974	1.10054	1.08745	1.07020
65	1.13258	1.12691	1.11734	1.10372	1.08577
66	1.14996	1.14407	1.13414	1.11998	1.10132
67	1.16734	1.16123	1.15092	1.13621	1.11683
68	1.18472	1.17838	1.16768	1.15243	1.13231
69	1.20210	1.19553	1.18444	1.16863	1.14776
70	1.21948	1.21267	1.20119	1.18481	1.16318
71	1.23685	1.22981	1.21794	1.20098	1.17857
72	1.25423	1.24695	1.23467	1.21713	1.19394
73	1.27160	1.26408	1.25139	1.23326	1.20928
74	1.28897	1.28121	1.26811	1.24938	1.22459
75	1.30635	1.29834	1.28481	1.26548	1.23989
76	1.32372	1.31546	1.30152	1.28158	1.25516
77	1.34109	1.33258	1.31821	1.29766	1.27041
78	1.35846	1.34970	1.33490	1.31372	1.28565
79	1.37583	1.36681	1.35158	1.32978	1.30086
80	1.39320	1.38393	1.36826	1.34583	1.31606
81	1.41056	1.40104	1.38493	1.36186	1.33124
82	1.42793	1.41814	1.40160	1.37789	1.34641
83	1.44530	1.43525	1.41826	1.39392	1.36157
84	1.46267	1.45236	1.43492	1.40993	1.37672
85	1.48003	1.46946	1.45158	1.42594	1.39186
86	1.49740	1.48656	1.46823	1.44195	1.40699
87	1.51476	1.50367	1.48488	1.45795	1.42211
88	1.53213	1.52077	1.50153	1.47395	1.43723
89	1.54950	1.53787	1.51818	1.48995	1.45235
90	1.56686	1.55497	1.53483	1.50594	1.46746

Table D. (*Continued*)

Second Kind, $D(\phi, k)$

$k \rightarrow$

ϕ^0	0.6	0.7	0.8	0.9
0	0.00000	0.00000	0.00000	0.00000
1	0.01745	0.01745	0.01745	0.01745
2	0.03490	0.03490	0.03490	0.03490
3	0.05235	0.05235	0.05234	0.05234
4	0.06979	0.06979	0.06978	0.06977
5	0.08723	0.08721	0.08720	0.08718
6	0.10465	0.10463	0.10460	0.10456
7	0.12206	0.12202	0.12198	0.12193
8	0.13946	0.13940	0.13934	0.13926
9	0.15685	0.15676	0.15667	0.15656
10	0.17422	0.17410	0.17397	0.17382
11	0.19156	0.19141	0.19123	0.19103
12	0.20889	0.20869	0.20846	0.20820
13	0.22620	0.22595	0.22565	0.22532
14	0.24348	0.24316	0.24280	0.24239
15	0.26073	0.26035	0.25990	0.25939
16	0.27796	0.27749	0.27695	0.27633
17	0.29516	0.29460	0.29395	0.29320
18	0.31233	0.31166	0.31089	0.31001
19	0.32946	0.32868	0.32777	0.32673
20	0.34656	0.34565	0.34459	0.34338
21	0.36362	0.36257	0.36134	0.35994
22	0.38065	0.37944	0.37803	0.37642
23	0.39764	0.39625	0.39464	0.39280
24	0.41458	0.41301	0.41118	0.40909
25	0.43149	0.42971	0.42765	0.42529
26	0.44835	0.44635	0.44403	0.44138
27	0.46516	0.46293	0.46034	0.45736
28	0.48193	0.47945	0.47656	0.47324
29	0.49866	0.49590	0.49269	0.48900
30	0.51533	0.51228	0.50873	0.50464
31	0.53195	0.52860	0.52468	0.52017
32	0.54853	0.54484	0.54053	0.53557
33	0.56505	0.56102	0.55629	0.55085
34	0.58152	0.57711	0.57195	0.56600
35	0.59793	0.59314	0.58751	0.58101
36	0.61429	0.60908	0.60297	0.59589
37	0.63060	0.62495	0.61832	0.61063
38	0.64684	0.64074	0.63356	0.62523
39	0.66303	0.65645	0.64870	0.63969
40	0.67917	0.67208	0.66372	0.65400
41	0.69524	0.68762	0.67863	0.66816
42	0.71125	0.70308	0.69343	0.68217
43	0.72721	0.71846	0.70812	0.69603
44	0.74310	0.73376	0.72268	0.70973

Table D. (*Continued*)

ϕ^0	0.6	0.7	0.8	0.9
45	0.75894	0.74896	0.73714	0.72327
46	0.77471	0.76409	0.75147	0.73665
47	0.79043	0.77912	0.76568	0.74987
48	0.80608	0.79407	0.77978	0.76293
49	0.82167	0.80893	0.79375	0.77582
50	0.83720	0.82371	0.80760	0.78855
51	0.85267	0.83840	0.82133	0.80111
52	0.86808	0.85300	0.83494	0.81350
53	0.88343	0.86751	0.84843	0.82572
54	0.89872	0.88194	0.86180	0.83777
55	0.91395	0.89628	0.87504	0.84964
56	0.92912	0.91054	0.88816	0.86135
57	0.94423	0.92471	0.90117	0.87288
58	0.95929	0.93880	0.91405	0.88425
59	0.97428	0.95280	0.92681	0.89544
60	0.98922	0.96672	0.93945	0.90646
61	1.00411	0.98056	0.95198	0.91731
62	1.01894	0.99432	0.96439	0.92799
63	1.03371	1.00800	0.97669	0.93850
64	1.04844	1.02161	0.98888	0.94884
65	1.06311	1.03514	1.00095	0.95902
66	1.07773	1.04859	1.01292	0.96904
67	1.09230	1.06197	1.02478	0.97889
68	1.10683	1.07529	1.03653	0.98859
69	1.12131	1.08853	1.04819	0.99813
70	1.13575	1.10171	1.05975	1.00752
71	1.15014	1.11482	1.07121	1.01676
72	1.16449	1.12788	1.08258	1.02585
73	1.17881	1.14087	1.09386	1.03481
74	1.19308	1.15381	1.10506	1.04363
75	1.20732	1.16669	1.11618	1.05231
76	1.22153	1.17953	1.12722	1.06088
77	1.23571	1.19231	1.13819	1.06932
78	1.24985	1.20506	1.14909	1.07766
79	1.26397	1.21776	1.15992	1.08588
80	1.27806	1.23042	1.17070	1.09401
81	1.29213	1.24304	1.18142	1.10205
82	1.30618	1.25564	1.19209	1.11001
83	1.32021	1.26820	1.20272	1.11789
84	1.33422	1.28075	1.21331	1.12570
85	1.34822	1.29326	1.22387	1.13345
86	1.36221	1.30577	1.23440	1.14116
87	1.37618	1.31825	1.24491	1.14883
88	1.39015	1.33073	1.25540	1.15647
89	1.40412	1.34320	1.26588	1.16409
90	1.41808	1.35566	1.27635	1.17170

Table E. Jacobi's Zeta and Epsilon Functions

$$k = 0.1$$

x	$Z(x)$	$E(x)$
0.0	0.00000	0.00000
0.1	0.00050	0.10000
0.2	0.00097	0.19997
0.3	0.00141	0.29991
0.4	0.00180	0.39979
0.5	0.00211	0.49960
0.6	0.00233	0.59933
0.7	0.00247	0.69896
0.8	0.00251	0.79850
0.9	0.00244	0.89794
1.0	0.00228	0.99728
1.1	0.00203	1.09653
1.2	0.00170	1.19569
1.3	0.00130	1.29480
1.4	0.00086	1.39385
1.5	0.00037	1.49286

$$K = 1.57475$$

$$k = 0.2$$

x	$Z(x)$	$E(x)$
0.0	0.00000	0.00000
0.1	0.00200	0.09999
0.2	0.00391	0.19989
0.3	0.00568	0.29965
0.4	0.00722	0.39917
0.5	0.00847	0.49842
0.6	0.00939	0.59733
0.7	0.00994	0.69587
0.8	0.01010	0.79402
0.9	0.00987	0.89178
1.0	0.00925	0.98915
1.1	0.00827	1.08616
1.2	0.00698	1.18285
1.3	0.00541	1.27928
1.4	0.00364	1.37549
1.5	0.00172	1.47157

$$K = 1.58687$$

Table E. (*Continued*)

k = 0.3

x	Z(x)	E(x)
0.0	0.00000	0.00000
0.1	0.00452	0.09997
0.2	0.00887	0.19976
0.3	0.01286	0.29921
0.4	0.01636	0.39815
0.5	0.01921	0.49645
0.6	0.02132	0.59400
0.7	0.02261	0.69074
0.8	0.02303	0.78661
0.9	0.02258	0.88160
1.0	0.02127	0.97574
1.1	0.01916	1.06908
1.2	0.01634	1.16170
1.3	0.01291	1.25372
1.4	0.00901	1.34527
1.5	0.00477	1.43647
1.6	0.00036	1.52751

K = 1.60805

k = 0.4

x	Z(x)	E(x)
0.0	0.00000	0.00000
0.1	0.00812	0.09995
0.2	0.01593	0.19958
0.3	0.02311	0.29859
0.4	0.02941	0.39671
0.5	0.03458	0.49370
0.6	0.03843	0.58939
0.7	0.04085	0.68363
0.8	0.04175	0.77635
0.9	0.04112	0.86755
1.0	0.03900	0.95726
1.1	0.03549	1.04557
1.2	0.03072	1.13263
1.3	0.02487	1.21861
1.4	0.01817	1.30373
1.5	0.01083	1.38822
1.6	0.00313	1.47234

K = 1.64000

Table E. (*Continued*)

k = 0.5

x	Z(x)	E(x)
0.0	0.00000	0.00000
0.1	0.01287	0.09992
0.2	0.02524	0.19934
0.3	0.03665	0.29780
0.4	0.04667	0.39487
0.5	0.05495	0.49020
0.6	0.06120	0.58351
0.7	0.06524	0.67460
0.8	0.06697	0.76337
0.9	0.06635	0.84981
1.0	0.06348	0.93399
1.1	0.05847	1.01604
1.2	0.05155	1.09616
1.3	0.04295	1.17462
1.4	0.03299	1.25170
1.5	0.02199	1.32775
1.6	0.01029	1.40311

K = 1.68575

k = 0.6

x	Z(x)	E(x)
0.0	0.00000	0.00000
0.1	0.01888	0.09988
0.2	0.03705	0.19905
0.3	0.05384	0.29684
0.4	0.06865	0.39264
0.5	0.08097	0.48596
0.6	0.09042	0.57641
0.7	0.09676	0.66374
0.8	0.09984	0.74782
0.9	0.09966	0.82864
1.0	0.09632	0.90631
1.1	0.09002	0.98100
1.2	0.08103	1.05301
1.3	0.06967	1.12265
1.4	0.05633	1.19031
1.5	0.04142	1.25639
1.6	0.02536	1.32134
1.7	0.00862	1.38559

K = 1.75075

Table E. (*Continued*)

$k = 0.7$

x	$Z(x)$	$E(x)$
0.0	0.00000	0.00000
0.1	0.02639	0.09984
0.2	0.05181	0.19871
0.3	0.07536	0.29571
0.4	0.09622	0.39002
0.5	0.11375	0.48100
0.6	0.12746	0.56816
0.7	0.13701	0.65116
0.8	0.14229	0.72989
0.9	0.14329	0.80434
1.0	0.14016	0.87466
1.1	0.13317	0.94112
1.2	0.12265	1.00405
1.3	0.10901	1.06386
1.4	0.09270	1.12099
1.5	0.07417	1.17592
1.6	0.05392	1.22912
1.7	0.03245	1.28110
1.8	0.01025	1.33235

$$K = 1.84569$$

$k = 0.8$

x	$Z(x)$	$E(x)$
0.0	0.00000	0.00000
0.1	0.03582	0.09979
0.2	0.07038	0.19832
0.3	0.10250	0.29441
0.4	0.13116	0.38703
0.5	0.15552	0.47536
0.6	0.17499	0.55879
0.7	0.18921	0.63698
0.8	0.19803	0.70978
0.9	0.20153	0.77724
1.0	0.19991	0.83958
1.1	0.19350	0.89714
1.2	0.18272	0.95034
1.3	0.16806	0.99964
1.4	0.15001	1.04556
1.5	0.12908	1.08859
1.6	0.10577	1.12926
1.7	0.08060	1.16805
1.8	0.05405	1.20547
1.9	0.02659	1.24197

$$K = 1.99530$$

Table E. (*Continued*)

$$k = 0.9$$

x	$Z(x)$	$E(x)$
0.0	0.00000	0.00000
0.1	0.04835	0.09973
0.2	0.09512	0.19787
0.3	0.13881	0.29294
0.4	0.17815	0.38367
0.5	0.21216	0.46905
0.6	0.24013	0.54840
0.7	0.26169	0.62133
0.8	0.27671	0.68774
0.9	0.28535	0.74775
1.0	0.28789	0.80167
1.1	0.28479	0.84994
1.2	0.27655	0.89308
1.3	0.26372	0.93163
1.4	0.24688	0.96617
1.5	0.22655	0.99722
1.6	0.20327	1.02532
1.7	0.17752	1.05094
1.8	0.14974	1.07454
1.9	0.12034	1.09652
2.0	0.08969	1.11725
2.1	0.05815	1.13709
2.2	0.02605	1.15637

$$K = 2.28055$$

For values of x not in the tables,
use the identities:

$Z(x + 2K) = Z(x);$
$Z(2K - x) = -Z(x).$

$E(x + 2K) = E(x) + 2E;$
$E(2K - x) = 2E - E(x).$

Table F. Sigma Function $\omega_1^{-1}\sigma(u,\omega_1,i\kappa\omega_1)$

$\kappa\rightarrow$

u/ω_1	0.2	0.4	0.6	0.8	1.0
0.0	0.00000	0.00000	0.00000	0.00000	0.00000
0.1	0.09980	0.09999	0.10000	0.10000	0.10000
0.2	0.19422	0.19959	0.19992	0.19997	0.19998
0.3	0.26388	0.29706	0.29939	0.29978	0.29988
0.4	0.28383	0.38843	0.39749	0.39910	0.39950
0.5	0.24715	0.46767	0.49262	0.49730	0.49846
0.6	0.17500	0.52775	0.58246	0.59340	0.59616
0.7	0.10086	0.56233	0.66412	0.68604	0.69169
0.8	0.04732	0.56762	0.73430	0.77343	0.78375
0.9	0.01808	0.54356	0.78968	0.85335	0.87059
1.0	0.00562	0.49411	0.82719	0.92314	0.94990
1.1	0.00142	0.42643	0.84438	0.97970	1.01867
1.2	0.00029	0.34935	0.83956	1.01942	1.07304
1.3	0.00005	0.27152	0.81192	1.03812	1.10807
1.4	0.00001	0.19991	0.76143	1.03089	1.11748
1.5	0.00000	0.13898	0.68859	0.99185	1.09326
1.6	0.00000	0.09056	0.59411	0.91385	1.02524
1.7	0.00000	0.05433	0.47848	0.78807	0.90049
1.8	0.00000	0.02864	0.34164	0.60352	0.70266
1.9	0.00000	0.01126	0.18273	0.34649	0.41112
2.0	0.00000	0.00000	0.00000	0.00000	0.00000

$\kappa\rightarrow$

u/ω_1	1.2	1.4	1.6	1.8	2.0
0.0	0.00000	0.00000	0.00000	0.00000	0.00000
0.1	0.10000	0.10000	0.10000	0.10000	0.10000
0.2	0.19999	0.19999	0.19999	0.19999	0.19999
0.3	0.29991	0.29991	0.29992	0.29992	0.29992
0.4	0.39960	0.39963	0.39964	0.39964	0.39964
0.5	0.49878	0.49887	0.49889	0.49890	0.49890
0.6	0.59692	0.59713	0.59719	0.59721	0.59722
0.7	0.69325	0.69369	0.69381	0.69385	0.69386
0.8	0.78662	0.78743	0.78766	0.78772	0.78774
0.9	0.87542	0.87679	0.87718	0.87729	0.87732
1.0	0.95746	0.95960	0.96021	0.96038	0.96043
1.1	1.02977	1.03294	1.03384	1.03409	1.03416
1.2	1.08847	1.09287	1.09413	1.09448	1.09458
1.3	1.12841	1.13423	1.13589	1.13636	1.13650
1.4	1.14293	1.15024	1.15232	1.15292	1.15309
1.5	1.12340	1.13209	1.13456	1.13527	1.13547
1.6	1.05873	1.06841	1.07117	1.07196	1.07218
1.7	0.93469	0.94460	0.94744	0.94825	0.94848
1.8	0.73318	0.74206	0.74460	0.74533	0.74553
1.9	0.43125	0.43713	0.43881	0.43930	0.43943
2.0	0.00000	0.00000	0.00000	0.00000	0.00000

Table F. (*Continued*)

Sigma Function $\sigma_1(u, \omega_1, i\kappa\omega_1)$

$\kappa \rightarrow$

u/ω_1	0.2	0.4	0.6	0.8	1.0
0.0	1.00000	1.00000	1.00000	1.00000	1.00000
0.1	0.90230	0.97438	0.98716	0.99047	0.99139
0.2	0.66283	0.90124	0.94920	0.96188	0.96543
0.3	0.39642	0.79082	0.88771	0.91418	0.92164
0.4	0.19301	0.65738	0.80509	0.84725	0.85926
0.5	0.07649	0.51598	0.70409	0.76079	0.77714
0.6	0.02465	0.37959	0.58747	0.65420	0.67374
0.7	0.00643	0.25712	0.45755	0.52653	0.54706
0.8	0.00133	0.15302	0.31594	0.37632	0.39463
0.9	0.00019	0.06790	0.16341	0.20164	0.21345
1.0	0.00000	0.00000	0.00000	0.00000	0.00000
1.1	−0.00002	−0.05327	−0.17473	−0.23149	−0.24976
1.2	−0.00001	−0.09418	−0.36123	−0.49601	−0.54030
1.3	−0.00000	−0.12415	−0.55939	−0.79674	−0.87638
1.4	−0.00000	−0.14379	−0.76798	−1.13651	−1.26289
1.5	−0.00000	−0.15334	−0.98419	−1.51737	−1.70449
1.6	−0.00000	−0.15326	−1.20333	−1.94002	−2.20515
1.7	−0.00000	−0.14465	−1.41875	−2.40320	−2.76754
1.8	−0.00000	−0.12932	−1.62211	−2.90298	−3.39211
1.9	−0.00000	−0.10969	−1.80385	−3.43187	−4.07582
2.0	−0.00000	−0.08832	−1.95391	−3.97791	−4.81048

$\kappa \rightarrow$

u/ω_1	1.2	1.4	1.6	1.8	2.0
0.0	1.0000	1.00000	1.00000	1.00000	1.00000
0.1	0.99165	0.99173	0.99175	0.99176	0.99176
0.2	0.96643	0.96672	0.96680	0.96682	0.96683
0.3	0.92376	0.92436	0.92453	0.92458	0.92460
0.4	0.86268	0.86365	0.86393	0.86401	0.86403
0.5	0.78181	0.78314	0.78352	0.78363	0.78366
0.6	0.67934	0.68094	0.68139	0.68152	0.68156
0.7	0.55298	0.55467	0.55515	0.55529	0.55532
0.8	0.39993	0.40145	0.40188	0.40200	0.40204
0.9	0.21689	0.21788	0.21816	0.21824	0.21826
1.0	0.00000	0.00000	0.00000	0.00000	0.00000
1.1	−0.25514	−0.25668	−0.25712	−0.25725	−0.25728
1.2	−0.55340	−0.55717	−0.55825	−0.55856	−0.55864
1.3	−0.90009	−0.90692	−0.90887	−0.90943	−0.90959
1.4	−1.30074	−1.31167	−1.31479	−1.31568	−1.31594
1.5	−1.76089	−1.77720	−1.78186	−1.78319	−1.78357
1.6	−2.28562	−2.30894	−2.31561	−2.31752	−2.31806
1.7	−2.87899	−2.91136	−2.92063	−2.92327	−2.92402
1.8	−3.54305	−3.58701	−3.59960	−3.60319	−3.60421
1.9	−4.27655	−4.33516	−4.35196	−4.35675	−4.35812
2.0	−5.07293	−5.14981	−5.17187	−5.17816	−5.17996

Table F. (*Continued*)

Sigma Function $\sigma_2(u, \omega_1, i\kappa\omega_1)$

$\kappa \rightarrow$

u/ω_1	0.2	0.4	0.6	0.8	1.0
0.0	1.00000	1.00000	1.00000	1.00000	1.00000
0.1	0.90230	0.97487	0.99011	0.99660	1.00002
0.2	0.66283	0.90336	0.96140	0.98686	1.00039
0.3	0.39643	0.79615	0.91670	0.97218	1.00199
0.4	0.19304	0.66828	0.86045	0.95484	1.00630
0.5	0.07655	0.53590	0.79836	0.93800	1.01538
0.6	0.02474	0.41324	0.73697	0.92553	1.03189
0.7	0.00655	0.31049	0.68320	0.92203	1.05906
0.8	0.00145	0.23297	0.64383	0.93262	1.10070
0.9	0.00029	0.18141	0.62502	0.96285	1.16116
1.0	0.00007	0.15292	0.63195	1.01850	1.24534
1.1	0.00002	0.14232	0.66832	1.10541	1.35866
1.2	0.00001	0.14339	0.73612	1.22924	1.50697
1.3	0.00000	0.14992	0.83526	1.39522	1.69659
1.4	0.00000	0.15654	0.96342	1.60788	1.93424
1.5	0.00000	0.15926	1.11596	1.87080	2.22702
1.6	0.00000	0.15581	1.28608	2.18638	2.58251
1.7	0.00000	0.14562	1.46508	2.55566	3.00882
1.8	0.00000	0.12963	1.64297	2.97836	3.51497
1.9	0.00000	0.10975	1.80924	3.45310	4.11131
2.0	0.00000	0.08832	1.95391	3.97791	4.81048

$\kappa \rightarrow$

u/ω_1	1.2	1.4	1.6	1.8	2.0
0.0	1.00000	1.00000	1.00000	1.00000	1.00000
0.1	1.00190	1.00293	1.00348	1.00378	1.00394
0.2	1.00782	1.01186	1.01405	1.01523	1.01586
0.3	1.01837	1.02730	1.03212	1.03472	1.03611
0.4	1.03461	1.05004	1.05838	1.06286	1.06526
0.5	1.05804	1.08127	1.09381	1.10054	1.10415
0.6	1.09060	1.12252	1.13973	1.14896	1.15390
0.7	1.13471	1.17575	1.19783	1.20967	1.21599
0.8	1.19331	1.24338	1.27024	1.28461	1.29229
0.9	1.26988	1.32832	1.35956	1.37622	1.38512
1.0	1.36848	1.43409	1.46897	1.48752	1.49740
1.1	1.49379	1.56489	1.60237	1.62221	1.63275
1.2	1.65123	1.72568	1.76448	1.78487	1.79566
1.3	1.84698	1.92244	1.96106	1.98116	1.99172
1.4	2.08818	2.16227	2.19918	2.21807	2.22792
1.5	2.38304	2.45374	2.48751	2.50434	2.51298
1.6	2.74115	2.80725	2.83681	2.85089	2.85792
1.7	3.17385	3.23556	3.26051	3.27149	3.27668
1.8	3.69478	3.75452	3.77554	3.78360	3.78700
1.9	4.32074	4.38410	4.40343	4.40957	4.41164
2.0	5.07293	5.14981	5.17187	5.17816	5.17996

Table F. (*Continued*)

Sigma Function $\sigma_3(u, \omega_1, i\kappa\omega_1)$

$\kappa \rightarrow$

u/ω_1	0.2	0.4	0.6	0.8	1.0
0.0	1.00000	1.00000	1.00000	1.00000	1.00000
0.1	1.19519	1.05075	1.02273	1.01293	1.00858
0.2	1.66316	1.19520	1.08938	1.05126	1.03418
0.3	2.11006	1.41084	1.19538	1.11361	1.07636
0.4	2.23754	1.66317	1.33338	1.19774	1.13444
0.5	1.94259	1.91041	1.49363	1.30063	1.20750
0.6	1.37468	2.11010	1.66463	1.41863	1.29448
0.7	0.79219	2.22659	1.83386	1.54762	1.39425
0.8	0.37169	2.23773	1.98878	1.68328	1.50576
0.9	0.14198	2.13894	2.11792	1.82140	1.62819
1.0	0.04416	1.94336	2.21193	1.95825	1.76118
1.1	0.01118	1.67804	2.26465	2.09108	1.90513
1.2	0.00230	1.37726	2.27389	2.21864	2.06155
1.3	0.00039	1.07511	2.24201	2.34186	2.23356
1.4	0.00005	0.79932	2.17610	2.46451	2.42645
1.5	0.00001	0.56774	2.08783	2.59407	2.64839
1.6	0.00000	0.38776	1.99294	2.74256	2.91135
1.7	0.00000	0.25805	1.91047	2.92747	3.23215
1.8	0.00000	0.17150	1.86167	3.17274	3.63368
1.9	0.00000	0.11829	1.86886	3.50969	4.14649
2.0	0.00000	0.08832	1.95391	3.97791	4.81048

$\kappa \rightarrow$

u/ω_1	1.2	1.4	1.6	1.8	2.0
0.0	1.00000	1.00000	1.00000	1.00000	1.00000
0.1	1.00644	1.00535	1.00477	1.00447	1.00430
0.2	1.02575	1.02142	1.01915	1.01795	1.01731
0.3	1.05788	1.04835	1.04335	1.04071	1.03930
0.4	1.10275	1.08636	1.07775	1.07319	1.07077
0.5	1.16032	1.13580	1.12290	1.11606	1.11242
0.6	1.23060	1.19720	1.17957	1.17021	1.16524
0.7	1.31374	1.27129	1.24881	1.23686	1.23050
0.8	1.41009	1.35913	1.33201	1.31757	1.30987
0.9	1.52039	1.46217	1.43099	1.41434	1.40545
1.0	1.64586	1.58241	1.54814	1.52977	1.51994
1.1	1.78846	1.72257	1.68656	1.66714	1.65672
1.2	1.95119	1.88634	1.85028	1.83066	1.82009
1.3	2.13838	2.07865	2.04451	2.02569	2.01549
1.4	2.35625	2.30612	2.27605	2.25910	2.24981
1.5	2.61342	2.57750	2.55366	2.53966	2.53182
1.6	2.92169	2.90435	2.88873	2.87861	2.87271
1.7	3.29699	3.30187	3.29598	3.29043	3.28678
1.8	3.76054	3.78998	3.79451	3.79373	3.79241
1.9	4.34032	4.39468	4.40910	4.41259	4.41325
2.0	5.07293	5.14981	5.17187	5.17816	5.17996

Table G. Weierstrass's Zeta Function $\omega_1 \zeta(u, \omega_1, i\kappa\omega_1)$

$\kappa \rightarrow$

u/ω_1	0.2	0.4	0.6	0.8	1.0
0.1	9.92012	9.99479	9.99896	9.99964	9.99980
0.2	4.45111	4.96006	4.99180	4.99714	4.99842
0.3	1.82787	3.20720	3.30649	3.32381	3.32801
0.4	−0.34130	2.22553	2.43891	2.47779	2.48737
0.5	−2.42076	1.51459	1.88645	1.95748	1.97525
0.6	−4.48176	0.91381	1.48121	1.59478	1.62366
0.7	−6.53893	0.36072	1.15160	1.31694	1.35963
0.8	−8.59530	−0.17127	0.86236	1.08682	1.14559
0.9	−10.65152	−0.69407	0.59413	0.88280	0.95927
1.0	−12.70769	−1.21341	0.33492	0.69038	0.78540
1.1	−14.76387	−1.73275	0.07570	0.49796	0.61153
1.2	−16.82008	−2.25555	−0.19253	0.29393	0.42521
1.3	−18.87646	−2.78754	−0.48177	0.06381	0.21117
1.4	−20.93363	−3.34063	−0.81137	−0.21403	−0.05286
1.5	−22.99463	−3.94141	−1.21662	−0.57673	−0.40445
1.6	−25.07409	−4.65235	−1.76908	−1.09704	−0.91657
1.7	−27.24326	−5.63402	−2.63666	−1.94305	−1.75722
1.8	−29.86650	−7.38688	−4.32197	−3.61639	−3.42763
1.9	−35.33551	−12.42161	−9.32912	−8.61888	−8.42901

$\kappa \rightarrow$

u/ω_1	1.2	1.4	1.6	1.8	2.0
0.1	9.99985	9.99986	9.99986	9.99986	9.99986
0.2	4.99877	4.99887	4.99890	4.99890	4.99891
0.3	3.32915	3.32947	3.32957	3.32959	3.32960
0.4	2.48998	2.49071	2.49092	2.49098	2.49100
0.5	1.98012	1.98149	1.98188	1.98199	1.98202
0.6	1.63160	1.63384	1.63448	1.63466	1.63471
0.7	1.37143	1.37477	1.37571	1.37598	1.37606
0.8	1.16191	1.16653	1.16784	1.16821	1.16831
0.9	0.98058	0.98662	0.98833	0.98882	0.98895
1.0	0.81196	0.81948	0.82162	0.82223	0.82240
1.1	0.64333	0.65234	0.65491	0.65563	0.65584
1.2	0.46201	0.47243	0.47540	0.47624	0.47648
1.3	0.25248	0.26419	0.26752	0.26847	0.26874
1.4	−0.00768	0.00512	0.00876	0.00979	0.01009
1.5	−0.35620	−0.34253	−0.33864	−0.33754	−0.33722
1.6	−0.86606	−0.85175	−0.84769	−0.84653	−0.84620
1.7	−1.70524	−1.69051	−1.68633	−1.68514	−1.68480
1.8	−3.37485	−3.35991	−3.35566	−3.35445	−3.35411
1.9	−8.37593	−8.36090	−8.35663	−8.35541	−8.35507

Table H. Weierstrass's Function $\omega_1^2 \mathscr{P}(u, \omega_1, i\kappa\omega_1)$

$$\kappa \rightarrow$$

u/ω_1	0.1	0.2	0.3	0.4	0.5
0.02	2501.6	2500.1	2500.0	2500.0	2500.0
0.04	631.11	625.40	625.08	625.03	625.01
0.06	290.56	278.66	277.96	277.83	277.80
0.08	176.91	157.78	156.56	156.35	156.29
0.10	128.84	102.31	100.48	100.15	100.06
0.12	106.09	72.640	70.123	69.665	69.536
0.14	94.690	55.178	51.925	51.317	51.144
0.16	88.808	44.227	40.215	39.444	39.222
0.18	85.726	37.054	32.284	31.340	31.065
0.20	84.097	32.209	26.702	25.577	25.245
0.22	83.232	28.866	22.656	21.346	20.953
0.24	82.772	26.522	19.656	18.160	17.704
0.26	82.527	24.860	17.392	15.710	15.190
0.28	82.396	23.672	15.658	13.794	13.208
0.30	82.326	22.819	14.315	12.275	11.623
0.32	82.289	22.202	13.266	11.057	10.338
0.34	82.269	21.756	12.441	10.070	9.2851
0.36	82.259	21.431	11.787	9.2636	8.4141
0.38	82.253	21.196	11.268	8.6007	7.6877
0.40	82.250	21.024	10.853	8.0525	7.0777
0.42	82.249	20.899	10.521	7.5969	6.5624
0.44	82.248	20.808	10.255	7.2167	6.1247
0.46	82.247	20.741	10.040	6.8983	5.7512
0.48	82.247	20.693	9.8676	6.6308	5.4314
0.50	82.247	20.658	9.7284	6.4056	5.1564
0.52	82.247	20.632	9.6160	6.2155	4.9194
0.54	82.247	20.613	9.5251	6.0549	4.7145
0.56	82.247	20.599	9.4517	5.9188	4.5370
0.58	82.247	20.589	9.3922	5.8035	4.3830
0.60	82.247	20.582	9.3441	5.7057	4.2492
0.62	82.247	20.576	9.3052	5.6226	4.1327
0.64	82.247	20.572	9.2736	5.5519	4.0312
0.66	82.247	20.569	9.2481	5.4919	3.9428
0.68	82.247	20.567	9.2274	5.4408	3.8657
0.70	82.247	20.566	9.2106	5.3974	3.7985
0.72	82.247	20.565	9.1970	5.3605	3.7400
0.74	82.247	20.564	9.1860	5.3292	3.6890
0.76	82.247	20.563	9.1771	5.3026	3.6447
0.78	82.247	20.563	9.1699	5.2800	3.6063
0.80	82.247	20.563	9.1641	5.2610	3.5731
0.82	82.247	20.562	9.1595	5.2450	3.5445
0.84	82.247	20.562	9.1557	5.2316	3.5202
0.86	82.247	20.562	9.1527	5.2205	3.4995
0.88	82.247	20.562	9.1503	5.2113	3.4823
0.90	82.247	20.562	9.1485	5.2039	3.4682
0.92	82.247	20.562	9.1470	5.1981	3.4569
0.94	82.247	20.562	9.1460	5.1938	3.4484
0.96	82.247	20.562	9.1453	5.1907	3.4424
0.98	82.247	20.562	9.1449	5.1889	3.4388
1.00	82.247	20.562	9.1447	5.1883	3.4376

Table H. (*Continued*)

$\kappa \rightarrow$

u/ω_1	0.6	0.7	0.8	0.9	1.0
0.02	2500.0	2500.0	2500.0	2500.0	2500.0
0.04	625.01	625.00	625.00	625.00	625.00
0.06	277.79	277.78	277.78	277.78	277.78
0.08	156.27	156.26	156.26	156.25	156.25
0.10	100.03	100.02	100.01	100.01	100.01
0.12	69.489	69.469	69.460	69.455	69.453
0.14	51.081	51.054	51.041	51.035	51.032
0.16	39.141	39.106	39.090	39.082	39.078
0.18	30.963	30.919	30.899	30.889	30.883
0.20	25.121	25.068	25.043	25.030	25.024
0.22	20.806	20.743	20.713	20.697	20.690
0.24	17.532	17.457	17.422	17.404	17.395
0.26	14.992	14.905	14.864	14.843	14.833
0.28	12.983	12.884	12.837	12.814	12.801
0.30	11.370	11.258	11.205	11.178	11.164
0.32	10.057	9.9317	9.8719	9.8418	9.8263
0.34	8.9748	8.8364	8.7699	8.7363	8.7190
0.36	8.0748	7.9226	7.8491	7.8121	7.7929
0.38	7.3193	7.1532	7.0727	7.0320	7.0109
0.40	6.6804	6.5001	6.4124	6.3680	6.3450
0.42	6.1364	5.9419	5.8470	5.7988	5.7738
0.44	5.6704	5.4617	5.3595	5.3075	5.2805
0.46	5.2692	5.0464	4.9369	4.8811	4.8520
0.48	4.9222	4.6854	4.5686	4.5089	4.4778
0.50	4.6209	4.3702	4.2461	4.1826	4.1495
0.52	4.3583	4.0940	3.9627	3.8954	3.8603
0.54	4.1287	3.8512	3.7128	3.6417	3.6046
0.56	3.9275	3.6370	3.4917	3.4168	3.3777
0.58	3.7508	3.4477	3.2955	3.2170	3.1759
0.60	3.5953	3.2800	3.1211	3.0390	2.9960
0.62	3.4582	3.1311	2.9658	2.8802	2.8353
0.64	3.3372	2.9989	2.8273	2.7383	2.6915
0.66	3.2304	2.8813	2.7037	2.6114	2.5629
0.68	3.1360	2.7767	2.5933	2.4978	2.4476
0.70	3.0527	2.6836	2.4947	2.3962	2.3443
0.72	2.9791	2.6009	2.4068	2.3054	2.2519
0.74	2.9142	2.5274	2.3284	2.2243	2.1693
0.76	2.8571	2.4623	2.2587	2.1521	2.0957
0.78	2.8070	2.4049	2.1970	2.0879	2.0303
0.80	2.7631	2.3543	2.1425	2.0312	1.9723
0.82	2.7250	2.3101	2.0946	1.9814	1.9214
0.84	2.6921	2.2717	2.0530	1.9379	1.8770
0.86	2.6641	2.2387	2.0172	1.9005	1.8386
0.88	2.6404	2.2108	1.9868	1.8687	1.8060
0.90	2.6208	2.1877	1.9616	1.8422	1.7789
0.92	2.6052	2.1691	1.9412	1.8209	1.7570
0.94	2.5932	2.1549	1.9256	1.8045	1.7402
0.96	2.5847	2.1448	1.9145	1.7929	1.7283
0.98	2.5797	2.1388	1.9079	1.7859	1.7212
1.00	2.5780	2.1368	1.9057	1.7836	1.7188

Table H. (*Continued*)

$\kappa \rightarrow$

u/ω_1	1.1	1.2	1.3	1.4	1.5
0.02	2500.0	2500.0	2500.0	2500.0	2500.0
0.04	625.00	625.00	625.00	625.00	625.00
0.06	277.78	277.78	277.78	277.78	277.78
0.08	156.25	156.25	156.25	156.25	156.25
0.10	100.01	100.00	100.00	100.00	100.00
0.12	69.452	69.451	69.451	69.451	69.450
0.14	51.030	51.029	51.029	51.029	51.029
0.16	39.075	39.074	39.074	39.073	39.073
0.18	30.881	30.879	30.878	30.878	30.878
0.20	25.020	25.019	25.018	25.017	25.017
0.22	20.686	20.684	20.682	20.682	20.682
0.24	17.390	17.388	17.387	17.386	17.385
0.26	14.827	14.824	14.823	14.822	14.822
0.28	12.795	12.792	12.790	12.789	12.789
0.30	11.157	11.153	11.151	11.150	11.150
0.32	9.8181	9.8138	9.8115	9.8103	9.8097
0.34	8.7099	8.7051	8.7026	8.7012	8.7005
0.36	7.7828	7.7775	7.7747	7.7732	7.7724
0.38	6.9998	6.9940	6.9909	6.9892	6.9883
0.40	6.3329	6.3265	6.3231	6.3213	6.3204
0.42	5.7607	5.7537	5.7500	5.7480	5.7470
0.44	5.2663	5.2588	5.2548	5.2527	5.2515
0.46	4.8368	4.8287	4.8244	4.8221	4.8209
0.48	4.4615	4.4528	4.4482	4.4458	4.4445
0.50	4.1321	4.1229	4.1179	4.1153	4.1139
0.52	3.8418	3.8320	3.8267	3.8240	3.8225
0.54	3.5850	3.5746	3.5690	3.5661	3.5645
0.56	3.3570	3.3461	3.3402	3.3371	3.3355
0.58	3.1542	3.1426	3.1365	3.1333	3.1315
0.60	2.9732	2.9612	2.9547	2.9513	2.9495
0.62	2.8115	2.7989	2.7922	2.7886	2.7867
0.64	2.6668	2.6536	2.6467	2.6429	2.6409
0.66	2.5371	2.5235	2.5162	2.5123	2.5103
0.68	2.4210	2.4068	2.3993	2.3952	2.3931
0.70	2.3168	2.3022	2.2944	2.2903	2.2880
0.72	2.2236	2.2085	2.2005	2.1962	2.1939
0.74	2.1402	2.1247	2.1164	2.1120	2.1097
0.76	2.0658	2.0499	2.0414	2.0369	2.0345
0.78	1.9996	1.9833	1.9747	1.9700	1.9676
0.80	1.9411	1.9244	1.9156	1.9108	1.9083
0.82	1.8895	1.8726	1.8635	1.8587	1.8561
0.84	1.8446	1.8273	1.8181	1.8132	1.8106
0.86	1.8057	1.7882	1.7789	1.7739	1.7713
0.88	1.7727	1.7550	1.7455	1.7405	1.7378
0.90	1.7452	1.7273	1.7178	1.7127	1.7099
0.92	1.7231	1.7050	1.6953	1.6902	1.6875
0.94	1.7060	1.6878	1.6781	1.6729	1.6701
0.96	1.6939	1.6756	1.6659	1.6607	1.6579
0.98	1.6867	1.6683	1.6586	1.6533	1.6506
1.00	1.6843	1.6659	1.6561	1.6509	1.6481

Table H. (*Continued*)

$\kappa \rightarrow$

u/ω_1	1.6	1.7	1.8	1.9	2.0
0.02	2500.0	2500.0	2500.0	2500.0	2500.0
0.04	625.00	625.00	625.00	625.00	625.00
0.06	277.78	277.78	277.78	277.78	277.78
0.08	156.25	156.25	156.25	156.25	156.25
0.10	100.00	100.00	100.00	100.00	100.00
0.12	69.450	69.450	69.450	69.450	69.450
0.14	51.029	51.028	51.028	51.028	51.028
0.16	39.073	39.073	39.073	39.073	39.073
0.18	30.878	30.878	30.878	30.878	30.878
0.20	25.017	25.017	25.017	25.017	25.017
0.22	20.681	20.681	20.681	20.681	20.681
0.24	17.385	17.385	17.385	17.385	17.385
0.26	14.821	14.821	14.821	14.821	14.821
0.28	12.788	12.788	12.788	12.788	12.788
0.30	11.149	11.149	11.149	11.149	11.149
0.32	9.8093	9.8091	9.8090	9.8090	9.8090
0.34	8.7001	8.6999	8.6998	8.6997	8.6997
0.36	7.7719	7.7717	7.7716	7.7715	7.7715
0.38	6.9878	6.9876	6.9875	6.9874	6.9874
0.40	6.3198	6.3196	6.3194	6.3193	6.3193
0.42	5.7464	5.7461	5.7460	5.7459	5.7459
0.44	5.2509	5.2506	5.2505	5.2504	5.2503
0.46	4.8202	4.8199	4.8197	4.8196	4.8195
0.48	4.4438	4.4434	4.4432	4.4431	4.4430
0.50	4.1132	4.1128	4.1126	4.1125	4.1124
0.52	3.8217	3.8213	3.8210	3.8209	3.8209
0.54	3.5637	3.5632	3.5630	3.5629	3.5628
0.56	3.3346	3.3341	3.3339	3.3337	3.3337
0.58	3.1306	3.1301	3.1298	3.1297	3.1296
0.60	2.9485	2.9480	2.9477	2.9476	2.9475
0.62	2.7857	2.7851	2.7849	2.7847	2.7846
0.64	2.6399	2.6393	2.6390	2.6388	2.6388
0.66	2.5092	2.5086	2.5083	2.5081	2.5080
0.68	2.3920	2.3914	2.3910	2.3909	2.3908
0.70	2.2869	2.2862	2.2859	2.2857	2.2856
0.72	2.1927	2.1920	2.1917	2.1915	2.1914
0.74	2.1084	2.1078	2.1074	2.1072	2.1071
0.76	2.0332	2.0325	2.0321	2.0319	2.0318
0.78	1.9663	1.9656	1.9652	1.9650	1.9649
0.80	1.9070	1.9062	1.9059	1.9057	1.9056
0.82	1.8548	1.8540	1.8536	1.8534	1.8533
0.84	1.8092	1.8085	1.8081	1.8078	1.8077
0.86	1.7698	1.7691	1.7687	1.7685	1.7684
0.88	1.7364	1.7356	1.7352	1.7350	1.7349
0.90	1.7085	1.7077	1.7073	1.7071	1.7070
0.92	1.6860	1.6852	1.6848	1.6846	1.6845
0.94	1.6687	1.6679	1.6675	1.6672	1.6671
0.96	1.6564	1.6556	1.6552	1.6550	1.6548
0.98	1.6491	1.6483	1.6479	1.6476	1.6475
1.00	1.6466	1.6458	1.6454	1.6452	1.6451

Table I. Weierstrass's Function $\omega_1^2 \mathscr{P}(u + \omega_3, \omega_1, i\kappa\omega_1)$

$\kappa \rightarrow$

u/ω_1	0.1	0.2	0.3	0.4	0.5
0.00	−164.49	−41.123	−18.277	−10.281	−6.5803
0.02	−141.66	−39.626	−17.979	−10.186	−6.5414
0.04	−87.972	−35.415	−17.109	−9.9065	−6.4261
0.06	−30.705	−29.225	−15.740	−9.4555	−6.2378
0.08	13.833	−21.993	−13.978	−8.8537	−5.9823
0.10	43.057	−14.595	−11.948	−8.1275	−5.6669
0.12	60.507	−7.6763	−9.7747	−7.3062	−5.3006
0.14	70.401	−1.6114	−7.5706	−6.4201	−4.8929
0.16	75.855	3.4582	−5.4273	−5.4983	−4.4537
0.18	78.816	7.5488	−3.4117	−4.5670	−3.9929
0.20	80.410	10.764	−1.5669	−3.6487	−3.5197
0.22	81.265	13.243	0.0848	−2.7616	−3.0427
0.24	81.723	15.127	1.5370	−1.9191	−2.5694
0.26	81.967	16.543	2.7950	−1.1310	−2.1061
0.28	82.097	17.600	3.8717	−0.4030	−1.6581
0.30	82.167	18.384	4.7841	0.2623	−1.2294
0.32	82.204	18.964	5.5509	0.8644	−0.8230
0.34	82.224	19.390	6.1913	1.4050	−0.4409
0.36	82.235	19.704	6.7230	1.8870	−0.0843
0.38	82.240	19.934	7.1627	2.3142	0.2464
0.40	82.243	20.103	7.5248	2.6908	0.5513
0.42	82.245	20.226	7.8223	3.0214	0.8310
0.44	82.246	20.316	8.0661	3.3104	1.0863
0.46	82.246	20.382	8.2655	3.5624	1.3185
0.48	82.246	20.431	8.4283	3.7813	1.5288
0.50	82.247	20.466	8.5611	3.9711	1.7188
0.52	82.247	20.492	8.6693	4.1353	1.8899
0.54	82.247	20.511	8.7573	4.2771	2.0435
0.56	82.247	20.524	8.8289	4.3993	2.1812
0.58	82.247	20.534	8.8871	4.5046	2.3042
0.60	82.247	20.542	8.9345	4.5951	2.4140
0.62	82.247	20.547	8.9729	4.6728	2.5118
0.64	82.247	20.551	9.0041	4.7395	2.5988
0.66	82.247	20.554	9.0294	4.7967	2.6759
0.68	82.247	20.556	9.0500	4.8457	2.7442
0.70	82.247	20.558	9.0666	4.8875	2.8046
0.72	82.247	20.559	9.0801	4.9233	2.8578
0.74	82.247	20.559	9.0911	4.9539	2.9047
0.76	82.247	20.560	9.1000	4.9799	2.9458
0.78	82.247	20.561	9.1071	5.0020	2.9817
0.80	82.247	20.561	9.1129	5.0207	3.0103
0.82	82.247	20.561	9.1176	5.0364	3.0400
0.84	82.247	20.561	9.1214	5.0497	3.0632
0.86	82.247	20.561	9.1244	5.0607	3.0830
0.88	82.247	20.561	9.1267	5.0697	3.0995
0.90	82.247	20.561	9.1286	5.0771	3.1131
0.92	82.247	20.562	9.1300	5.0828	3.1240
0.94	82.247	20.562	9.1311	5.0872	3.1323
0.96	82.247	20.562	9.1318	5.0902	3.1381
0.98	82.247	20.562	9.1322	5.0920	3.1416
1.00	82.247	20.562	9.1323	5.0925	3.1427

Table I. (*Continued*)

u/ω_1	$\kappa \to$				
	0.6	0.7	0.8	0.9	1.0
0.00	−4.5724	−3.3672	−2.5942	−2.0761	−1.7188
0.02	−4.5536	−3.3571	−2.5883	−2.0725	−1.7164
0.04	−4.4978	−3.3269	−2.5706	−2.0615	−1.7094
0.06	−4.4061	−3.2772	−2.5415	−2.0434	−1.6977
0.08	−4.2804	−3.2087	−2.5012	−2.0184	−1.6814
0.10	−4.1235	−3.1224	−2.4502	−1.9866	−1.6607
0.12	−3.9384	−3.0198	−2.3892	−1.9483	−1.6358
0.14	−3.7287	−2.9022	−2.3188	−1.9040	−1.6068
0.16	−3.4982	−2.7714	−2.2398	−1.8540	−1.5740
0.18	−3.2510	−2.6292	−2.1531	−1.7988	−1.5376
0.20	−2.9911	−2.4773	−2.0597	−1.7388	−1.4979
0.22	−2.7222	−2.3176	−1.9604	−1.6747	−1.4551
0.24	−2.4482	−2.1520	−1.8562	−1.6068	−1.4097
0.26	−2.1723	−1.9823	−1.7481	−1.5358	−1.3618
0.28	−1.8977	−1.8101	−1.6369	−1.4621	−1.3119
0.30	−1.6271	−1.6370	−1.5237	−1.3864	−1.2602
0.32	−1.3627	−1.4644	−1.4092	−1.3090	−1.2070
0.34	−1.1064	−1.2936	−1.2943	−1.2306	−1.1527
0.36	−0.8597	−1.1258	−1.1797	−1.1516	−1.0976
0.38	−0.6239	−0.9618	−1.0661	−1.0724	−1.0420
0.40	−0.3997	−0.8026	−0.9541	−0.9935	−0.9861
0.42	−0.1877	−0.6489	−0.8442	−0.9152	−0.9302
0.44	0.0118	−0.5010	−0.7370	−0.8380	−0.8746
0.46	0.1987	−0.3595	−0.6328	−0.7621	−0.8196
0.48	0.3731	−0.2247	−0.5319	−0.6878	−0.7653
0.50	0.5352	−0.0967	−0.4347	−0.6154	−0.7120
0.52	0.6853	0.0244	−0,3414	−0.5452	−0.6598
0.54	0.8240	0.1385	−0.2521	−0.4772	−0.6089
0.56	0.9518	0.2457	−0.1670	−0.4118	−0.5595
0.58	1.0691	0.3460	−0.0862	−0.3489	−0.5117
0.60	1.1765	0.4397	−0.0096	−0.2888	−0.4656
0.62	1.2746	0.5269	0.0626	−0.2316	−0.4214
0.64	1.3640	0.6078	0.1304	−0.1772	−0.3791
0.66	1.4453	0.6826	0.1940	−0.1258	−0.3388
0.68	1.5190	0.7516	0.2533	−0.0774	−0.3007
0.70	1.5856	0.8149	0.3085	−0.0320	−0.2646
0.72	1.6456	0.8729	0.3596	0.0104	−0.2308
0.74	1.6995	0.9257	0.4066	0.0498	−0.1992
0.76	1.7477	0.9737	0.4497	0.0861	−0.1698
0.78	1.7906	1.0169	0.4889	0.1194	−0.1428
0.80	1.8286	1.0556	0.5244	0.1497	−0.1181
0.82	1.8621	1.0901	0.5562	0.1771	−0.0957
0.84	1.8912	1.1204	0.5844	0.2014	−0.0756
0.86	1.9163	1.1468	0.6091	0.2229	−0.0579
0.88	1.9376	1.1694	0.6304	0.2414	−0.0425
0.90	1.9553	1.1883	0.6482	0.2571	−0.0295
0.92	1.9695	1.2036	0.6628	0.2699	−0.0189
0.94	1.9805	1.2154	0.6740	0.2798	−0.0106
0.96	1.9882	1.2238	0.6821	0.2868	−0.0047
0.98	1.9928	1.2288	0.6869	0.2911	−0.0012
1.00	1.9944	1.2304	0.6885	0.2925	0.0000

Table I. (*Continued*)

$\kappa \rightarrow$

u/ω_1	1.1	1.2	1.3	1.4	1.5
0.00	-1.4677	-1.2890	-1.1608	-1.0684	-1.0014
0.02	-1.4661	-1.2879	-1.1601	-1.0679	-1.0011
0.04	-1.4614	-1.2847	-1.1578	-1.0663	-0.9999
0.06	-1.4536	-1.2793	-1.1541	-1.0636	-0.9981
0.08	-1.4427	-1.2719	-1.1489	-1.0600	-0.9955
0.10	-1.4288	-1.2624	-1.1423	-1.0553	-0.9921
0.12	-1.4121	-1.2509	-1.1343	-1.0496	-0.9881
0.14	-1.3926	-1.2375	-1.1249	-1.0430	-0.9834
0.16	-1.3704	-1.2222	-1.1142	-1.0355	-0.9780
0.18	-1.3458	-1.2052	-1.1023	-1.0270	-0.9720
0.20	-1.3188	-1.1866	-1.0892	-1.0177	-0.9653
0.22	-1.2897	-1.1663	-1.0750	-1.0076	-0.9581
0.24	-1.2586	-1.1447	-1.0598	-0.9968	-0.9503
0.26	-1.2257	-1.1217	-1.0435	-0.9852	-0.9420
0.28	-1.1912	-1.0976	-1.0264	-0.9730	-0.9333
0.30	-1.1553	-1.0724	-1.0086	-0.9603	-0.9241
0.32	-1.1183	-1.0463	-0.9900	-0.9469	-0.9145
0.34	-1.0802	-1.0193	-0.9708	-0.9331	-0.9046
0.36	-1.0414	-0.9917	-0.9510	-0.9190	-0.8943
0.38	-1.0020	-0.9636	-0.9308	-0.9044	-0.8838
0.40	-0.9622	-0.9351	-0.9103	-0.8896	-0.8731
0.42	-0.9222	-0.9063	-0.8896	-0.8746	-0.8622
0.44	-0.8821	-0.8774	-0.8686	-0.8594	-0.8511
0.46	-0.8422	-0.8485	-0.8476	-0.8441	-0.8400
0.48	-0.8026	-0.8196	-0.8266	-0.8288	-0.8288
0.50	-0.7635	-0.7910	-0.8057	-0.8135	-0.8177
0.52	-0.7250	-0.7627	-0.7849	-0.7983	-0.8066
0.54	-0.6872	-0.7349	-0.7645	-0.7833	-0.7956
0.56	-0.6504	-0.7075	-0.7443	-0.7685	-0.7847
0.58	-0.6145	-0.6809	-0.7246	-0.7539	-0.7740
0.60	-0.5797	-0.6549	-0.7053	-0.7397	-0.7635
0.62	-0.5462	-0.6297	-0.6866	-0.7258	-0.7533
0.64	-0.5139	-0.6054	-0.6684	-0.7124	-0.7434
0.66	-0.4830	-0.5821	-0.6510	-0.6994	-0.7338
0.68	-0.4536	-0.5598	-0.6342	-0.6869	-0.7245
0.70	-0.4257	-0.5385	-0.6182	-0.6750	-0.7157
0.72	-0.3994	-0.5184	-0.6031	-0.6637	-0.7073
0.74	-0.3747	-0.4995	-0.5888	-0.6530	-0.6993
0.76	-0.3517	-0.4818	-0.5754	-0.6429	-0.6918
0.78	-0.3304	-0.4654	-0.5629	-0.6336	-0.6849
0.80	-0.3109	-0.4503	-0.5514	-0.6249	-0.6784
0.82	-0.2931	-0.4365	-0.5409	-0.6170	-0.6725
0.84	-0.2772	-0.4242	-0.5315	-0.6099	-0.6672
0.86	-0.2630	-0.4132	-0.5231	-0.6036	-0.6624
0.88	-0.2508	-0.4036	-0.5158	-0.5980	-0.6583
0.90	-0.2404	-0.3955	-0.5096	-0.5933	-0.6548
0.92	-0.2318	-0.3888	-0.5044	-0.5895	-0.6519
0.94	-0.2252	-0.3836	-0.5005	-0.5864	-0.6496
0.96	-0.2204	-0.3799	-0.4976	-0.5843	-0.6480
0.98	-0.2176	-0.3777	-0.4959	-0.5830	-0.6470
1.00	-0.2166	-0.3769	-0.4953	-0.5825	-0.6467

Table I. (*Continued*)

			$\kappa \rightarrow$		
u/ω_1	1.6	1.7	1.8	1.9	2.0
0.00	−0.9529	−0.9175	−0.8918	−0.8731	−0.8594
0.02	−0.9526	−0.9173	−0.8917	−0.8730	−0.8593
0.04	−0.9518	−0.9168	−0.8913	−0.8727	−0.8591
0.06	−0.9504	−0.9158	−0.8906	−0.8722	−0.8587
0.08	−0.9486	−0.9144	−0.8896	−0.8715	−0.8582
0.10	−0.9462	−0.9127	−0.8883	−0.8706	−0.8576
0.12	−0.9433	−0.9106	−0.8868	−0.8695	−0.8568
0.14	−0.9399	−0.9082	−0.8851	−0.8682	−0.8558
0.16	−0.9360	−0.9054	−0.8830	−0.8667	−0.8548
0.18	−0.9317	−0.9023	−0.8808	−0.8651	−0.8536
0.20	−0.9269	−0.8988	−0.8783	−0.8633	−0.8523
0.22	−0.9217	−0.8951	−0.8756	−0.8613	−0.8508
0.24	−0.9161	−0.8910	−0.8726	−0.8591	−0.8493
0.26	−0.9102	−0.8867	−0.8695	−0.8569	−0.8476
0.28	−0.9038	−0.8821	−0.8662	−0.8545	−0.8459
0.30	−0.8972	−0.8773	−0.8627	−0.8519	−0.8440
0.32	−0.8903	−0.8723	−0.8590	−0.8493	−0.8421
0.34	−0.8831	−0.8671	−0.8553	−0.8465	−0.8401
0.36	−0.8757	−0.8617	−0.8513	−0.8437	−0.8380
0.38	−0.8680	−0.8562	−0.8473	−0.8407	−0.8359
0.40	−0.8603	−0.8505	−0.8432	−0.8377	−0.8337
0.42	−0.8523	−0.8447	−0.8390	−0.8347	−0.8314
0.44	−0.8443	−0.8389	−0.8347	−0.8316	−0.8292
0.46	−0.8362	−0.8330	−0.8304	−0.8284	−0.8269
0.48	−0.8281	−0.8271	−0.8261	−0.8253	−0.8246
0.50	−0.8199	−0.8211	−0.8217	−0.8221	−0.8223
0.52	−0.8118	−0.8152	−0.8174	−0.8189	−0.8200
0.54	−0.8037	−0.8093	−0.8131	−0.8158	−0.8176
0.56	−0.7958	−0.8035	−0.8088	−0.8126	−0.8154
0.58	−0.7879	−0.7977	−0.8046	−0.8096	−0.8131
0.60	−0.7802	−0.7921	−0.8005	−0.8065	−0.8109
0.62	−0.7727	−0.7865	−0.7964	−0.8036	−0.8087
0.64	−0.7654	−0.7812	−0.7925	−0.8007	−0.8066
0.66	−0.7583	−0.7760	−0.7887	−0.7979	−0.8046
0.68	−0.7515	−0.7709	−0.7850	−0.7952	−0.8026
0.70	−0.7500	−0.7661	−0.7815	−0.7926	−0.8007
0.72	−0.7388	−0.7616	−0.7781	−0.7901	−0.7989
0.74	−0.7329	−0.7572	−0.7749	−0.7878	−0.7972
0.76	−0.7273	−0.7531	−0.7719	−0.7856	−0.7955
0.78	−0.7221	−0.7493	−0.7691	−0.7835	−0.7940
0.80	−0.7174	−0.7458	−0.7665	−0.7816	−0.7926
0.82	−0.7130	−0.7425	−0.7641	−0.7798	−0.7913
0.84	−0.7090	−0.7396	−0.7619	−0.7783	−0.7902
0.86	−0.7055	−0.7370	−0.7600	−0.7768	−0.7891
0.88	−0.7024	−0.7347	−0.7583	−0.7756	−0.7882
0.90	−0.6998	−0.7328	−0.7569	−0.7746	−0.7875
0.92	−0.6976	−0.7312	−0.7557	−0.7737	−0.7868
0.94	−0.6960	−0.7299	−0.7548	−0.7730	−0.7863
0.96	−0.6948	−0.7290	−0.7541	−0.7725	−0.7860
0.98	−0.6940	−0.7285	−0.7537	−0.7722	−0.7858
1.00	−0.6938	−0.7283	−0.7536	−0.7721	−0.7857

Table J. Weierstrass's Function $\omega_1^2 \mathcal{P}(u, \omega_1 . \frac{1}{2}\omega_1(1 + iv))$

$v \rightarrow$

u/ω_1	0.2	0.4	0.6	0.8	1.0
0.02	2500.1	2500.0	2500.0	2500.0	2500.0
0.04	625.40	625.02	625.00	625.00	625.00
0.06	278.66	277.83	277.78	277.77	277.77
0.08	157.777	156.34	156.25	156.23	156.23
0.10	102.308	100.14	99.992	99.971	99.976
0.12	72.639	69.642	69.432	69.402	69.410
0.14	55.177	51.285	51.003	50.962	50.974
0.16	44.226	39.401	39.038	38.986	39.002
0.18	37.053	31.284	30.831	30.767	30.788
0.20	32.208	25.505	24.955	24.879	24.906
0.22	28.864	21.256	20.603	20.513	20.547
0.24	26.520	18.047	17.286	17.183	17.225
0.26	24.858	15.571	14.697	14.582	14.634
0.28	23.669	13.625	12.634	12.508	12.571
0.30	22.815	12.070	10.961	10.825	10.900
0.32	22.197	10.810	9.5812	9.4368	9.5256
0.34	21.748	9.7738	8.4257	8.2754	8.3802
0.36	21.421	8.9105	7.4444	7.2911	7.4138
0.38	21.181	8.1810	6.5996	6.4468	6.5895
0.40	21.004	7.5550	5.8625	5.7144	5.8794
0.42	20.872	7.0089	5.2111	5.0724	5.2621
0.44	20.771	6.5234	4.6278	4.5041	4.7211
0.46	20.690	6.0828	4.0989	3.9963	4.2432
0.48	20.623	5.6735	3.6132	3.5387	3.8181
0.50	20.562	5.2841	3.1616	3.1230	3.4376
0.52	20.501	4.9043	2.7369	2.7424	3.0950
0.54	20.434	4.5247	2.3330	2.3916	2.7850
0.56	20.354	4.1367	1.9450	2.0662	2.5030
0.58	20.253	3.7324	1.5688	1.7628	2.2457
0.60	20.123	3.3039	1.2012	1.4786	2.0099
0.62	19.949	2.8442	0.83997	1.2113	1.7933
0.64	19.715	2.3464	0.48316	0.95915	1.5939
0.66	19.398	1.8043	0.12968	0.72092	1.4101
0.68	18.969	1.2127	−0.22099	0.49562	1.2406
0.70	18.389	0.56714	−0.56877	0.28258	1.0842
0.72	17.603	−0.13499	−0.91302	0.08140	0.94005
0.74	16.546	−0.89439	−1.2525	−0.10810	0.80752
0.76	15.128	−1.7091	−1.5855	−0.28591	0.68603
0.78	13.244	−2.5741	−1.9096	−0.45187	0.57512
0.80	10.765	−3.4803	−2.2223	−0.60568	0.47447
0.82	7.5495	−4.4146	−2.5204	−0.74695	0.38382
0.84	3.4587	−5.3593	−2.8004	−0.87523	0.30299
0.86	−1.6110	−6.2922	−3.0590	−0.99002	0.23182
0.88	−7.6759	−7.1874	−3.2924	−1.0908	0.17025
0.90	−14.594	−8.0161	−3.4970	−1.1771	0.11820
0.92	−21.993	−8.7481	−3.6696	−1.2484	0.07564
0.94	−29.224	−9.3542	−3.8073	−1.3044	0.04254
0.96	−35.414	−9.8084	−3.9074	−1.3446	0.01891
0.98	−39.626	−10.090	−3.9683	−1.3688	0.00473
1.00	−41.123	−10.185	−3.9887	−1.3769	0.00000

Table J. (*Continued*)

u/ω_1	$v \to$ 1.2	1.4	1.6	1.8	2.0
0.02	2500.0	2500.0	2500.0	2500.0	2500.0
0.04	625.00	625.00	625.00	625.00	625.00
0.06	277.77	277.78	277.78	277.78	277.78
0.08	156.24	156.25	156.25	156.25	156.25
0.10	99.986	99.993	99.998	100.00	100.00
0.12	69.424	69.435	69.442	69.446	69.448
0.14	50.993	51.008	51.017	51.022	51.025
0.16	39.027	39.046	39.058	39.065	39.069
0.18	30.820	30.843	30.858	30.867	30.872
0.20	24.945	24.975	24.993	25.004	25.010
0.22	20.595	20.631	20.653	20.666	20.673
0.24	17.283	17.325	17.352	17.367	17.375
0.26	14.702	14.752	14.782	14.800	14.810
0.28	12.650	12.708	12.743	12.764	12.775
0.30	10.992	11.058	11.098	11.121	11.134
0.32	9.6310	9.7059	9.7516	9.7776	9.7920
0.34	8.5000	8.5843	8.6355	8.6647	8.6808
0.36	7.5490	7.6433	7.7003	7.7327	7.7505
0.38	6.7412	6.8459	6.9089	6.9446	6.9643
0.40	6.0486	6.1641	6.2334	6.2725	6.2941
0.42	5.4498	5.5766	5.6523	5.6950	5.7184
0.44	4.9282	5.0667	5.1489	5.1952	5.2206
0.46	4.4707	4.6213	4.7102	4.7602	4.7875
0.48	4.0670	4.2300	4.3257	4.3794	4.4087
0.50	3.7088	3.8843	3.9870	4.0444	4.0758
0.52	3.3893	3.5777	3.6874	3.7486	3.7820
0.54	3.1030	3.3045	3.4213	3.4862	3.5216
0.56	2.8456	3.0603	3.1841	3.2527	3.2901
0.58	2.6133	2.8412	2.9720	3.0444	3.0838
0.60	2.4030	2.6442	2.7820	2.8581	2.8995
0.62	2.2122	2.4667	2.6114	2.6912	2.7344
0.64	2.0388	2.3065	2.4580	2.5413	2.5864
0.66	1.8811	2.1617	2.3199	2.4066	2.4536
0.68	1.7375	2.0308	2.1955	2.2856	2.3343
0.70	1.6067	1.9125	2.0834	2.1768	2.2273
0.72	1.4878	1.8057	1.9826	2.0791	2.1312
0.74	1.3799	1.7093	1.8920	1.9915	2.0452
0.76	1.2821	1.6226	1.8108	1.9131	1.9682
0.78	1.1939	1.5449	1.7383	1.8432	1.8997
0.80	1.1148	1.4756	1.6738	1.7812	1.8390
0.82	1.0443	1.4142	1.6168	1.7265	1.7854
0.84	0.98199	1.3603	1.5669	1.6786	1.7386
0.86	0.92765	1.3134	1.5237	1.6372	1.6982
0.88	0.88100	1.2734	1.4868	1.6020	1.6638
0.90	0.84184	1.2399	1.4561	1.5726	1.6351
0.92	0.81001	1.2127	1.4312	1.5488	1.6119
0.94	0.78537	1.1918	1.4120	1.5305	1.5941
0.96	0.76784	1.1769	1.3984	1.5175	1.5814
0.98	0.75735	1.1680	1.3903	1.5098	1.5739
1.00	0.75385	1.1651	1.3876	1.5072	1.5713

Table J. (*Continued*)

$v \rightarrow$

u/ω_1	2.2	2.4	2.6	2.8	3.0
0.02	2500.0	2500.0	2500.0	2500.0	2500.0
0.04	625.00	625.00	625.00	625.00	625.00
0.06	277.78	277.78	277.78	277.78	277.78
0.08	156.25	156.25	156.25	156.25	156.25
0.10	100.00	100.00	100.00	100.00	100.00
0.12	69.449	69.450	69.450	69.450	69.450
0.14	51.027	51.027	51.028	51.028	51.028
0.16	39.071	39.072	39.072	39.073	39.073
0.18	30.874	30.876	30.877	30.877	30.877
0.20	25.013	25.014	25.015	25.016	25.016
0.22	20.677	20.679	20.680	20.680	20.681
0.24	17.380	17.382	17.348	17.384	17.385
0.26	14.815	14.818	14.819	14.820	14.821
0.28	12.781	12.784	12.786	12.787	12.787
0.30	11.141	11.145	11.147	11.148	11.148
0.32	9.7999	9.8041	9.8063	9.8075	9.8082
0.34	8.6895	8.6942	8.6968	8.6981	8.6988
0.36	7.7602	7.7654	7.7682	7.7697	7.7705
0.38	6.9750	6.9807	6.9838	6.9854	6.9863
0.40	6.3057	6.3120	6.3154	6.3172	6.3181
0.42	5.7311	5.7379	5.7416	5.7436	5.7446
0.44	5.2344	5.2417	5.2457	5.2478	5.2490
0.46	4.8023	4.8103	4.8146	4.8168	4.8181
0.48	4.4246	4.4331	4.4377	4.4402	4.4415
0.50	4.0928	4.1019	4.1067	4.1094	4.1107
0.52	3.8000	3.8097	3.8148	3.8176	3.8191
0.54	3.5407	3.5509	3.5564	3.5594	3.5609
0.56	3.3103	3.3211	3.3269	3.3300	3.3317
0.58	3.1051	3.1164	3.1225	3.1258	3.1275
0.60	2.9217	2.9337	2.9401	2.9435	2.9453
0.62	2.7577	2.7702	2.7769	2.7804	2.7823
0.64	2.6107	2.6237	2.6307	2.6344	2.6364
0.66	2.4789	2.4924	2.4996	2.5035	2.5056
0.68	2.3605	2.3746	2.3821	2.3861	2.3882
0.70	2.2544	2.2689	2.2766	2.2808	2.2830
0.72	2.1592	2.1741	2.1821	2.1864	2.1887
0.74	2.0739	2.0893	2.0976	2.1020	2.1043
0.76	1.9978	2.0136	2.0221	2.0266	2.0290
0.78	1.9300	1.9462	1.9548	1.9595	1.9619
0.80	1.8699	1.8865	1.8953	1.9000	1.9025
0.82	1.8170	1.8339	1.8429	1.8477	1.8503
0.84	1.7708	1.7879	1.7971	1.8020	1.8046
0.86	1.7308	1.7482	1.7576	1.7625	1.7652
0.88	1.6968	1.7145	1.7239	1.7290	1.7316
0.90	1.6685	1.6864	1.6959	1.7010	1.7037
0.92	1.6456	1.6637	1.6733	1.6784	1.6812
0.94	1.6280	1.6462	1.6559	1.6611	1.6638
0.96	1.6156	1.6338	1.6436	1.6488	1.6515
0.98	1.6081	1.6264	1.6362	1.6414	1.6442
1.00	1.6056	1.6240	1.6337	1.6390	1.6417

Table J. (*Continued*)

$v \rightarrow$

u/ω_1	3.2	3.4	3.6	3.8	4.0
0.02	2500.0	2500.0	2500.0	2500.0	2500.0
0.04	625.00	625.00	625.00	625.00	625.00
0.06	277.78	277.78	277.78	277.78	277.78
0.08	156.25	156.25	156.25	156.25	156.25
0.10	100.00	100.00	100.00	100.00	100.00
0.12	69.450	69.450	69.450	69.450	69.450
0.14	51.028	51.028	51.028	51.028	51.028
0.16	39.073	39.073	39.073	39.073	39.073
0.18	30.877	30.877	30.877	30.877	30.878
0.20	25.016	25.016	25.016	25.016	25.016
0.22	20.681	20.681	20.681	20.681	20.681
0.24	17.385	17.385	17.385	17.385	17.385
0.26	14.821	14.821	14.821	14.821	14.821
0.28	12.788	12.788	12.788	12.788	12.788
0.30	11.149	11.149	11.149	11.149	11.149
0.32	9.8085	9.8087	9.8088	9.8089	9.8089
0.34	8.6992	8.6994	8.6995	8.6996	8.6996
0.36	7.7710	7.7712	7.7713	7.7714	7.7714
0.38	6.9868	6.9870	6.9872	6.9872	6.9873
0.40	6.3187	6.3189	6.3191	6.3192	6.3192
0.42	5.7452	5.7455	5.7456	5.7457	5.7457
0.44	5.2496	5.2499	5.2501	5.2501	5.2502
0.46	4.8187	4.8191	4.8192	4.8193	4.8194
0.48	4.4422	4.4425	4.4427	4.4428	4.4429
0.50	4.1115	4.1119	4.1121	4.1122	4.1123
0.52	3.8199	3.8203	3.8205	3.8206	3.8207
0.54	3.5618	3.5622	3.5625	3.5626	3.5626
0.56	3.3326	3.3330	3.3333	3.3334	3.3335
0.58	3.1285	3.1290	3.1292	3.1294	3.1294
0.60	2.9463	2.9468	2.9471	2.9472	2.9473
0.62	2.7834	2.7839	2.7842	2.7843	2.7844
0.64	2.6374	2.6380	2.6383	2.6385	2.6386
0.66	2.5067	2.5072	2.5076	2.5077	2.5078
0.68	2.3894	2.3900	2.3903	2.3905	2.3906
0.70	2.2842	2.2848	2.2851	2.2853	2.2854
0.72	2.1899	2.1906	2.1909	2.1911	2.1912
0.74	2.1056	2.1062	2.1066	2.1068	2.1069
0.76	2.0303	2.0309	2.0313	2.0315	2.0316
0.78	1.9632	1.9639	1.9643	1.9645	1.9646
0.80	1.9039	1.9046	1.9050	1.9052	1.9053
0.82	1.8516	1.8524	1.8528	1.8530	1.8531
0.84	1.8060	1.8067	1.8071	1.8074	1.8075
0.86	1.7666	1.7674	1.7678	1.7680	1.7681
0.88	1.7331	1.7338	1.7343	1.7345	1.7346
0.90	1.7052	1.7059	1.7064	1.7066	1.7067
0.92	1.6826	1.6834	1.6838	1.6841	1.6842
0.94	1.6653	1.6661	1.6665	1.6667	1.6668
0.96	1.6530	1.6538	1.6542	1.6544	1.6546
0.98	1.6457	1.6465	1.6469	1.6471	1.6472
1.00	1.6432	1.6440	1.6445	1.6447	1.6448

Table K. Stationary Values and Invariants of $\mathscr{P}(u, \omega_1, i\kappa\omega_1)$

κ	$\omega_1^2 e_1$	$\omega_1^2 e_2$	$\omega_1^2 e_3$	$\omega_1^4 g_2$	$\omega_1^6 g_3$	G
0.1	82.247	82.247	−164.49	81174	−4450900	1.00000
0.2	20.562	20.562	−41.123	5073.4	−69545	1.00000
0.3	9.1447	9.1323	−18.277	1002.2	−6105.5	1.00000
0.4	5.1883	5.0925	−10.281	317.10	−1086.6	1.00026
0.5	3.4376	3.1427	−6.5803	129.99	−284.36	1.00605
0.6	2.5780	1.9944	−4.5724	63.060	−94.035	1.05033
0.7	2.1368	1.2304	−3.3672	34.835	−35.411	1.24856
0.8	1.9057	0.68845	−2.5942	21.671	−13.614	2.03368
0.9	1.7836	0.29249	−2.0761	15.154	−4.3324	6.86723
1.0	1.7188	0.00000	−1.7188	11.817	0.0000	∞
1.1	1.6843	−0.21663	−1.4677	10.076	2.1420	8.25712
1.2	1.6659	−0.37693	−1.2890	9.1578	3.2376	2.71371
1.3	1.6561	−0.49530	−1.1608	8.6712	3.8089	1.66452
1.4	1.6509	−0.58253	−1.0684	8.4125	4.1099	1.30546
1.5	1.6481	−0.64668	−1.0014	8.2748	4.2694	1.15126
1.6	1.6466	−0.69378	−0.95286	8.2013	4.3542	1.07766
1.7	1.6458	−0.72831	−0.91753	8.1622	4.3993	1.04060
1.8	1.6454	−0.75361	−0.89181	8.1413	4.4234	1.02143
1.9	1.6452	−0.77213	−0.87307	8.1302	4.4362	1.01137
2.0	1.6451	−0.78567	−0.85940	8.1242	4.4431	1.00605

Stationary Values and Invariants of $\mathscr{P}(u, \omega_1, \frac{1}{2}\omega_1(1 + iv))$

v	$\omega_1^2 e_1$	$\omega_1^2 e_2, \omega_1^2 e_3$	$\omega_1^4 g_2$	$\omega_1^6 g_3$	G
0.2	−41.123	20.562 ± 0.19157i	5073.2	−69550	0.99974
0.4	−10.185	5.0925 ± 2.4270i	287.65	−1296.5	0.52439
0.6	−3.9887	1.9944 ± 3.9154i	−13.592	−308.06	−0.00098
0.8	−1.3769	0.68845 ± 3.9979i	−58.245	−90.640	−0.89080
1.0	0.00000	0.00000 ± 3.4376i	−47.268	0.00000	−∞
1.2	0.75385	−0.37693 ± 2.7300i	−28.107	22.902	−1.5679
1.4	1.1651	−0.58253 ± 2.0834i	−13.289	21.809	−0.18277
1.6	1.3876	−0.69378 ± 1.5574i	−3.9257	16.133	−0.00861
1.8	1.5072	−0.75361 ± 1.1516i	1.5105	11.419	0.00098
2.0	1.5713	−0.78567 ± 0.84665i	4.5401	8.3854	0.04929
2.2	1.6056	−0.80282 ± 0.62056i	6.1939	6.6128	0.20126
2.4	1.6240	−0.81198 ± 0.45410i	7.0869	5.6223	0.41704
2.6	1.6337	−0.81687 ± 0.33201i	7.5664	5.0810	0.62146
2.8	1.6390	−0.81948 ± 0.24263i	7.8231	4.7885	0.77335
3.0	1.6417	−0.82087 ± 0.17727i	7.9603	4.6314	0.87096
3.2	1.6432	−0.82162 ± 0.12949i	8.0336	4.5473	0.92865
3.4	1.6440	−0.82201 ± 0.09459i	8.0727	4.5024	0.96119
3.6	1.6445	−0.82223 ± 0.06909i	8.0936	4.4784	0.97908
3.8	1.6447	−0.82234 ± 0.05047i	8.1047	4.4655	0.98878
4.0	1.6448	−0.82240 ± 0.03686i	8.1106	4.4587	0.99399

Table L. Weierstrass's Function $\omega_1^2 \mathcal{P}(u|g_2, g_3)$ $(G = g_2^3/27g_3^2)$

$g_3 > 0$

$G \rightarrow$

u/ω_1	-6.0	-5.0	-4.0	-3.0	-2.0
0.02	2500.0	2500.0	2500.0	2500.0	2500.0
0.04	625.00	625.00	625.00	625.00	625.00
0.06	277.77	277.77	277.77	277.77	277.77
0.08	156.24	156.24	156.24	156.24	156.24
0.10	99.982	99.982	99.983	99.984	99.985
0.12	69.418	69.419	69.420	69.421	69.423
0.14	50.985	50.986	50.987	50.989	50.991
0.16	39.016	39.017	39.019	39.021	39.025
0.18	30.806	30.807	30.809	30.812	30.817
0.20	24.928	24.930	24.932	24.936	24.942
0.22	20.574	20.576	20.580	20.584	20.591
0.24	17.258	17.261	17.264	17.270	17.278
0.26	14.672	14.676	14.680	14.686	14.696
0.28	12.616	12.620	12.625	12.632	12.643
0.30	10.952	10.957	10.963	10.971	10.984
0.32	9.5857	9.5909	9.5977	9.6072	9.6218
0.34	8.4487	8.4546	8.4623	8.4731	8.4896
0.36	7.4914	7.4980	7.5068	7.5189	7.5373
0.38	6.6768	6.6843	6.6940	6.7076	6.7282
0.40	5.9771	5.9854	5.9963	6.0113	6.0342
0.42	5.3709	5.3801	5.3921	5.4087	5.4339
0.44	4.8415	4.8516	4.8648	4.8831	4.9108
0.46	4.3759	4.3870	4.4015	4.4215	4.4517
0.48	3.9638	3.9759	3.9917	4.0135	4.0464
0.50	3.5968	3.6100	3.6272	3.6508	3.6865
0.52	3.2683	3.2826	3.3012	3.3267	3.3652
0.54	2.9729	2.9883	3.0083	3.0358	3.0772
0.56	2.7061	2.7227	2.7442	2.7736	2.8180
0.58	2.4643	2.4820	2.5050	2.5365	2.5838
0.60	2.2444	2.2633	2.2878	2.3214	2.3717
0.62	2.0439	2.0640	2.0901	2.1257	2.1791
0.64	1.8608	1.8822	1.9097	1.9475	2.0039
0.66	1.6934	1.7160	1.7451	1.7849	1.8443
0.68	1.5403	1.5640	1.5946	1.6365	1.6989
0.70	1.4001	1.4250	1.4572	1.5011	1.5664
0.72	1.2720	1.2981	1.3317	1.3776	1.4458
0.74	1.1552	1.1824	1.2174	1.2652	1.3361
0.76	1.0489	1.0771	1.1136	1.1632	1.2368
0.78	0.95253	0.98185	1.0196	1.0709	1.1471
0.80	0.86571	0.89600	0.93499	0.98799	1.0665
0.82	0.78802	0.81923	0.85938	0.91391	0.99467
0.84	0.71916	0.75121	0.79242	0.84838	0.93117
0.86	0.65887	0.69168	0.73386	0.79110	0.87573
0.88	0.60696	0.64044	0.68347	0.74185	0.82811
0.90	0.56327	0.59733	0.64109	0.70045	0.78811
0.92	0.52767	0.56221	0.60659	0.66675	0.75558
0.94	0.50006	0.53499	0.57985	0.64065	0.73040
0.96	0.48040	0.51560	0.56080	0.62207	0.71248
0.98	0.46861	0.50398	0.54940	0.61094	0.70174
1.00	0.46469	0.50011	0.54560	0.60723	0.69817

Table L. (Continued)

$$g_3 > 0$$
$$G \rightarrow$$

u/ω_1	-1.8	-1.6	-1.4	-1.2	-1.0
0.02	2500.0	2500.0	2500.0	2500.0	2500.0
0.04	625.00	625.00	625.00	625.00	625.00
0.06	277.77	277.77	277.77	277.77	277.77
0.08	156.24	156.24	156.24	156.24	156.24
0.10	99.986	99.986	99.986	99.987	99.988
0.12	69.424	69.424	69.425	69.426	69.427
0.14	50.992	50.993	50.994	50.995	50.996
0.16	39.026	39.027	39.028	39.030	39.031
0.18	30.818	30.819	30.821	30.823	30.825
0.20	24.943	24.945	24.947	24.949	24.952
0.22	20.593	20.595	20.597	20.600	20.603
0.24	17.280	17.283	17.285	17.289	17.293
0.26	14.698	14.701	14.705	14.709	14.713
0.28	12.646	12.650	12.654	12.658	12.664
0.30	10.987	10.991	10.996	11.001	11.007
0.32	9.6257	9.6302	9.6353	9.6413	9.6484
0.34	8.4940	8.4991	8.5049	8.5116	8.5196
0.36	7.5423	7.5480	7.5545	7.5621	7.5710
0.38	6.7338	6.7401	6.7473	6.7557	6.7657
0.40	6.0404	6.0474	6.0554	6.0647	6.0757
0.42	5.4407	5.4484	5.4573	5.4675	5.4796
0.44	4.9183	4.9267	4.9364	4.9476	4.9609
0.46	4.4599	4.4691	4.4797	4.4919	4.5063
0.48	4.0553	4.0653	4.0767	4.0900	4.1057
0.50	3.6961	3.7069	3.7193	3.7336	3.7505
0.52	3.3756	3.3873	3.4006	3.4160	3.4342
0.54	3.0883	3.1009	3.1152	3.1317	3.1512
0.56	2.8299	2.8433	2.8586	2.8762	2.8970
0.58	2.5965	2.6108	2.6271	2.6459	2.6680
0.60	2.3852	2.4004	2.4176	2.4376	2.4611
0.62	2.1934	2.2094	2.2277	2.2488	2.2736
0.64	2.0189	2.0359	2.0552	2.0775	2.1036
0.66	1.8602	1.8780	1.8983	1.9217	1.9491
0.68	1.7155	1.7342	1.7555	1.7800	1.8088
0.70	1.5838	1.6034	1.6256	1.6512	1.6812
0.72	1.4639	1.4843	1.5075	1.5342	1.5654
0.74	1.3550	1.3762	1.4003	1.4280	1.4604
0.76	1.2563	1.2783	1.3033	1.3320	1.3655
0.78	1.1673	1.1900	1.2158	1.2454	1.2801
0.80	1.0874	1.1108	1.1373	1.1678	1.2035
0.82	1.0161	1.0401	1.0674	1.0987	1.1353
0.84	0.95312	0.97775	1.0057	1.0378	1.0752
0.86	0.89816	0.92332	0.95184	0.98460	1.0229
0.88	0.85097	0.87660	0.90564	0.93899	0.97794
0.90	0.81133	0.83737	0.86686	0.90073	0.94026
0.92	0.77911	0.80547	0.83535	0.86964	0.90966
0.94	0.75416	0.78079	0.81096	0.84558	0.88599
0.96	0.73641	0.76323	0.79361	0.82847	0.86916
0.98	0.72578	0.75271	0.78322	0.81823	0.85908
1.00	0.72224	0.74921	0.77976	0.81482	0.85573

Table L. (*Continued*)

$$g_3 > 0$$
$$G \rightarrow$$

u/ω_1	−0.8	−0.6	−0.4	−0.2	0
0.02	2500.0	2500.0	2500.0	2500.0	2500.0
0.04	625.00	625.00	625.00	625.00	625.00
0.06	277.77	277.77	277.77	277.78	277.78
0.08	156.24	156.24	156.24	156.25	156.25
0.10	99.989	99.990	99.991	99.993	100.00
0.12	69.428	69.430	69.432	69.435	69.445
0.14	50.998	51.000	51.003	51.007	51.021
0.16	39.034	39.036	39.040	39.045	39.063
0.18	30.828	30.831	30.836	30.843	30.865
0.20	24.955	24.960	24.965	24.974	25.001
0.22	20.607	20.613	20.619	20.630	20.662
0.24	17.298	17.304	17.312	17.324	17.363
0.26	14.719	14.726	14.736	14.750	14.795
0.28	12.670	12.679	12.690	12.706	12.758
0.30	11.015	11.024	11.037	11.056	11.115
0.32	9.6570	9.6678	9.6822	9.7035	9.7704
0.34	8.5293	8.5415	8.5577	8.5816	8.6566
0.36	7.5819	7.5955	7.6136	7.6403	7.7237
0.38	6.7778	6.7929	6.8130	6.8425	6.9348
0.40	6.0890	6.1058	6.1279	6.1604	6.2617
0.42	5.4943	5.5127	5.5369	5.5726	5.6832
0.44	4.9769	4.9970	5.0235	5.0623	5.1825
0.46	4.5238	4.5456	4.5744	4.6165	4.7464
0.48	4.1246	4.1482	4.1793	4.2248	4.3646
0.50	3.7709	3.7964	3.8299	3.8788	4.0286
0.52	3.4561	3.4835	3.5194	3.5718	3.7317
0.54	3.1747	3.2039	3.2423	3.2982	3.4683
0.56	2.9221	2.9533	2.9942	3.0536	3.2339
0.58	2.6946	2.7278	2.7712	2.8341	3.0246
0.60	2.4893	2.5244	2.5703	2.6368	2.8372
0.62	2.3034	2.3405	2.3888	2.4589	2.6693
0.64	2.1350	2.1739	2.2248	2.2983	2.5184
0.66	1.9821	2.0229	2.0762	2.1531	2.3829
0.68	1.8433	1.8860	1.9416	2.0219	2.2609
0.70	1.7172	1.7617	1.8197	1.9032	2.1513
0.72	1.6028	1.6491	1.7094	1.7960	2.0527
0.74	1.4993	1.5473	1.6097	1.6993	1.9643
0.76	1.4057	1.4553	1.5198	1.6123	1.8852
0.78	1.3215	1.3727	1.4391	1.5343	1.8146
0.80	1.2461	1.2988	1.3670	1.4647	1.7519
0.82	1.1791	1.2331	1.3030	1.4031	1.6965
0.84	1.1200	1.1752	1.2467	1.3489	1.6481
0.86	1.0685	1.1248	1.1977	1.3018	1.6062
0.88	1.0244	1.0817	1.1558	1.2616	1.5706
0.90	0.98744	1.0455	1.1207	1.2279	1.5408
0.92	0.95741	1.0162	1.0922	1.2007	1.5167
0.94	0.93419	0.99352	1.0702	1.1796	1.4982
0.96	0.91768	0.97740	1.0546	1.1647	1.4850
0.98	0.90781	0.96776	1.0453	1.1557	1.4772
1.00	0.90452	0.96455	1.0421	1.1528	1.4746

Table L. (*Continued*)

$$g_3 > 0$$
$$G \rightarrow$$

u/ω_1	0.2	0.4	0.6	0.8	1.0
0.02	2500.0	2500.0	2500.0	2500.0	2500.0
0.04	625.00	625.00	625.00	625.00	625.00
0.06	277.78	277.78	277.78	277.78	277.78
0.08	156.25	156.25	156.25	156.25	156.25
0.10	100.00	100.00	100.00	100.00	100.00
0.12	69.449	69.450	69.450	69.450	69.450
0.14	51.027	51.027	51.028	51.028	51.028
0.16	39.071	39.072	39.072	39.073	39.073
0.18	30.874	30.876	30.877	30.877	30.878
0.20	25.013	25.014	25.015	25.016	25.016
0.22	20.677	20.679	20.680	20.681	20.681
0.24	17.380	17.382	17.383	17.384	17.385
0.26	14.815	14.818	14.819	14.820	14.821
0.28	12.781	12.784	12.786	12.787	12.788
0.30	11.141	11.144	11.146	11.148	11.149
0.32	9.7998	9.8038	9.8061	9.8077	9.8089
0.34	8.6895	8.6940	8.6965	8.6983	8.6997
0.36	7.7602	7.7651	7.7680	7.7700	7.7714
0.38	6.9749	6.9804	6.9835	6.9857	6.9873
0.40	6.3057	6.3117	6.3151	6.3175	6.3192
0.42	5.7311	5.7376	5.7413	5.7439	5.7458
0.44	5.2343	5.2413	5.2454	5.2481	5.2502
0.46	4.8023	4.8099	4.8142	4.8172	4.8195
0.48	4.4245	4.4327	4.4373	4.4405	4.4430
0.50	4.0927	4.1014	4.1063	4.1097	4.1123
0.52	3.7999	3.8091	3.8144	3.8180	3.8208
0.54	3.5406	3.5504	3.5560	3.5598	3.5627
0.56	3.3102	3.3205	3.3264	3.3305	3.3336
0.58	3.1050	3.1158	3.1220	3.1263	3.1295
0.60	2.9216	2.9330	2.9395	2.9440	2.9474
0.62	2.7576	2.7695	2.7763	2.7810	2.7845
0.64	2.6106	2.6230	2.6301	2.6350	2.6387
0.66	2.4787	2.4916	2.4990	2.5041	2.5079
0.68	2.3604	2.3738	2.3814	2.3867	2.3907
0.70	2.2542	2.2681	2.2760	2.2814	2.2855
0.72	2.1590	1.1733	2.1814	2.1871	2.1913
0.74	2.0738	2.0885	2.0968	2.1026	2.1070
0.76	1.9977	2.0127	2.0213	2.0272	2.0317
0.78	1.9299	1.9453	1.9541	1.9602	1.9647
0.80	1.8698	1.8855	1.8945	1.9007	1.9054
0.82	1.8168	1.8329	1.8421	1.8484	1.8532
0.84	1.7706	1.7870	1.7963	1.8027	1.8076
0.86	1.7307	1.7473	1.7567	1.7633	1.7682
0.88	1.6967	1.7135	1.7231	1.7297	1.7347
0.90	1.6684	1.6854	1.6951	1.7018	1.7068
0.92	1.6455	1.6626	1.6725	1.6792	1.6843
0.94	1.6279	1.6452	1.6550	1.6619	1.6670
0.96	1.6154	1.6328	1.6427	1.6495	1.6547
0.98	1.6080	1.6254	1.6353	1.6422	1.6474
1.00	1.6055	1.6229	1.6329	1.6398	1.6449

Table L. (*Continued*)

$$g_3 > 0$$
$$G \rightarrow$$

u/ω_1	1.2	1.4	1.6	1.8	2.0
0.02	2500.0	2500.0	2500.0	2500.0	2500.0
0.04	625.00	625.00	625.00	625.00	625.00
0.06	277.78	277.78	277.78	277.78	277.78
0.08	156.25	156.25	156.25	156.25	156.25
0.10	100.00	100.00	100.00	100.00	100.00
0.12	69.450	69.451	69.451	69.451	69.451
0.14	51.029	51.029	51.029	51.029	51.029
0.16	39.073	39.073	39.074	39.074	39.074
0.18	30.878	30.878	30.878	30.879	30.879
0.20	25.017	25.017	25.017	25.018	25.018
0.22	20.682	20.682	20.682	20.683	20.683
0.24	17.386	17.386	17.386	17.387	17.387
0.26	14.822	14.822	14.823	14.823	14.823
0.28	12.789	12.789	12.790	12.790	12.791
0.30	11.150	11.151	11.151	11.152	11.152
0.32	9.8099	9.8107	9.8113	9.8119	9.8124
0.34	8.7007	8.7016	8.7023	8.7030	8.7035
0.36	7.7726	7.7736	7.7744	7.7751	7.7758
0.38	6.9886	6.9897	6.9906	6.9914	6.9921
0.40	6.3207	6.3218	6.3228	6.3237	6.3244
0.42	5.7473	5.7486	5.7497	5.7506	5.7514
0.44	5.2519	5.2533	5.2545	5.2555	5.2564
0.46	4.8213	4.8227	4.8240	4.8251	4.8260
0.48	4.4449	4.4465	4.4478	4.4490	4.4500
0.50	4.1144	4.1161	4.1175	4.1188	4.1198
0.52	3.8230	3.8248	3.8263	3.8276	3.8288
0.54	3.5650	3.5669	3.5686	3.5700	3.5712
0.56	3.3360	3.3380	3.3397	3.3412	3.3425
0.58	3.1321	3.1342	3.1360	3.1375	3.1389
0.60	2.9501	2.9523	2.9542	2.9558	2.9572
0.62	2.7873	2.7896	2.7916	2.7933	2.7948
0.64	2.6416	2.6440	2.6461	2.6478	2.6494
0.66	2.5110	2.5135	2.5156	2.5174	2.5190
0.68	2.3938	2.3964	2.3986	2.4005	2.4022
0.70	2.2888	2.2915	2.2937	2.2957	2.2974
0.72	2.1947	2.1974	2.1998	2.2018	2.2036
0.74	2.1104	2.1133	2.1157	2.1178	2.1196
0.76	2.0353	2.0382	2.0407	2.0428	2.0447
0.78	1.9684	1.9714	1.9739	1.9761	1.9780
0.80	1.9091	1.9122	1.9148	1.9170	1.9190
0.82	1.8570	1.8601	1.8628	1.8650	1.8670
0.84	1.8115	1.8146	1.8173	1.8196	1.8217
0.86	1.7721	1.7754	1.7781	1.7804	1.7825
0.88	1.7387	1.7420	1.7447	1.7471	1.7492
0.90	1.7108	1.7141	1.7169	1.7193	1.7215
0.92	1.6884	1.6917	1.6945	1.6969	1.6991
0.94	1.6711	1.6744	1.6773	1.6797	1.6818
0.96	1.6588	1.6622	1.6650	1.6675	1.6696
0.98	1.6515	1.6549	1.6577	1.6602	1.6623
1.00	1.6490	1.6524	1.6553	1.6578	1.6599

Table L. (*Continued*)

$$g_3 > 0$$
$$G \rightarrow$$

u/ω_1	3.0	4.0	5.0	6.0
0.02	2500.0	2500.0	2500.0	2500.0
0.04	625.00	625.00	625.00	625.00
0.06	277.78	277.78	277.78	277.78
0.08	156.25	156.25	156.25	156.25
0.10	100.00	100.00	100.00	100.00
0.12	69.451	69.451	69.451	69.452
0.14	51.030	51.030	51.030	51.030
0.16	39.074	39.075	39.075	39.075
0.18	30.879	30.880	30.880	30.880
0.20	25.019	25.019	25.020	25.020
0.22	20.684	20.684	20.685	20.685
0.24	17.388	17.389	17.389	17.390
0.26	14.825	14.826	14.826	14.827
0.28	12.792	12.793	12.794	12.794
0.30	11.154	11.155	11.156	11.156
0.32	9.8142	9.8154	9.8163	9.8170
0.34	8.7056	8.7069	8.7079	8.7087
0.36	7.7780	7.7795	7.7806	7.7814
0.38	6.9946	6.9962	6.9974	6.9983
0.40	6.3272	6.3290	6.3303	6.3313
0.42	5.7544	5.7564	5.7578	5.7589
0.44	5.2596	5.2617	5.2632	5.2644
0.46	4.8295	4.8318	4.8334	4.8347
0.48	4.4537	4.4561	4.4579	4.4593
0.50	4.1238	4.1264	4.1283	4.1297
0.52	3.8330	3.8357	3.8377	3.8393
0.54	3.5756	3.5785	3.5807	3.5823
0.56	3.3472	3.3503	3.3525	3.3542
0.58	3.1438	3.1471	3.1494	3.1512
0.60	2.9624	2.9658	2.9682	2.9701
0.62	2.8002	2.8037	2.8063	2.8083
0.64	2.6550	2.6587	2.6614	2.6634
0.66	2.5249	2.5287	2.5315	2.5336
0.68	2.4082	2.4122	2.4151	2.4173
0.70	2.3037	2.3078	2.3108	2.3131
0.72	2.2100	2.2143	2.2173	2.2197
0.74	2.1263	2.1306	2.1338	2.1362
0.76	2.0515	2.0560	2.0592	2.0617
0.78	1.9850	1.9896	1.9929	1.9955
0.80	1.9261	1.9308	1.9342	1.9368
0.82	1.8743	1.8791	1.8825	1.8852
0.84	1.8291	1.8339	1.8374	1.8401
0.86	1.7900	1.7949	1.7985	1.8012
0.88	1.7568	1.7618	1.7654	1.7682
0.90	1.7292	1.7342	1.7378	1.7407
0.92	1.7068	1.7119	1.7156	1.7184
0.94	1.6897	1.6948	1.6985	1.7013
0.96	1.6775	1.6826	1.6864	1.6892
0.98	1.6702	1.6754	1.6791	1.6820
1.00	1.6678	1.6730	1.6767	1.6796

Table L. (*Continued*)

$$g_3 < 0$$
$$G \rightarrow$$

u/ω_1	-6.0	-5.0	-4.0	-3.0	-2.0
0.02	2500.0	2500.0	2500.0	2500.0	2500.0
0.04	625.00	625.00	625.00	625.00	625.00
0.06	277.77	277.77	277.77	277.77	277.77
0.08	156.23	156.23	156.23	156.23	156.23
0.10	99.972	99.972	99.972	99.971	99.971
0.12	69.404	69.404	69.404	69.403	69.402
0.14	50.966	50.965	50.965	50.964	50.963
0.16	38.991	38.990	38.989	38.988	38.987
0.18	30.773	30.772	30.771	30.770	30.768
0.20	24.888	24.886	24.885	24.883	24.881
0.22	20.525	20.523	20.521	20.519	20.516
0.24	17.198	17.197	17.194	17.191	17.188
0.26	14.601	14.599	14.596	14.593	14.588
0.28	12.532	12.530	12.526	12.522	12.516
0.30	10.855	10.851	10.847	10.842	10.835
0.32	9.4729	9.4691	9.4643	9.4580	9.4495
0.34	8.3192	8.3147	8.3090	8.3016	8.2914
0.36	7.3437	7.3385	7.3318	7.3231	7.3109
0.38	6.5096	6.5034	6.4957	6.4854	6.4710
0.40	5.7885	5.7815	5.7726	5.7607	5.7438
0.42	5.1595	5.1514	5.1412	5.1274	5.1077
0.44	4.6056	4.5965	4.5848	4.5690	4.5461
0.46	4.1140	4.1037	4.0903	4.0723	4.0459
0.48	3.6743	3.6626	3.6475	3.6270	3.5968
0.50	3.2781	3.2650	3.2480	3.2249	3.1905
0.52	2.9189	2.9043	2.8853	2.8592	2.8203
0.54	2.5914	2.5751	2.5539	2.5247	2.4809
0.56	2.2910	2.2730	2.2495	2.2170	2.1679
0.58	2.0144	1.9946	1.9686	1.9326	1.8778
0.60	1.7587	1.7369	1.7083	1.6686	1.6079
0.62	1.5215	1.4977	1.4664	1.4228	1.3557
0.64	1.3009	1.2750	1.2410	1.1933	1.1197
0.66	1.0955	1.0675	1.0305	0.97868	0.89822
0.68	0.90406	0.87386	0.83396	0.77782	0.69033
0.70	0.72568	0.69326	0.65036	0.58985	0.49519
0.72	0.55965	0.52500	0.47909	0.41418	0.31226
0.74	0.40545	0.36859	0.31967	0.25036	0.14116
0.76	0.26273	0.22369	0.17180	0.09814	-0.01827
0.78	0.13126	0.09008	0.03529	-0.04262	-0.16611
0.80	0.01090	-0.03233	-0.08991	-0.17193	-0.30227
0.82	-0.09838	-0.14356	-0.20380	-0.28973	-0.42662
0.84	-0.19657	-0.24356	-0.30629	-0.39590	-0.53892
0.86	-0.28358	-0.33226	-0.39727	-0.49025	-0.63894
0.88	-0.35934	-0.40951	-0.47658	-0.57259	-0.72638
0.90	-0.42371	-0.47519	-0.54405	-0.64271	-0.80096
0.92	-0.47658	-0.52916	-0.59953	-0.70041	-0.86241
0.94	-0.51784	-0.57129	-0.64285	-0.74550	-0.91048
0.96	-0.54739	-0.60147	-0.67390	-0.77783	-0.94497
0.98	-0.56515	-0.61961	-0.69256	-0.79727	-0.96573
1.00	-0.57107	-0.62567	-0.69879	-0.80376	-0.97266

Table L. (*Continued*)

$$g_3 < 0$$
$$G \rightarrow$$

u/ω_1	-1.8	-1.6	-1.4	-1.2	-1.0
0.02	2500.0	2500.0	2500.0	2500.0	2500.0
0.04	625.00	625.00	625.00	625.00	625.00
0.06	277.77	277.77	277.77	277.77	277.77
0.08	156.23	156.23	156.23	156.23	156.23
0.10	99.971	99.971	99.971	99.971	99.971
0.12	69.402	69.402	69.402	69.402	69.402
0.14	50.963	50.962	50.962	50.962	50.962
0.16	38.987	38.986	38.986	38.986	38.986
0.18	30.768	30.767	30.767	30.767	30.767
0.20	24.880	24.880	24.879	24.879	24.879
0.22	20.515	20.515	20.514	20.514	20.513
0.24	17.187	17.186	17.185	17.184	17.183
0.26	14.587	14.586	14.585	14.584	14.583
0.28	12.515	12.513	12.512	12.510	12.509
0.30	10.833	10.832	10.830	10.828	10.826
0.32	9.4474	9.4452	9.4429	9.4405	9.4380
0.34	8.2889	8.2862	8.2833	8.2802	8.2771
0.36	7.3078	7.3046	7.3011	7.2973	7.2933
0.38	6.4674	6.4634	6.4592	6.4545	6.4496
0.40	5.7395	5.7348	5.7296	5.7240	5.7179
0.42	5.1026	5.0971	5.0909	5.0842	5.0767
0.44	4.5402	4.5337	4.5264	4.5184	4.5094
0.46	4.0391	4.0315	4.0230	4.0135	4.0028
0.48	3.5889	3.5801	3.5702	3.5591	3.5464
0.50	3.1814	3.1713	3.1599	3.1469	3.1321
0.52	2.8100	2.7984	2.7853	2.7703	2.7530
0.54	2.4692	2.4560	2.4410	2.4239	2.4039
0.56	2.1547	2.1398	2.1228	2.1033	2.0805
0.58	1.8630	1.8463	1.8272	1.8051	1.7791
0.60	1.5914	1.5727	1.5513	1.5264	1.4971
0.62	1.3375	1.3167	1.2929	1.2651	1.2321
0.64	1.0995	1.0766	1.0502	1.0193	0.98248
0.66	0.87610	0.85090	0.82180	0.78769	0.74689
0.68	0.66619	0.63864	0.60676	0.56929	0.52433
0.70	0.46899	0.43903	0.40430	0.36338	0.31412
0.72	0.28396	0.25155	0.21392	0.16948	0.11584
0.74	0.11076	0.07589	0.03534	-0.01265	-0.07072
0.76	-0.05077	-0.08808	-0.13155	-0.18309	-0.24559
0.78	-0.20066	-0.24038	-0.28672	-0.34175	-0.40864
0.80	-0.33882	-0.38088	-0.43001	-0.48844	-0.55960
0.82	-0.46507	-0.50937	-0.56117	-0.62287	-0.69814
0.84	-0.57918	-0.62559	-0.67991	-0.74469	-0.82383
0.86	-0.68085	-0.72921	-0.78586	-0.85349	-0.93622
0.88	-0.76979	-0.81990	-0.87865	-0.94886	-1.0348
0.90	-0.84567	-0.89733	-0.95792	-1.0304	-1.1192
0.92	-0.90822	-0.96117	-1.0233	-1.0977	-1.1889
0.94	-0.95717	-1.0112	-1.0745	-1.1504	-1.2436
0.96	-0.99230	-1.0470	-1.1113	-1.1883	-1.2829
0.98	-1.01344	-1.0686	-1.1335	-1.2111	-1.3065
1.00	-1.02050	-1.0758	-1.1409	-1.2187	-1.3144

Table L. (*Continued*)

$$g_3 < 0$$
$$G \rightarrow$$

u/ω_1	-0.8	-0.6	-0.4	-0.2	0.0
0.02	2500.0	2500.0	2500.0	2500.0	2500.0
0.04	625.00	625.00	625.00	625.00	625.00
0.06	277.77	277.77	277.77	277.77	277.78
0.08	156.23	156.23	156.23	156.23	156.25
0.10	99.971	99.971	99.972	99.974	99.999
0.12	69.402	69.402	69.403	69.406	69.442
0.14	50.962	50.963	50.964	50.968	51.016
0.16	38.986	38.986	38.988	38.993	39.054
0.18	30.767	30.767	30.769	30.775	30.851
0.20	24.879	24.879	24.881	24.888	24.980
0.22	20.513	20.513	20.516	20.523	20.632
0.24	17.183	17.183	17.186	17.194	17.320
0.26	14.582	14.582	14.584	14.593	14.736
0.28	12.508	12.508	12.509	12.518	12.679
0.30	10.824	10.824	10.825	10.834	11.011
0.32	9.4358	9.4343	9.4350	9.4438	9.6361
0.34	8.2741	8.2717	8.2714	8.2794	8.4855
0.36	7.2893	7.2858	7.2842	7.2910	7.5088
0.38	6.4444	6.4395	6.4363	6.4412	6.6681
0.40	5.7113	5.7048	5.6995	5.7019	5.9346
0.42	5.0686	5.0600	5.0523	5.0513	5.2861
0.44	4.4994	4.4885	4.4778	4.4726	4.7049
0.46	3.9906	3.9771	3.9628	3.9525	4.1771
0.48	3.5319	3.5153	3.4968	3.4804	3.6913
0.50	3.1148	3.0947	3.0715	3.0479	3.2384
0.52	2.7328	2.7088	2.6802	2.6481	2.8108
0.54	2.3804	2.3520	2.3173	2.2755	2.4022
0.56	2.0532	2.0201	1.9786	1.9258	2.0075
0.58	1.7479	1.7095	1.6606	1.5954	1.6226
0.60	1.4616	1.4175	1.3603	1.2815	1.2440
0.62	1.1920	1.1418	1.0758	0.98197	0.86923
0.64	0.93754	0.88075	0.80521	0.69511	0.49632
0.66	0.69682	0.63310	0.54741	0.41978	0.12406
0.68	0.46890	0.39793	0.30153	0.15524	-0.24811
0.70	0.25316	0.17465	0.06708	-0.09886	-0.62008
0.72	0.04920	-0.03705	-0.15616	-0.34257	-0.99107
0.74	-0.14311	-0.23725	-0.36815	-0.57564	-1.3596
0.76	-0.32374	-0.42580	-0.56861	-0.79756	-1.7237
0.78	-0.49249	-0.60242	-0.75711	-1.0076	-2.0806
0.80	-0.64904	-0.76667	-0.93305	-1.2048	-2.4271
0.82	-0.79294	-0.91801	-1.0957	-1.3882	-2.7594
0.84	-0.92372	-1.0558	-1.2443	-1.5567	-3.0734
0.86	-1.0408	-1.1795	-1.3780	-1.7089	-3.3648
0.88	-1.1437	-1.2883	-1.4960	-1.8439	-3.6290
0.90	-1.2318	-1.3817	-1.5974	-1.9603	-3.8615
0.92	-1.3047	-1.4590	-1.6815	-2.0572	-4.0584
0.94	-1.3619	-1.5197	-1.7477	-2.1336	-4.2157
0.96	-1.4030	-1.5634	-1.7953	-2.1887	-4.3305
0.98	-1.4278	-1.5898	-1.8241	-2.2220	-4.4003
1.00	-1.4361	-1.5986	-1.8337	-2.2332	-4.4238

Table L. (*Continued*)

$$g_3 < 0$$
$$G \rightarrow$$

u/ω_1	0.2	0.4	0.6	0.8	1.0
0.02	2500.0	2500.0	2500.0	2500.0	∞
0.04	625.01	625.02	625.03	625.04	
0.06	277.80	277.82	277.84	277.87	
0.08	156.30	156.32	156.35	156.41	
0.10	100.07	100.11	100.16	100.25	
0.12	69.543	69.601	69.672	69.798	
0.14	51.152	51.229	51.326	51.493	
0.16	39.230	39.330	39.453	39.667	
0.18	31.071	31.195	31.348	31.612	
0.20	25.247	25.398	25.583	25.900	
0.22	20.950	21.129	21.347	21.720	
0.24	17.692	17.900	18.154	18.584	
0.26	15.164	15.403	15.693	16.181	
0.28	13.164	13.435	13.762	14.310	
0.30	11.555	11.858	12.223	12.830	
0.32	10.239	10.575	10.978	11.644	
0.34	9.1480	9.5163	9.9574	10.682	
0.36	8.2296	8.6307	9.1096	9.8915	
0.38	7.4456	7.8792	8.3953	9.2336	
0.40	6.7662	7.2316	7.7843	8.6777	
0.42	6.1681	6.6644	7.2528	8.1998	
0.44	5.6328	6.1586	6.7815	7.7804	
0.46	5.1448	5.6985	6.3544	7.4034	
0.48	4.6917	5.2710	5.9581	7.0550	
0.50	4.2626	4.8647	5.5807	6.7230	
0.52	3.8482	4.4699	5.2117	6.3965	
0.54	3.4406	4.0776	4.8417	6.0650	
0.56	3.0325	3.6797	4.4616	5.7187	
0.58	2.6177	3.2689	4.0630	5.3478	
0.60	2.1904	2.8382	3.6377	4.9424	
0.62	1.7458	2.3815	3.1779	4.4925	
0.64	1.2796	1.8929	2.6759	3.9879	
0.66	0.78860	1.3674	2.1249	3.4183	
0.68	0.27022	0.80110	1.5185	2.7738	
0.70	−0.27679	0.19092	0.85135	2.0447	
0.72	−0.85240	−0.46453	0.11981	1.2226	
0.74	−1.4550	−1.1648	−0.67782	0.30131	
0.76	−2.0813	−1.9070	−1.5404	−0.72252	
0.78	−2.7260	−2.6858	−2.4632	−1.8477	
0.80	−3.3815	−3.4925	−3.4375	−3.0672	
0.82	−4.0384	−4.3151	−4.4492	−4.3665	
0.84	−4.6848	−5.1382	−5.4794	−5.7224	
0.86	−5.3074	−5.9433	−6.5034	−7.1020	
0.88	−5.8913	−6.7089	−7.4919	−8.4629	
0.90	−6.4207	−7.4120	−8.4119	−9.7546	
0.92	−6.8802	−8.0288	−9.2285	−10.921	
0.94	−7.2552	−8.5366	−9.9073	−11.904	
0.96	−7.5330	−8.9154	−10.417	−12.652	
0.98	−7.7038	−9.1494	−10.734	−13.119	
1.00	−7.7615	−9.2286	−10.841	−13.278	

Table L. (*Continued*)

$$g_3 < 0$$
$$G \rightarrow$$

u/ω_1	1.2	1.4	1.6	1.8	2.0
0.02	2500.0	2500.0	2500.0	2500.0	2500.0
0.04	625.00	625.00	625.00	625.00	625.00
0.06	277.78	277.78	277.78	277.78	277.78
0.08	156.26	156.26	156.26	156.26	156.26
0.10	100.02	100.01	100.01	100.01	100.01
0.12	69.471	69.465	69.463	69.461	69.460
0.14	51.057	51.049	51.045	51.043	51.042
0.16	39.110	39.099	39.095	39.092	39.090
0.18	30.924	30.911	30.905	30.901	30.899
0.20	25.073	25.057	25.050	25.046	25.043
0.22	20.749	20.730	20.722	20.716	20.713
0.24	17.465	17.443	17.433	17.427	17.423
0.26	14.914	14.888	14.876	14.869	14.865
0.28	12.895	12.865	12.851	12.843	12.838
0.30	11.270	11.236	11.221	11.212	11.206
0.32	9.9448	9.9070	9.8899	9.8797	9.8728
0.34	8.8510	8.8089	8.7899	8.7785	8.7709
0.36	7.9386	7.8923	7.8712	7.8587	7.8503
0.38	7.1707	7.1200	7.0969	7.0832	7.0739
0.40	6.5191	6.4640	6.4388	6.4239	6.4138
0.42	5.9625	5.9028	5.8756	5.8594	5.8485
0.44	5.4838	5.4196	5.3903	5.3729	5.3611
0.46	5.0700	5.0013	4.9699	4.9512	4.9386
0.48	4.7106	4.6374	4.6039	4.5839	4.5704
0.50	4.3969	4.3193	4.2837	4.2625	4.2481
0.52	4.1223	4.0402	4.0025	3.9800	3.9648
0.54	3.8809	3.7944	3.7547	3.7310	3.7150
0.56	3.6682	3.5775	3.5358	3.5109	3.4940
0.58	3.4803	3.3854	3.3417	3.3156	3.2979
0.60	3.3140	3.2150	3.1694	3.1421	3.1237
0.62	3.1665	3.0636	3.0161	2.9877	2.9685
0.64	3.0355	2.9288	2.8796	2.8501	2.8301
0.66	2.9192	2.8088	2.7578	2.7273	2.7066
0.68	2.8157	2.7019	2.6492	2.6177	2.5963
0.70	2.7238	2.6066	2.5524	2.5199	2.4978
0.72	2.6421	2.5218	2.4661	2.4326	2.4099
0.74	2.5696	2.4464	2.3892	2.3549	2.3316
0.76	2.5055	2.3795	2.3210	2.2859	2.2620
0.78	2.4489	2.3203	2.2606	2.2247	2.2003
0.80	2.3991	2.2682	2.2073	2.1707	2.1459
0.82	2.3556	2.2225	2.1606	2.1234	2.0981
0.84	2.3178	2.1829	2.1200	2.0823	2.0566
0.86	2.2855	2.1488	2.0851	2.0468	2.0208
0.88	2.2581	2.1199	2.0555	2.0168	1.9904
0.90	2.2354	2.0959	2.0309	1.9918	1.9652
0.92	2.2171	2.0766	2.0111	1.9717	1.9449
0.94	2.2031	2.0618	1.9959	1.9563	1.9293
0.96	2.1932	2.0514	1.9852	1.9454	1.9183
0.98	2.1873	2.0451	1.9788	1.9388	1.9117
1.00	2.1854	2.0430	1.9766	1.9367	1.9095

Table L. (*Continued*)

$$g_3 < 0$$
$$G \rightarrow$$

u/ω_1	3.0	4.0	5.0	6.0
0.02	2500.0	2500.0	2500.0	2500.0
0.04	625.00	625.00	625.00	625.00
0.06	277.78	277.78	277.78	277.78
0.08	156.26	156.26	156.26	156.25
0.10	100.01	100.01	100.01	100.01
0.12	69.458	69.457	69.456	69.456
0.14	51.038	51.037	51.036	51.036
0.16	39.086	39.084	39.083	39.082
0.18	30.894	30.891	30.890	30.889
0.20	25.036	25.033	25.032	25.031
0.22	20.705	20.701	20.700	20.698
0.24	17.413	17.409	17.407	17.405
0.26	14.853	14.849	14.846	14.845
0.28	12.825	12.820	12.817	12.815
0.30	11.191	11.185	11.182	11.180
0.32	9.8565	9.8497	9.8459	9.8434
0.34	8.7527	8.7452	8.7409	8.7381
0.36	7.8302	7.8218	7.8171	7.8140
0.38	7.0518	7.0427	7.0375	7.0341
0.40	6.3897	6.3798	6.3741	6.3704
0.42	5.8223	5.8115	5.8054	5.8013
0.44	5.3329	5.3213	5.3146	5.3103
0.46	4.9083	4.8958	4.8887	4.8840
0.48	4.5381	4.5247	4.5171	4.5121
0.50	4.2137	4.1994	4.1913	4.1860
0.52	3.9284	3.9133	3.9047	3.8990
0.54	3.6765	3.6605	3.6515	3.6455
0.56	3.4535	3.4367	3.4271	3.4208
0.58	3.2555	3.2378	3.2278	3.2212
0.60	3.0793	3.0608	3.0503	3.0434
0.62	2.9222	2.9029	2.8920	2.8848
0.64	2.7820	2.7619	2.7505	2.7431
0.66	2.6567	2.6359	2.6241	2.6163
0.68	2.5447	2.5232	2.5110	2.5029
0.70	2.4446	2.4224	2.4098	2.4015
0.72	2.3552	2.3324	2.3194	2.3108
0.74	2.2755	2.2520	2.2387	2.2299
0.76	2.2045	2.1805	2.1668	2.1578
0.78	2.1415	2.1169	2.1030	2.0937
0.80	2.0859	2.0608	2.0465	2.0371
0.82	2.0371	2.0115	1.9970	1.9874
0.84	1.9945	1.9686	1.9538	1.9441
0.86	1.9579	1.9316	1.9166	1.9067
0.88	1.9268	1.9002	1.8850	1.8750
0.90	1.9009	1.8740	1.8587	1.8486
0.92	1.8801	1.8530	1.8375	1.8273
0.94	1.8641	1.8368	1.8212	1.8110
0.96	1.8527	1.8253	1.8097	1.7994
0.98	1.8460	1.8185	1.8028	1.7925
1.00	1.8437	1.8162	1.8005	1.7902

Table M. Stationary Values and Invariants of $\mathscr{P}(u|g_2,g_3)$
$$g_3 > 0$$

G	ν	$\omega_1^2 e_1$	$\omega_1^2 e_2$ and $\omega_1^2 e_3$	$\omega_1^4 g_2$	$\omega_1^6 g_3$
−6.0	1.1086	0.46469	$-0.23234 \mp 3.0536i$	−36.650	17.432
−5.0	1.1185	0.50011	$-0.25005 \mp 3.0181i$	−35.685	18.347
−4.0	1.1317	0.54560	$-0.27280 \mp 2.9709i$	−34.412	19.425
−3.0	1.1505	0.60723	$-0.30362 \mp 2.9041i$	−32.628	20.709
−2.0	1.1803	0.69817	$-0.34909 \mp 2.7988i$	−29.870	22.216
−1.8	1.1887	0.72224	$-0.36112 \mp 2.7695i$	−29.115	22.535
−1.6	1.1983	0.74921	$-0.37461 \mp 2.7359i$	−28.256	22.852
−1.4	1.2096	0.77976	$-0.38988 \mp 2.6968i$	−27.267	23.158
−1.2	1.2230	0.81482	$-0.40741 \mp 2.6506i$	−26.110	23.439
−1.0	1.2394	0.85573	$-0.42786 \mp 2.5946i$	−24.732	23.670
−0.8	1.2600	0.90452	$-0.45226 \mp 2.5250i$	−23.047	23.807
−0.6	1.2874	0.96455	$-0.48227 \mp 2.4343i$	−20.913	23.761
−0.4	1.3265	1.0421	$-0.52107 \mp 2.3082i$	−18.054	23.342
−0.2	1.3918	1.1528	$-0.57638 \mp 2.1075i$	−13.780	22.012
0.0	1.7321	1.4746	$-0.73729 \mp 1.2770i$	0	12.825
0.2	2.1987	1.6055	$-0.80274 \mp 0.62180i$	6.1862	6.6212
0.4	2.3846	1.6229	$-0.81146 \mp 0.46516i$	7.0361	5.6792
0.6	2.5767	1.6329	$-0.81645 \mp 0.34435i$	7.5247	5.1284
0.8	2.8457	1.6398	$-0.81988 \mp 0.22585i$	7.8624	4.7436
1.0	$+\infty$	1.6449	$-0.82247 \mp 0i$	8.1174	4.4509
	κ				
1.2	1.4594	1.6490	−0.62301 −1.0260	8.3205	4.2165
1.4	1.3638	1.6524	−0.55399 −1.0984	8.4880	4.0222
1.6	1.3123	1.6553	−0.50756 −1.1477	8.6299	3.8571
1.8	1.2783	1.6578	−0.47261 −1.1852	8.7523	3.7142
2.0	1.2536	1.6599	−0.44478 −1.2152	8.8596	3.5886
3.0	1.1863	1.6678	−0.35776 −1.3101	9.2517	3.1267
4.0	1.1540	1.6730	−0.30915 −1.3638	9.5088	2.8215
5.0	1.1341	1.6767	−0.27674 −1.4000	9.6958	2.5984
6.0	1.1202	1.6796	−0.25305 −1.4266	9.8403	2.4253

Table M. (*Continued*)

$$g_3 < 0$$

G	v	$\omega_1^2 e_1$	$\omega_1^2 e_2$ and $\omega_1^2 e_3$	$\omega_1^4 g_2$	$\omega_1^6 g_3$
−6.0	0.90206	−0.57107	0.28554 ∓ 3.7527i	−55.352	−32.355
−5.0	0.89405	−0.62567	0.31283 ∓ 3.7758i	−55.852	−35.924
−4.0	0.88361	−0.69879	0.34940 ∓ 3.8051i	−56.450	−40.812
−3.0	0.86919	−0.80376	0.40188 ∓ 3.8440i	−57.167	−48.026
−2.0	0.84723	−0.97266	0.48633 ∓ 3.8991i	−57.975	−60.071
−1.8	0.84127	−1.0205	0.51025 ∓ 3.9132i	−58.128	−63.571
−1.6	0.83450	−1.0758	0.53792 ∓ 3.9286i	−58.263	−67.663
−1.4	0.82674	−1.1409	0.57043 ∓ 3.9456i	−58.367	−72.527
−1.2	0.81767	−1.2187	0.60937 ∓ 3.9645i	−58.412	−78.430
−1.0	0.80686	−1.3144	0.65721 ∓ 3.9855i	−58.352	−85.783
−0.8	0.79363	−1.4361	0.71804 ∓ 4.0088i	−58.095	−95.276
−0.6	0.77677	−1.5986	0.79929 ∓ 4.0345i	−57.443	−108.17
−0.4	0.75387	−1.8337	0.91687 ∓ 4.0615i	−55.896	−127.16
−0.2	0.71847	−2.2332	1.1166 ∓ 4.0827i	−51.713	−160.03
0	0.57735	−4.4238	2.2119 ∓ 3.8311i	0	−346.29
0.2	0.45481	−7.7615	3.8807 ∓ 3.0060i	144.58	−748.09
0.4	0.41936	−9.2286	4.6143 ∓ 2.6451i	227.51	−1044.2
0.6	0.38809	−10.841	5.4207 ∓ 2.2863i	331.70	−1501.0
0.8	0.35141	−13.278	6.6392 ∓ 1.8289i	515.57	−2518.9
	κ				
1.2	0.68520	2.1854	1.3270 −3.5124	37.747	−40.742
1.4	0.73325	2.0430	1.0304 −3.0734	29.363	−25.880
1.6	0.76200	1.9766	0.87412 −2.8508	25.596	−19.702
1.8	0.78227	1.9367	0.77230 −2.7090	23.371	−16.207
2.0	0.79773	1.9095	0.69892 −2.6084	21.877	−13.924
3.0	0.84294	1.8437	0.50350 −2.3472	18.325	−8.7159
4.0	0.86656	1.8162	0.41170 −2.2279	16.863	−6.6633
5.0	0.88179	1.8005	0.35591 −2.1564	16.037	−5.5274
6.0	0.89268	1.7902	0.31754 −2.1077	15.496	−4.7926

Bibliography

1. Belyakov, V.M. (1965) *Tables of Elliptical Integrals* (trans. Basu, P.). Pergamon Press, Oxford.
2. Bellman, R. (1961) *Brief Introduction to Theta Functions.* Holt, Rinehart and Winston, New York.
3. Bowman, F. (1953) *Introduction to Elliptic Functions, with Applications.* English Universities Press, London.
4. Byrd, P.F., Friedman, M.D. (1971) *Handbook of Elliptic Integrals for Engineers and Scientists.* Springer-Verlag, New York.
5. Cayley, A. (1961) *Elementary Treatise on Elliptic Functions.* Dover, New York.
6. Copson, E.T. (1935) *Theory of Functions of a Complex Variable.* Oxford University Press, London, pp. 345–443.
7. Dixon, A.C. (1894) *Elliptic Functions.* Macmillan, London.
8. Greenhill, A.G. (1892) *Applications of Elliptic Functions.* Macmillan, London.
9. Hancock, H. (1910) *Lectures on the Theory of Elliptic Functions.* Wiley, New York.
10. MacRobert, T.M. (1947) *Functions of a Complex Variable.* Macmillan, London.
11. Milne-Thomson, L.M. (1950) *Jacobian Elliptic Function Tables.* Dover, New York.
12. Neville, E.H. (1944) *Jacobian Elliptic Functions.* Oxford University Press, London.
13. Neville, E.H. (1971) *Elliptic Functions: A Primer* (Ed. Langford, W.J.). Pergamon Press, Oxford.
14. Spenceley, G.W., Spenceley, R.M. (1947) *Smithsonian Elliptic Functions Tables.* Smithsonian Institution, Washington, D.C.
15. Tannery, J., Molk, J. (1902) *Fonctions Elliptiques.* Gauthier-Villars, Paris.
16. du Val, P. (1973) *Elliptic Functions and Elliptic Curves.* Cambridge University Press, London.
17. Whittaker, E.T., Watson, G.N. (1950) *A Course of Modern Analysis.* Cambridge University Press, London, pp. 429–535.

Index

Applied Mathematical Sciences

Printed in the USA
CPSIA information can be obtained
at www.ICGtesting.com
LVHW010620200224
772311LV00003B/206

9 780387 969657